The publisher gratefully acknowledges

the generous contribution to this book

provided by the Richard and Rhoda Goldman Fund.

TREES AND SHRUBS OF CALIFORNIA

CALIFORNIA NATURAL HISTORY GUIDES

Phyllis Faber and Robert Ornduff, General Editors

LATEST TITLES IN THE SERIES

59. *Glaciers of California: Modern and Ice Age Glaciers, the Origin of Yosemite Valley, and a Glacier Tour in the Sierra Nevada*, by Bill Guyton
60. *Sierra East: Edge of the Great Basin*, edited by Genny Smith
61. *Natural History of the Islands of California*, by Allan A. Schoenherr, C. Robert Feldmeth, and Michael J. Emerson
62. *Trees and Shrubs of California*, by John D. Stuart and John O. Sawyer

The University of California Press wishes to dedicate this book to the memory of Robert Ornduff, 1932–2000.

CALIFORNIA NATURAL HISTORY GUIDES 62

TREES AND SHRUBS OF CALIFORNIA

JOHN D. STUART
JOHN O. SAWYER

ILLUSTRATED BY
ANDREA J. PICKART

UNIVERSITY OF CALIFORNIA PRESS

Berkeley Los Angeles London

University of California Press
Berkeley and Los Angeles, California

University of California Press, Ltd.
London, England

Photo credits: Plates 3, 6, 7, 8, 9, 10, 15, 17, 18, 22, 29, 30, 31, 38, 39, 40—John D. Stuart; Plates 1, 2, 4, 5, 12, 13, 16, 19, 20, 21, 23, 24, 25, 27, 28, 33, 34, 35, 37—John O. Sawyer; Plates 11, 14, 26, 32, 36—Andrea J. Pickart

Library of Congress Cataloging-in-Publication Data

Stuart, John David.
Trees and shrubs of California / John D. Stuart, John O. Sawyer ; illustrated by Andrea J. Pickart.
p. cm. — (California natural history guides ; 62)
Includes bibliographical references (p.) and index.
ISBN 0-520-22109-5 (cloth : alk. paper) — ISBN 0-520-22110-9 (pbk. : alk. paper)
1. Trees—California—Identification. 2. Shrubs—California—Identification. 3. Trees—California—Pictorial works. 4. Shrubs—California—Pictorial works. I. Sawyer, John O., 1939–. II. Title. III. Series.

QK149 S73 2001
582.16'09794—dc21 00-025834

Manufactured in China

10 09 08 07 06 05 04 03 02 01
10 9 8 7 6 5 4 3 2 1

The paper used in this publication meets the minimum requirements of ANSI/NISO Z39.48-1992 (R 1997) (*Permanence of Paper*). ⊗

To Mary, Jane, Pete, Danny, Robbie, and David

CONTENTS

List of Plates ix
Preface xi
Introduction 1
How the Book Is Organized 1
Classification 1
Nomenclature 2
How to Use the Book to Identify Trees and Shrubs 4
California's Forests and Woodlands 5

CONIFERS 19

KEY TO GENERA 19
DESCRIPTIONS OF GENERA AND SPECIES 22

BROADLEAVED TREES AND SHRUBS 111

KEY TO GROUPS 111
KEY TO GENERA 112
DESCRIPTIONS OF GENERA AND SPECIES 130

Appendix A: Genera Grouped by Distinctive Morphological Features 417
Conifers 417
Broadleaved Trees and Shrubs 419

Appendix B: Checklist of Trees and Shrubs 425
Gnetophytes 425
Conifers 425
Broadleaved Trees and Shrubs 427

Glossary 439
References 451
Index 455

PLATES

following page 214

1. Bristlecone fir, *Abies bracteata*
2. Green ephedra, *Ephedra viridis*
3. Mountain juniper, *Juniperus occidentalis* var. *australis*
4. Brewer spruce, *Picea breweriana*
5. Foxtail pine, *Pinus balfouriana*
6. Western white pine, *Pinus monticola*
7. Monterey pine, *Pinus radiata*
8. Ghost pine, *Pinus sabiniana*
9. Redwood, *Sequoia sempervirens*
10. Giant sequoia, *Sequoiadendron giganteum*
11. Mountain hemlock, *Tsuga mertensiana*
12. California buckeye, *Aesculus californica*
13. Sitka alder, *Alnus viridis* ssp. *sinuata*
14. Bearberry, *Arctostaphylos uva-ursi*
15. Whiteleaf manzanita, *Arctostaphylos viscida*
16. Oregon-grape, *Berberis aquifolium*
17. Spice bush, *Calycanthus occidentalis*
18. Blue blossom, *Ceanothus thyrsiflorus*
19. Western redbud, *Cercis occidentalis*
20. Curlleaf mountain-mahogany, *Cercocarpus ledifolius*

21. Mountain dogwood, *Cornus nuttallii*
22. California hazel, *Corylus cornuta* var. *californica*
23. Fremont silk tassel, *Garrya fremontii*
24. California black walnut, *Juglans californica*
25. Tanoak, *Lithocarpus densiflorus*
26. Twinberry, *Lonicera involucrata*
27. Black cottonwood, *Populus balsamifera* ssp. *trichocarpa*
28. Cliffrose, *Purshia mexicana* var. *stansburyana*
29. Blue oak, *Quercus douglasii*
30. California black oak, *Quercus kelloggii*
31. California black oak, *Quercus kelloggii*
32. Cascara, *Rhamnus purshiana*
33. Pacific rhododendron, *Rhododendron macrophyllum*
34. Skunkbush, *Rhus trilobata*
35. Gummy gooseberry, *Ribes lobbii*
36. Red flowering currant, *Ribes sanguineum*
37. Black sage, *Salvia mellifera*
38. Douglas spiraea, *Spiraea douglasii*
39. California bay, *Umbellularia californica*
40. California fan palm, *Washingtonia filifera*

PREFACE

This manual can be used in the field, office, or home. Our intended audience is amateur and professional botanists, natural resource professionals, students, and other people who have an interest in the trees and shrubs of California. Our purpose was to produce an easy-to-use, portable manual of California trees and shrubs. Essentially, all native California tree species and most common shrub species are in this book. We have not, however, included every shrub species, as there are simply too many for a compact guide. In some instances, we treat 1 or 2 species as representative of a genus, rather than describe every species. Among large genera, we have included the more common species; for example, 13 of the 56 manzanita species in California, 17 of the 43 ceanothus species, 13 of the 30 currant and gooseberry species, and 13 of the 30 willows. In general, the book has relatively complete coverage of the common shrubs found in forests, woodlands, and chaparrals, but less complete coverage of the desert scrubs.

We have tried to make our book a user-friendly manual with minimal technical nomenclature, one that employs more familiar terms whenever possible in the dichotomous keys and descriptions. Unavoidable jargon is defined in the glossary.

ACKNOWLEDGMENTS

We are pleased to recognize Jim André, Nona Chiariello, R. Jane Cole, Peter Jain, Jim Rorabaugh, Mary Stuart, Danny Stuart,

Robbie Stuart, and Robin Wills for their help in collecting plant specimens; Marian Perry, Maralyn Renner, and Mary Stuart for their reviews; and Jennifer Key, Tom Mahony, Steve Steinberg, and Tom Voorhees for help with computer-generated maps. Our ecological sections and our maps are adapted from ecological-unit maps produced by the USDA Forest Service and the Natural Resource Conservation Service. In particular, we used data from the map entitled *Ecological Units of California: Subsections* (August 1994), compiled by C. B. Goudey and D. W. Smith and available from the Pacific Southwest Region of the USDA Forest Service in San Francisco, California.

INTRODUCTION

HOW THE BOOK IS ORGANIZED

Conifers and broadleaved trees and shrubs are treated separately in this book. Each group has its own set of keys to genera and species, as well as plant descriptions. Plant descriptions are organized alphabetically by genus and then by species. In a few cases, we have included separate subspecies or varieties. Genera in which we include more than one species have short generic descriptions and species keys. Detailed species descriptions follow the generic descriptions. A species description includes growth habit, distinctive characteristics, habitat, range (including a map), and remarks. Most species descriptions have an illustration showing leaves and either cones, flowers, or fruits. Illustrations were drawn from fresh specimens with the intent of showing diagnostic characteristics. Plant rarity is based on rankings derived from the California Native Plant Society and federal and state lists (Skinner and Pavlik 1994).

Two lists are presented in the appendixes. The first is a list of species grouped by distinctive morphological features. The second is a checklist of trees and shrubs indexed alphabetically by family, genus, species, and common name.

CLASSIFICATION

To classify is a natural human trait. It is our nature to place objects into similar groups and to place those groups into a hier-

TABLE 1 CLASSIFICATION HIERARCHY OF A CONIFER AND A BROADLEAVED TREE

Taxonomic rank	Conifer	Broadleaved tree
Kingdom	Plantae	Plantae
Division	Pinophyta	Magnoliophyta
Class	Pinopsida	Magnoliopsida
Order	Pinales	Sapindales
Family	Pinaceae	Aceraceae
Genus	*Abies*	*Acer*
Species epithet	*magnifica*	*glabrum*
Variety	*shastensis*	*torreyi*
Common name	Shasta red fir	mountain maple

archy. Biologists group plants by morphological and genetic characteristics. An example of a widely accepted taxonomic hierarchy is found in Table 1. This is not a static classification, and modifications are proposed based on new morphological and genetic information.

Taxonomists often use intermediate ranks such as tribe or subgenus to organize complexity in large families and genera. Our book has occasional references to ranks below genus. The classification of lodgepole pine (see Table 2) serves as a good example of ranks below genus.

NOMENCLATURE

The rules and procedures for naming plants can be found in *The International Code of Botanical Nomenclature* (Greuter and others 1994). The valid code is based in part on the principles that names reflect the earliest published description, that there can be only one correct name for a plant, and that scientific names are in Latin.

Species names are made up of two parts, the genus name and the species epithet. It is incorrect to use only the species epithet, since little information is gained from it alone. The species epithet *menziesii,* for example, is used for Douglas-fir (*Pseudotsuga menziesii*) and Pacific madrone (*Arbutus menziesii*). The genus name and species epithet are normally italicized or underlined. The first letter of the genus is always capitalized and the epithet, subspecies, and variety names are lowercased irrespective of

TABLE 2 SUBDIVISIONS OF LODGEPOLE PINE

Taxonomic rank	Lodgepole pine
Genus	*Pinus*
Subgenus	Pinus
Section	Pinus
Subsection	Contortae
Species epithet	*contorta*
Subspecies	*murrayana*

Source: Little and Critchfield 1969.

source. Roman type is used when writing the names of all ranks higher than genus, the author's name (i.e., the name of the botanist who first formally described the plant), and the abbreviations for subspecies (ssp. or subsp.) and variety (var.). When writing about a species, it is a convention to abbreviate that species' name after the first instance, giving the first initial of the genus name rather than rewriting the entire name, as long as the discussion includes no other genus that begins with the same first letter. The plural of *species* is *species* and the plural of *genus* is *genera*. *Species* and *specific* are appropriate adjective forms for *species,* and *generic* is the appropriate form for *genus.*

Plants are occasionally reclassified and renamed based on new interpretations of genetic and morphological evidence. Most changes are proposed for the ranks genus, species, and subspecies, but not all proposed changes become widely accepted. When you are in doubt about the proper name for a plant, we recommend that that you consult the most recent, authoritative book on regional flora.

Common names have fewer rules and conventions than scientific names have. They differ from scientific names in the following ways: they are often the only names known by many people; they are familiar in only one language; a species can have more than one common name (for example, California bay, Oregon myrtle, pepperwood, California laurel, and a few other names all refer to one species); and more than one species or genus can have the same common name (for example, sage). Rules for common names include lowercasing all words except for proper names (for example, sugar pine versus Torrey pine) and hyphenating names or making them into one word if the object is not "true" (for example, Douglas-fir is not a member of

Abies, the firs, and western redcedar is not a member of *Cedrus*, the cedars). As with scientific names, when you are in doubt about the common name, consult the most recent, authoritative book on regional flora. Organizations such as the U.S. Forest Service maintain lists of preferred common names.

HOW TO USE THE BOOK TO IDENTIFY TREES AND SHRUBS

Identifying plants from a guidebook usually entails a set of sequential steps. We recommend that the following procedures be adopted when using this book.

1. Determine whether the unknown specimen is a conifer or broadleaved plant and then turn to the appropriate key.
2. Address questions of growth habit before you begin the keying process:
   a. Is it a tree or a shrub? Little (1979) defined trees as woody plants "having one erect perennial stem or trunk at least 3 inches (7.5 centimeters) in diameter at breast height (4½ feet or 1.3 meters), a more or less definitely formed crown of foliage, and a height of at least 13 feet (4 meters)." Shrubs, in contrast, are smaller and generally multistemmed.
   b. Is it erect or prostrate?
   c. How tall is it?
   d. Does it have multiple stems?
3. Use the key to identify the genus of your plant. Work your way through the key by selecting between sequential pairs (couplets) of alternately indented opposing statements (dichotomies). The couplets describe a small set of the plant's morphological characteristics, and occasionally geographical or habitat characteristics are included. Choose the statement that best fits your plant. If the statement ends with a name, find the genus in the alphabetically arranged pages. If the statement does not end with a name, go to the next indented pair of statements and continue the process until you eventually arrive at a name.
4. Read the description of the genus to ensure that it fits your plant.

5. Next, use the species key in the same manner in order to identify the species. Genera in which we present only one California species do not have a species key.
6. Read the species description to be sure it fits your plant. In addition to the species' morphological characteristics, pay close attention to its growth habit, habitat, and range. References to largest individuals are included in some species descriptions. Largest individuals are determined using a combination of height, circumference, and crown spread (Cannon 1998). Diameters refer to trunks. The largest plant, therefore, is not necessarily the tallest.
7. Check to see if the habitat and range described in the book match the habitat and range of your plant. We use the terms *coastal, low-elevation, foothill, montane, subalpine,* and *alpine* to further define a species' habitat.

   The shaded parts of each range map correspond to the ecological regions in which the plant grows (Map 1) (Goudey and Smith 1994). This does not imply, however, that the plant grows everywhere inside the shaded areas. Combine the habitat description with the elevation range to better define the plant's natural range.

CALIFORNIA'S FORESTS AND WOODLANDS

Most of California's landscapes have characteristic trees and shrubs. The north coast, for example, is home to redwood and Douglas-fir forests; the central coast has isolated stands of Monterey cypress and Monterey pine; the Sierra Nevada is noted for its mixed conifer forests; southern California has extensive chaparral stands; desert regions are noted for Joshua trees, junipers, and pinyons; and the Central Valley is surrounded by foothills with blue oak woodlands.

Tree-dominated vegetation can be called either a forest or woodland. Forests typically have trees close enough that their crowns touch. Not all crowns, however, touch or overlap, as forests usually have large gaps in their canopies. The older the forest, generally speaking, the larger the gaps. Some forests have gaps amounting to as much as 75 to 80 percent of the area. Woodlands, in contrast, have widely spaced trees with grass or shrubs among

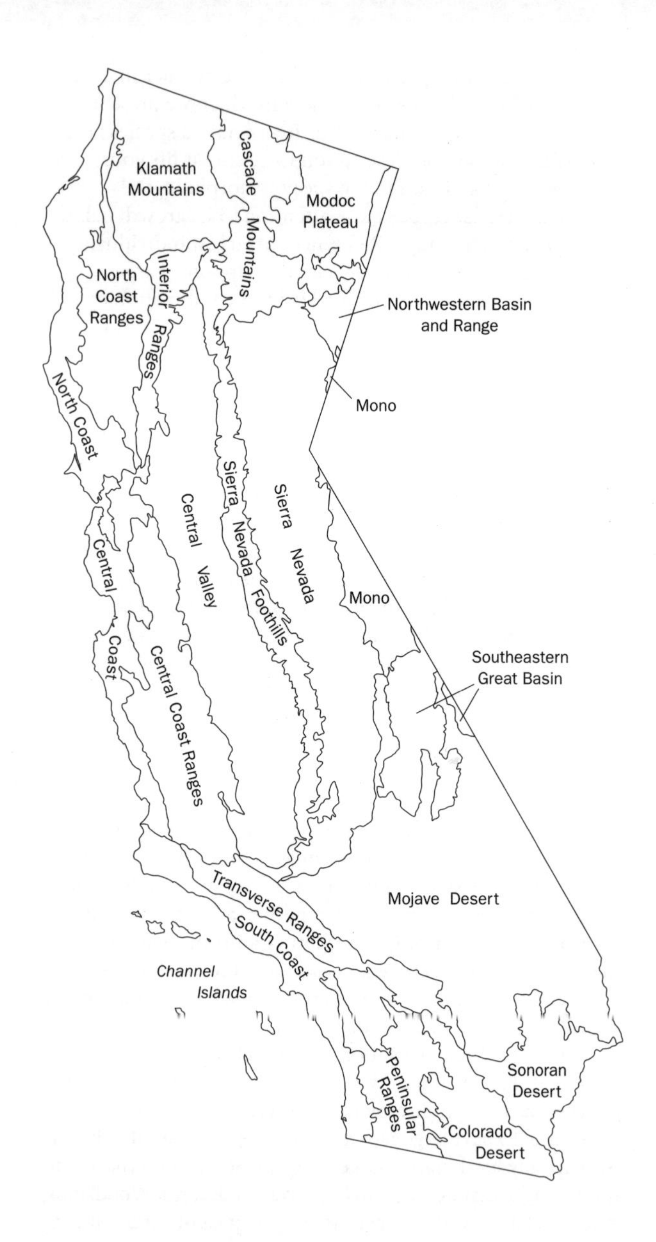

Map 1. Ecological sections of California

them. In general, forests have more than 20 percent canopy cover, while woodlands have less than 20 percent canopy cover.

Shrub-dominated areas have various names. *Chaparral* is a name used for shrublands composed of species with thickened, evergreen, leathery, sclerophyllous leaves. *Scrub* is a term applied to shrublands that have either widely scattered or drought-deciduous shrubs (e.g., southern California's coastal scrub). Occasionally, shrub types are named after a commonly occurring shrub species, such as big sagebrush in the Great Basin.

Forests and woodlands vary greatly throughout the state. Oaks dominate in many of the state's lower-elevation forests and woodlands, while conifers are generally more common on mid- and high-elevation mountains as well as in low-elevation coastal forests. In some areas, conifer forests and woodlands are distinct from those dominated by broadleaved trees. In other areas there is considerable mixing of conifers and broadleaved trees, as is the case with the so-called mixed evergreen forests.

The state's vegetation diversity is largely a function of varying climates, landforms, geological formations, and soils. Even though these factors vary continuously across landscapes, it is possible to characterize regions within the state as having similar ecological attributes. The Mojave Desert's vegetation, climate, landforms, geological formations, and soils are quite different from those found in, for example, the Klamath Mountains. Some tree and shrub species are found exclusively in a single ecological region (e.g., Torrey pine and Sadler oak), while others can be found in many regions (e.g., canyon live oak and wedgeleaf ceanothus). Knowing the ecological region(s) in which a plant grows is useful for identification and for understanding ecological requirements.

Increasingly, ecologists, botanists, natural resource professionals, geologists, and geographers characterize California's ecosystems (Bailey 1995), flora (Hickman 1993), and vegetation (Sawyer and Keeler-Wolf 1995) on the basis of region. The 20 ecological sections (see Map 1) we recognize are adapted from those described by Miles and Goudey in their book *Ecological Subregions of California: Section and Subsection Descriptions* (1997). For a detailed description of California's vegetation, refer to Barbour and Billings (1988), Barbour and Major (1988), Holland and Keil (1995), and Sawyer and Keeler-Wolf (1995). The ecological subregions of California described by Miles and Goudey (1997) are based on the *National Hierarchical Frame-*

TABLE 3 ECOLOGICAL UNITS OF CALIFORNIA

HUMID TEMPERATE DOMAIN
Mediterranean Division
California Coastal Chaparral Forest and Shrub Province
Central Coast Section
South Coast Section
California Dry Steppe Province
Central Valley Section
California Coastal Steppe–Mixed Forest–Redwood Forest Province
North Coast Section
Mediterranean Regime Mountains Division
Sierran Steppe–Mixed Forest–Coniferous Forest Province
Klamath Mountains Section
North Coast Ranges Section
Interior Ranges Section
Cascade Mountains Section
Sierra Nevada Foothills Section
Sierra Nevada Section
Modoc Plateau Section
California Coastal Range Open Woodland–Shrub–Coniferous Forest–Meadow Province
Central Coast Ranges Section
Transverse Ranges Section
Peninsular Ranges Section
DRY DOMAIN
Tropical/Subtropical Desert Division
American Semidesert and Desert Province
Mojave Desert Section
Sonoran Desert Section
Colorado Desert Section
Temperate Desert Division
Intermountain Semidesert and Desert Province
Mono Section
Southeastern Great Basin Section
Intermountain Semidesert Province
Northwestern Basin and Range Section

Source: Adapted from Miles and Goudey 1997.

work of Ecological Units (McNab and Avers 1994) and Bailey's *Description of the Ecoregions of the United States* (1995). ECOMAP (1993), which is a companion to *Ecological Subregions of California,* maps California using this classification.

The hierarchical levels of the national framework of ecoregions are domain, division, province, section, and subsection. Large climatic zones define domains and divisions, and broad-scale natural vegetation types define provinces. Sections and subsections are further delimited based on landforms, soils, local climates, and vegetation. California, for example, is divided into the humid temperate domain, characteristic of most of the state, and the dry domain, associated with the deserts. California's humid temperate domain has the Mediterranean division and the Mediterranean regime mountains division. California's dry domain is made up of the tropical/subtropical desert division and the temperate desert division (see Table 3).

HUMID TEMPERATE DOMAIN

Mediterranean Division

The Pacific Ocean greatly moderates the climate of the coastal regions and, to a lesser extent, of the inland valleys. Winters are moderate and typically rainy, while summers are warm to hot and dry.

CALIFORNIA COASTAL CHAPARRAL FOREST AND SHRUB PROVINCE

Central Coast Section This section extends from San Francisco Bay to San Luis Obispo Bay. Prominent landmarks include Mount Diablo, the San Francisco Peninsula, Santa Cruz Mountains, Salinas Valley, and Santa Lucia Mountains. Mountaintops typically vary between 450 m (1,500 ft) and 1,200 m (4,000 ft) in elevation. Annual precipitation ranges from 30 cm (12 in.) to 150 cm (60 in.). Summer and winter temperatures are moderate. Coastal slopes are covered with redwood forests or coast live oak and mixed oak woodlands. Blue oak and valley oak woodlands are more evident inland. Much of the ocean-facing and inland, low-elevation slopes are covered with chaparrals made up of ceanothus, chamise, and manzanita. At higher elevations, canyon live oak is common and stands of Coulter pine suggest the beginning of a montane zone on the highest peaks.

South Coast Section This section extends from Point Sal to San Diego. Prominent landmarks include the Santa Ynez Mountains, Santa Susana Mountains, Santa Monica Mountains, Channel Islands, Los Angeles Basin, and San Diego Bay. Elevations range from sea level to around 1,000 m (3,000 ft), and annual precipitation varies between 25 cm (10 in.) and 75 cm (30 in.). Summer temperatures are moderate to warm, while winters are moderate. The South Coast Section characteristically has woodlands of California walnut, coast live oak, and/or Engelmann oak punctuating slopes of coastal scrubs, chaparrals, and grasslands. Coastal scrubs, once extensive, are shrublands consisting of black sage, California buckwheat, California encelia, California sagebrush, and white sage. Chaparral is the most extensive shrubland type in southern California, and it varies greatly in species composition. Many slopes are covered with monotonous expanses of chamise or scrub oak, but others are covered by a mixture of shrubs, including many species of ceanothus, manzanita, and oaks. Bigcone Douglas-fir is found on steep north slopes.

CALIFORNIA DRY STEPPE PROVINCE

Central Valley Section This section lies between the Coast Ranges and the Sierra Nevada. Prominent landforms include the Sacramento and San Joaquin Valleys, Sutter Buttes, and Tulare Lake. The Sacramento River joins the San Joaquin River in the delta region west of Stockton, eventually draining into San Francisco Bay. Elevations range from sea level to 600 m (2,000 ft). Annual precipitation varies from 12 cm (5 in.) to 63 cm (25 in.). Summer temperatures are hot, while winter temperatures are moderate. Prior to settlement of the Sacramento Valley, valley oak woodlands were extensive among the grasslands and marshes. In the San Joaquin Valley, the woodlands were less extensive, and basins supported saltbush scrubs.

CALIFORNIA COASTAL STEPPE—MIXED FOREST—REDWOOD FOREST PROVINCE

North Coast Section This section stretches from San Francisco Bay to the Oregon border. The north coast is home to the Smith, Klamath, Mad, Van Duzen, Mattole, Eel, Noyo, Navarro, Gualala, Russian, and Napa Rivers. Other notable landmarks include Humboldt Bay, Cape Mendocino, and Point Reyes. The

Pacific Ocean significantly moderates the climate, producing relatively mild temperatures year round. Annual precipitation varies from 50 cm (20 in.) to 300 cm (120 in.) and elevations range from sea level to 900 m (3,000 ft). Conifers such as beach pine, Bishop pine, grand fir, pygmy cypress, and Sitka spruce dominate coastal forests. Adjacent to the coastal forests are more or less continuous redwood forests that contain the tallest trees in the world. Drier conditions farther inland support Douglas-fir–tanoak forests, Oregon white oak woodlands, chaparrals, and grasslands.

Mediterranean Regime Mountains Division

The Pacific Ocean's influence on climate is diminished for these interior mountainous regions. Winters are cool to cold and rainy to snowy. Summers are warm and dry.

SIERRAN STEPPE—MIXED FOREST—CONIFEROUS FOREST PROVINCE

Klamath Mountains Section The Klamath Mountains are found in northwestern California and separate the North Coast Section from the Cascade Mountains. The region is known for its mountains, steep canyons, and high biodiversity. The Siskiyou, Marble, Salmon, Scott, Trinity, and Yolla Bolly Mountains collectively make up the section, with elevations ranging from 60 m (200 ft) to 2,700 m (9,000 ft). Annual precipitation varies between 45 cm (18 in.) and 300 cm (120 in.). Summer temperatures are moderate to hot, while winter temperatures are cold. Low-elevation canyons are the realm of canyon live oak or Douglas-fir–tanoak forests intermingled. Chaparrals and blue oak woodlands are more extensive on the eastern side of the Klamath Mountains. The western montane elevations are home to Douglas-fir-dominated forests that, with increasing elevation, blend into white fir and then Shasta red fir forests. In the eastern ranges, the montane forests are dominated by Douglas-fir, Douglas-fir and ponderosa pine, mixed conifers like those of the Sierra Nevada, white fir, and Shasta red fir. Greenleaf manzanita, huckleberry oak, and tobacco brush are principal associates of the montane chaparrals. The highest ridges and north slopes support mountain hemlock forests and subalpine foxtail or whitebark pine woodlands.

North Coast Ranges Section This section is bounded by

the North Coast Section, the Interior Ranges Section, and the Klamath Mountains Section. South Fork Mountain, the Yolla Bolly Mountains, Round Valley, and Clear Lake are prominent landmarks. Elevations range from 100 m (300 ft) to 2,500 m (8,100 ft). Annual precipitation varies from 60 cm (25 in.) to 300 cm (120 in.) and summer temperatures are typically warm to hot. Lower-elevation forests are mixtures of Douglas-fir and tanoak. With increasing elevation, the forests become dominated by Douglas-fir and ponderosa pine and eventually first by white fir and then by red fir. The highest peaks support subalpine foxtail pine and Jeffrey pine woodlands. Blue oak or Oregon white oak woodlands, chaparrals, and grasslands are found in the interior lowlands.

Interior Ranges Section This section separates the North Coast Ranges Section from the Sacramento Valley. Annual precipitation varies between 38 cm (15 in.) and 100 cm (40 in.). Winters are cool and wet, while summers are hot and dry. The foothills are covered with grasslands or blue oak woodlands. Montane slopes are covered with chaparrals often dominated by chamise or black oak and ponderosa pine forests. White fir is more common at montane elevations. Only the highest peaks support stands of red fir.

Cascade Mountains Section This section is the southern edge of the Cascades, which extend from Oregon and Washington. Volcanic mountains—notably Mount Shasta and Mount Lassen—and broad valleys interspersed with volcanic ridges characterize the area. Elevations range between 600 m (2,000 ft) and 4,300 m (14,000 ft). Annual precipitation varies from 20 cm (8 in.) to 200 cm (80 in.). Summers are generally warm and dry, while winters are cold and snowy. The principal montane-elevation forest types are Jeffrey pine, mixed conifer, and ponderosa pine. Greenleaf manzanita is prominent in much of the montane chaparral. White fir and red fir dominate upper montane forests. The subalpine areas are home to lodgepole pine, western white pine, and whitebark pine forests.

Sierra Nevada Foothills Section This section extends from near Red Bluff in the north to the Tehachapi Mountains in the south. Much of this area has characteristically steep canyons separated by somewhat flattened ridges. Major waterways include the Feather, Sacramento, Yuba, American, Tuolumne, Stanislaus, Merced, San Joaquin, Kings, and Kern Rivers. Elevations range between 60 m (200 ft) and 1,500 m

(5,000 ft). Annual precipitation varies from 15 cm (6 in.) to 100 cm (40 in.). Winters are moderate to cool, and summers are typically warm to hot. Grasslands and blue oak woodlands characterize this area. Ghost pine and interior live oak are prominent associates of blue oak. Chaparrals and grasslands can be found on exposed slopes. Important riparian species include California sycamore, Fremont cottonwood, white alder, and a variety of willows.

Sierra Nevada Section The Sierra Nevada is the dominant mountain range in California. It separates the temperate desert regions from western central and northern California. Lake Tahoe, the High Sierra, Yosemite Valley, and Kings Canyon are internationally renowned landmarks. Elevations range from 300 m (1,000 ft) in canyon bottoms to the summit of Mount Whitney at 4,419 m (14,495 ft). Annual precipitation varies between 25 cm (10 in.) and 230 cm (90 in.). Summers are warm to cool and winters are cold and wet. Ponderosa pine forests mark the lower montane zone in the Sierra Nevada. Mixed conifer forests of Douglas-fir, incense-cedar, ponderosa pine, sugar pine, and white fir become extensive with increasing elevation. Groves of giant sequoia are sprinkled throughout the higher elevations within the mixed conifer forest in the central and southern regions. White fir and then red fir dominate the forests of the snowy upper montane. Still higher, the subalpine forests of lodgepole pine mix with extensive meadows. Near the tree line, whitebark pine and mountain hemlock forests are found in the north, and foxtail pine forests in the south.

Modoc Plateau Section This section lies to the east of the Cascade Mountains in northeastern California. Prominent features include the Warner Mountains, the Pit River, and Lower Klamath, Tule, and Goose Lakes. Elevations vary from 900 m (3,000 ft) to 3,000 m (9,900 ft). Annual precipitation ranges between 20 cm (8 in.) and 75 cm (30 in.). Winters are cold and summers warm. Broad expanses of grasslands, big sagebrush, and western juniper woodlands characterize valleys and lower slopes. With increasing elevation, pure or mixed forests of Jeffrey and ponderosa pine can be found. Higher elevations support aspen, lodgepole pine, and white fir stands. In the Warner Mountains, a montane white fir zone is well developed below subalpine whitebark pine woodlands. Bulrush and cattail marshes are extensive.

CALIFORNIA COASTAL RANGE OPEN WOODLAND–SHRUB–CONIFEROUS FOREST–MEADOW PROVINCE

Central Coast Ranges Section This section separates the Central Valley from the Central Coast Section. It extends from east of San Francisco Bay to the Transverse Ranges. The Diablo, Gabilan, interior Santa Lucia, Caliente, and Temblor Ranges are prominent within it. The Carrizo Plain is renowned. Elevations range from 30 m (100 ft) to 1,700 m (5,700 ft). Annual precipitation varies from 15 cm (6 in.) to 100 cm (40 in.). Winters are moderate and summers hot. Low-elevation slopes and valleys are covered by a mosaic of chaparrals, grasslands, forests, and woodlands. Pure and mixed forests and woodlands of blue oak, California bay, coast live oak, ghost pine, tanoak, and valley oak are extensive. Chaparrals and coastal scrubs exhibit a confusing mixture of species. With elevation, bigcone Douglas-fir and Coulter pine signify the beginning of the montane forests, of which canyon live oak is an important component. The highest peaks support stands of incense-cedar, Jeffrey pine, and ponderosa pine. The eastern slopes support blue oak or California juniper woodlands and chaparrals. Eastern foothills sustain grasslands or open stands of bladderpod, California ephedra, and various saltbush species.

Transverse Ranges Section This section extends in a generally east-west direction from the mountains north of Santa Barbara to the mountains north and east of Palm Springs. Prominent mountain ranges within this section include the San Rafael–Topatopa, San Gabriel, San Bernardino, and Little San Bernardino Mountains. Elevations range from 100 m (300 ft) to 3,500 m (11,500 ft). Precipitation varies between 15 cm (6 in.) and 100 cm (40 in.). Winters are cool to cold and summers moderate to warm. Chaparrals and coastal scrubs dominate lower elevations. Above this zone, Coulter pine, Jeffrey pine, ponderosa pine, incense-cedar, and black oak signify the beginning of the montane forests. Montane chaparrals can be found intermixed with these forests. The highest peaks harbor subalpine stands of limber pine, lodgepole pine, and mountain juniper. The low-elevation eastern mountain slopes are covered with big sagebrush and a desert chaparral composed of flannelbush, cupleaf ceanothus, and desert mountain-mahogany. California juniper, Joshua tree, and singleleaf pinyon woodlands may be found above these types.

Peninsular Ranges Section This section separates the South Coast Section from the Colorado Desert. The Santa Ana, San Jacinto, Santa Rosa, and Laguna Mountains are significant ranges within this section. Elevations range between 100 m (300 ft) and 3,300 m (10,800 ft). Precipitation varies from 15 cm (6 in.) to 80 cm (30 in.). Winters are moderate to cold and summers warm to hot. Chaparrals and coastal scrubs are found at lower elevations. With increasing elevation, Coulter pine mixed with California black, canyon, and interior live oaks signify the beginnings of montane elevations. At still higher elevations, the mixed conifer forests have a different mix than in the Sierra Nevada; for example, incense-cedar, Jeffrey pine, sugar pine, white fir, and live oak. The highest peaks harbor subalpine stands of lodgepole pine, Jeffrey pine, and limber pine. The steep eastern slope supports desert chaparrals and California juniper and singleleaf pinyon woodlands.

DRY DOMAIN

Tropical/Subtropical Desert Division

An interior, continental climate is dominant in these regions. Summers are hot to very hot and very dry. Winters are cool or warm, and dry.

AMERICAN SEMIDESERT AND DESERT PROVINCE

Mojave Desert Section This section is a vast expanse found south and east of the Sierra Nevada and east of the Transverse Ranges. Prominent landscape features include Death Valley, Owens Valley, Rogers Lake, the Kelso Dunes, New York Mountains, Granite Mountains, and Cima Dome. Elevations range between 86 m below sea level (–282 ft) and 2,400 m (7,900 ft). Annual precipitation is highly variable, from 7 cm (3 in.) to 20 cm (8 in.). Winters are moderate to cool and summers hot. The lower elevations of the Mojave Desert are covered by creosote bush scrubs punctuated with woodlands of catclaw acacia or smoke tree in the washes. Joshua tree, singleleaf pinyon, or California or Utah juniper woodlands can be found above the valleys. Some mountains in the eastern Mojave Desert are sufficiently high enough to support Great Basin bristlecone pine, limber pine, or white fir.

Sonoran Desert Section This section lies east of the Colorado Desert and south of the Mojave Desert. The Colorado

River borders its eastern edge. Prominent landmarks include the Chuckawalla and Chocolate Mountains and the Colorado River. Elevations range between 100 m (300 ft) and 1,300 m (4,400 ft). Annual precipitation is highly variable, from 7 cm (3 in.) to 15 cm (6 in.). Winters are mild and summers are very hot. Black bush is common at higher elevations in the isolated ranges in the Sonoran Desert. On upland surfaces, sparse groupings of brittlebush and creosote bush are the more conspicuous shrubs growing between patches of desert pavement. The broad washes support woodlands of blue palo verde, catclaw acacia, ironwood, and smoke tree. Mesquite and saltbush species are found on basin floors. The borders of the Colorado River are covered with tamarisk.

Colorado Desert Section This section separates the Peninsular Ranges from the Sonoran Desert. The Salton Sea, Algodones Dunes, and the Coachella and Imperial Valleys are dominant landscape features. Elevations range between 70 m below sea level (–230 ft) and 670 m (2,200 ft). Annual precipitation is highly variable, from 7 cm (3 in.) to 15 cm (6 in.). Winters are mild and summers are very hot. Creosote bush and many kinds of cactus dominate the upland desert scrubs on the valley bottoms. California fan palm and Fremont cottonwood can be found in the canyons, and blue palo verde, ironwood, and smoke tree in the washes. On the sand dunes colorful displays of herbs occur among scattered individual creosote bushes.

Temperate Desert Division

An interior, continental climate is dominant in these regions. Summers are hot to very hot and very dry. Winters are cold and dry.

INTERMOUNTAIN SEMIDESERT AND DESERT PROVINCE

Mono Section This section lies east of the Sierra Nevada and is the western extension the Basin and Range geomorphic province found in Nevada. The Sweetwater Mountains, the Bodie Hills, Mono Lake, and the White Mountains are prominent landmarks. Elevations range between 1,300 m (4,400 ft) and 4,300 m (14,200 ft). Annual precipitation varies from 12 cm (5 in.) to 75 cm (30 in.). Winter temperatures are cold to very cold and summers warm to cool. Shadscale and big sagebrush are extensive in basins and on lower slopes. Singleleaf pinyon and Utah juniper woodlands appear with increasing elevation. Above the woodlands can be found Jeffrey pine and white fir forests. With further elevation, big sagebrush is replaced by low sagebrush. On the eastern slopes of the Sierra Nevada, upper montane red fir forests are locally extensive. East of the Owens Valley, Great Basin bristlecone pine woodlands inhabit the subalpine zone. Watercourses harbor black cottonwood, water birch, and several willow species.

Southeastern Great Basin Section This section is bounded by the Sierra Nevada to the west, the Mono Section to the north, and the Mojave Desert to the south. The Inyo Mountains, Panamint Range, Saline Valley, and Eureka Valley are prominent features. Elevations range between 300 m (1,000 ft) and 3,300 m (11,000 ft). Annual precipitation varies from 10 cm (4 in.) to 50 cm (20 in.). Winters are cold to very cold and summers are warm to hot. Soils of the lowest elevations have salt accumulations that inhibit the growth of vascular plants. Around these playas grow saltbush species. Further up the slopes grow shrubs associated with the Mojave Desert, such as brittlebush and creosote bush. With higher elevation, the cover changes to one dominated by big sagebrush, black bush, and singleleaf pinyon and Utah juniper woodlands. The highest peaks support stands of Great Basin bristlecone pine and limber pine.

INTERMOUNTAIN SEMIDESERT PROVINCE

Northwestern Basin and Range Section This section lies to the east of the Modoc Plateau and extends into northwestern Nevada. Significant landmarks include the Surprise Valley, Madeline Plain, Skedaddle Mountains, and Honey Lake. Elevations range between 1,200 m (4,000 ft) and 2,400 m (8,000 ft). Annual precipitation varies from 10 cm (4 in.) to 50 cm (20 in.). Winters are cold and summers warm to hot. Shadscale and winter fat can be found in the basins. Big sagebrush, bitterbrush, and low sagebrush grow on higher slopes. The highest slopes and ridges support curlleaf mountain-mahogany and western juniper woodlands.

CONIFERS

KEY TO GENERA

1. Leaves are scalelike or awl-like 2
1. Leaves are needles or are linear 10
    2. Stems are ridged and jointed, resembling horsetails . . . . **ephedra** (*Ephedra*)*
    2. Stems are not ridged and jointed 3
3. Leaf sprays are flat 4
3. Leaf sprays are more or less round in cross section 7
    4. Scalelike leaves are much longer than broad and are 6 mm (.25 in.) to 25 mm (1 in.) long. The partially opened mature seed cones resemble duck bills . . . . **incense-cedar** (*Calocedrus decurrens*)
    4. Scalelike leaves are about as long as broad and are generally less than 6 mm (.25 in.) long. The cones do not resemble duck bills 5
5. Leaves are blue- or gray-green and have abundant, sticky resin . . . . **McNab cypress** (*Cupressus macnabiana*)
5. Leaves are green and have very little to no sticky resin . . . 6

* *Ephedra* belongs to the Gnetophyta and therefore is not a conifer. Like the conifers, *Ephedra* is a gymnosperm, and we include it here for convenience.

6. Leaves have stomatal bloom resembling a bow tie or butterfly underneath. The mature seed cones with their overlapping scales resemble rosebuds. These cones range from 10 mm (.4 in.) to 20 mm (.75 in.) long and are terminal and erect. Mature seed-cone scales are leathery .... **western redcedar (*Thuja plicata*)**
6. Leaves have either a white stomatal X pattern underneath or no stomatal bloom at all. Mature seed cones are more or less spherical and have central projections on peltate scales. The mature seed cones range from 6 mm (.25 in.) to 12 mm (.5 in.) in diameter and are pendant or erect. Mature seed cones are woody .... **whitecedar (*Chamaecyparis*)**
7. Seed cones are fleshy and berrylike .... **juniper (*Juniperus*)**
7. Seed cones are woody 8
8. Leaves are scalelike. Seed cones are more or less spherical. Immature cones resemble small soccer balls .... **cypress (*Cupressus*)**
8. Leaves are awl-like. Seed cones are barrel-like 9
9. Seed cones are 20 mm (.75 in.) to 25 mm (1 in.) long. Trees grow in the Coast Ranges .... **redwood (*Sequoia sempervirens*)**
9. Seed cones are 45 mm (1.75 in.) to 70 mm (2.75 in.) long. Trees grow in the Sierra Nevada .... **giant sequoia (*Sequoiadendron giganteum*)**
10. Leaves are needles and are set in bundles of 1 to 5 (usually 2, 3, or 5). Papery bundle sheaths are found at the bases of new leaves. Bundle sheaths may be persistent or deciduous on mature leaves .... **pine (*Pinus*)**
10. Leaves are linear, are attached singly, and lack sheaths at their bases 11
11. Leaf tips are sharp pointed to spiny and leaves are set in flat sprays. Seeds are surrounded by fleshy arils or are in barrel-like woody cones with peltate scales 12
11. Leaf tips are either sharp pointed or blunt (bristlecone fir has spiny leaf tips), and leaves are set in either flat or bottlebrush-like sprays. Seeds are in cones with overlapping scales. (This trait is most evident with green or wet cones.) 14
12. Leaves are 25 mm (1 in.) to 64 mm (2.5 in.) long and spine tipped. Seeds are not in cones but are surrounded by green to purple arils .... **California-nutmeg (*Torreya californica*)**

12. Leaves are less than 25 mm (1 in.) long and are sharp pointed but not spine tipped. Seeds are surrounded by red arils or are in barrel-like woody cones 13

13. Leaves have petioles and pale green stomatal bloom beneath. Seeds are not in cones but are surrounded by red, fleshy arils . . . . **Pacific yew** (***Taxus brevifolia***)

13. Leaves lack petioles and have whitish stomatal bloom beneath. Seed cones are barrel-like and have peltate scales . . . . **redwood** (***Sequoia sempervirens***)

14. Leaves arise from round and flat scars. Seed cones are erect and have deciduous scales, so cones are rarely found intact on the forest floor . . . . **fir** (***Abies***)

14. Leaves arise from pegs. Seed cones are pendant at maturity and have persistent scales, so cones fall as intact units 15

15. Leaves arise from minute pegs. Seed cones have 3-pronged bracts that are longer than the scales . . . . **Douglas-fir** (***Pseudotsuga***)

15. Leaves arise from prominent pegs. Seed cone scales are longer than the bracts 16

16. Pegs are more or less perpendicular to twigs. Leaf tips are sharp pointed or round. Seed-cone scales are thin and papery. Tops of trees are erect . . . . **spruce** (***Picea***)

16. Pegs are nearly parallel to twigs, with part of each peg fused with the twig. Leaf tips are round to blunt. Seed-cone scales are leathery. Tops of trees droop . . . . **hemlock** (***Tsuga***)

DESCRIPTIONS OF GENERA AND SPECIES

ABIES (FIR)

The genus *Abies* includes 43 species; 10 are native to North America and 7 of these are native to California. *Abies* trees are evergreen, and some attain heights of over 60 m (200 ft). Fir species growing at montane elevations have columnar stems and domelike crowns. Firs in the subalpine often have spirelike crowns and are considerably shorter than species growing at lower elevations. Several fir species develop a nearly prostrate, shrublike form when growing near the tree line.

Leaves are evergreen and linear and lack petioles. They are spirally arranged, although some species have flat sprays due to twisted leaf bases. Some species have evident stomatal bloom only on the undersides of their leaves, while other species have stomatal bloom on both surfaces. Erect seed cones are found on the highest branches, and droopy pollen cones in the mid to upper canopy. Mature seed cones have deciduous scales and bracts, and as a result intact cones are rarely found on the ground. Bracts are either longer or shorter than scales. Erect seed-cone axes resemble candles and persist into the following year after scale and seed dispersal. Twigs typically have round or egg-shaped, blunt, resinous buds. Leaf scars are round and flat. Younger trees have smooth, thin bark with abundant resin blisters. The bark of mature trees is slightly to deeply furrowed.

Firs typically grow in cool, moist environments. Several of California's firs are rare enough that you must hunt for them in order to see them. The bristlecone fir, for example, is found only in the Santa Lucia Range of Monterey and San Luis Obispo Counties. Silver fir and subalpine fir are restricted to California's Klamath Mountains of Siskiyou County but are extensive in western North American mountains outside the state. Other firs are more extensive within the state. *Abies* species are called true firs as opposed to the Douglas-firs of the genus *Pseudotsuga.*

1. Leaf tips are spiny. Seed cones have bristlelike, exserted bracts that are much longer than the scales. Trees grow only in the Santa Lucia Mountains . . . . . . **bristlecone fir** (*A. bracteata*)

1. Leaf tips are notched, round, or blunt. Seed cones lack bristlelike, exserted bracts. Trees do not grow in the Santa Lucia Mountains 2
   2. Leaves are more or less arranged in flat or V-shaped sprays. Leaves are twisted at their bases. Stomatal bloom can be on one or both sides 3
   2. Leaves are not arranged in flat sprays. Leaves are not twisted at their bases. Stomatal bloom can be found on both leaf surfaces 5
3. Stomatal bloom is on both surfaces. Leaf length is uniform and ranges between 4 cm (1.5 in.) and 7 cm (2.75 in.). Leaf tips are round or notched . . . . **white fir** (*A. concolor*)
3. Stomatal bloom is only on lower surfaces. Leaves vary in length but are less than 4 cm (1.5 in.) long. Leaf tips are usually notched 4
   4. Leaves are arranged in flat sprays and are usually 2.5 cm (1 in.) to 4 cm (1.5 in.) long. Trees grow at low elevations, including along some portions of the coast . . . . **grand fir** (*A. grandis*)
   4. Leaves appear to be arranged in flat sprays when viewed from below, but from above they appear to be clustered and they point toward the ends of twigs. Leaves are mostly less than 3 cm (1.25 in.) long. In California, trees grow only at high elevations in the Klamath Mountains . . . . **Pacific silver fir** (*A. amabilis*)
5. Leaves do not have bases that tend to parallel the twigs and they do not resemble hockey sticks. The most recent year's leaves have dull stomatal bloom on both surfaces. Seed cones are thin and less than 10 cm (4 in.) long. In California, trees are found only in the Klamath Mountains . . . . **subalpine fir** (*A. lasiocarpa*)
5. Leaves have bases that tend to parallel the twigs, making them resemble hockey sticks (this characteristic is prominent on upper branches or on branches with sun exposure). The most recent year's leaves have conspicuous stomatal bloom on both surfaces. Cones are thick and more than 10 cm (4 in.) long 6
   6. Cones lack exserted bracts. Leaves are ridged or flat but not grooved on their upper surfaces . . . . **red fir** (*A. magnifica* var. *magnifica*)
   6. Cones have exserted bracts. Most leaves are either ridged or grooved on their upper surfaces 7

Figure 1
Pacific silver fir,
Abies amabilis

7. Exserted bracts cover less than 50% of the cone surface. Bract tips are round and have a tail. Most leaves are ridged on their upper surfaces . . . . **Shasta red fir (*A. magnifica* var. *shastensis*)**
7. Exserted bracts cover more than 50% of the cone surface. Bract tips gradually taper to a point. Most leaves are grooved on their upper surfaces . . . . **noble fir (*A. procera*)**

PACIFIC SILVER FIR (Fig. 1) ***Abies amabilis***

DESCRIPTION: A tall, erect, single-stemmed tree. Mature trees are typically 24 m (80 ft) to 48 m (160 ft) tall and 30 cm (1 ft) to 120 cm (4 ft) in diameter. The largest grows in Forks, Washington, and is 62 m (203 ft) tall and 2.4 m (94 in.) in diameter. Younger trees have crowns similar to those of other true firs. Older ones often have spirelike or pyramidal crowns. Trees are long-lived, with the oldest surviving more than 400 years. **LEAVES** are linear and about 25 mm (1 in.) long. They appear to be arranged in flat sprays when viewed from below, but when viewed from above they appear to be clustered and tend to point toward the ends of twigs and diagonally upward. There is no stomatal bloom on the upper surfaces of leaves, but a silvery bloom is evident when viewed from below. The tips of the leaves are either notched or round. Leaves are grooved on their upper sides. **CONES** are 9 cm (3.5 in.) to 15 cm (6 in.) long and purple to green. Scales and bracts are deciduous. Bracts are shorter than the scales. **BARK** is thin and chalky white to ashy gray. Younger trees have many resin blisters.

HABITAT AND RANGE: Outside of California it grows in subalpine coniferous forests throughout the Cascades and coastal mountains of Oregon, Washington, British Columbia, and Alaska. In California it is rare and can be found growing only in the Marble and Siskiyou Mountains, between 1,700 m (5,600 ft) and 2,100 m (7,000 ft) elevation. The species normally grows in cool, moist habitats.

REMARKS: Silver fir is very tolerant of shade and susceptible to fire damage. It is initially slow growing and can grow in the understory for decades before it is released from competition. Silver fir can vegetatively regenerate new stems and roots when branches come into contact with the soil (this is known as layering). In the Pacific Northwest, silver fir is used as lumber, plywood, pulp, and Christmas trees and greens. Another common name for this species is lovely fir.

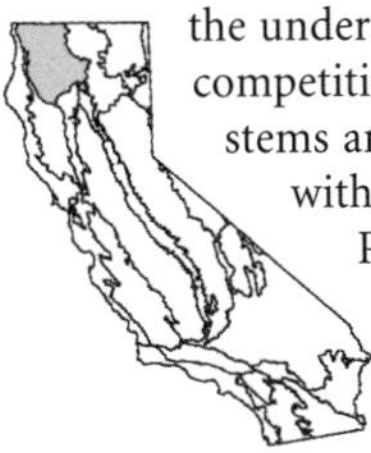

BRISTLECONE FIR (Fig. 2; Pl. 1) *Abies bracteata*

DESCRIPTION: A tall, erect, single-stemmed tree. Mature trees are typically 12 m (40 ft) to 30 m (100 ft) tall and 45 cm (18 in.) to 90 cm (36 in.) in diameter. The largest tree is 55 m (182 ft) tall and 1.3 m (52 in.) in diameter and grows in Los Padres National Forest. Species longevity is unknown. Trees have spirelike crowns with branches extending nearly to the ground. Smaller branches droop. **LEAVES** are linear, 3 cm (1.25 in.) to 5.5 cm (2.25 in.) long, and arranged in flat sprays. Individual leaves are flat and stiff, with sharply pointed tips. Stomatal bloom is found underneath. **CONES** are egg shaped and 6 cm (2.5 in.) to 10 cm (4 in.) long, with conspicuous exserted, bristle-tipped bracts. Bracts are 2.5 cm (1 in.) to 5 cm (2 in.) long. Scales and bracts are deciduous. **BUDS** are sharp pointed, tan, and 2 cm (.75 in.) to 2.5 cm (1 in.) long. **BARK** on mature trees is fissured.

HABITAT AND RANGE: The species is uncommon and grows

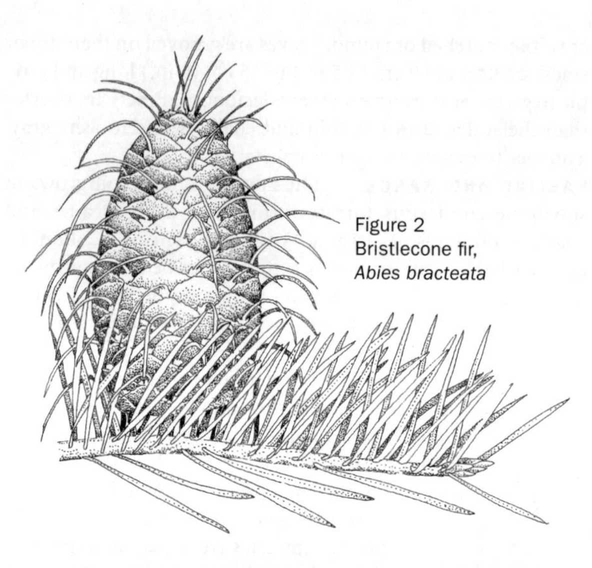

Figure 2
Bristlecone fir,
Abies bracteata

in coastal and montane forests in the Santa Lucia Mountains. Its most northerly stands are found near the headwaters of the Carmel and Little Sur Rivers, and the most southerly stands are found along Arroyo de la Cruz near San Simeon. It grows at elevations between 200 m (700 ft) and 1,600 m (5,200 ft), although it is usually found growing in steep, rocky, fire-resistant sites at elevations from 600 m (2,000 ft) to 1,500 m (5,000 ft).

REMARKS: Bristlecone fir is the rarest of North America's true firs but is locally common in its native range. The species is moderately shade tolerant and is easily killed by fire. Spanish missionaries made incense from its aromatic resin. Another common name for this species is Santa Lucia fir.

WHITE FIR (Fig. 3) ***Abies concolor***

DESCRIPTION: A tall, erect, single-stemmed tree. Mature trees are usually 40 m (130 ft) to 55 m (180 ft) tall and 1 m (40 in.) to 1.5 m (60 in.) in diameter. The largest is found in Yosemite

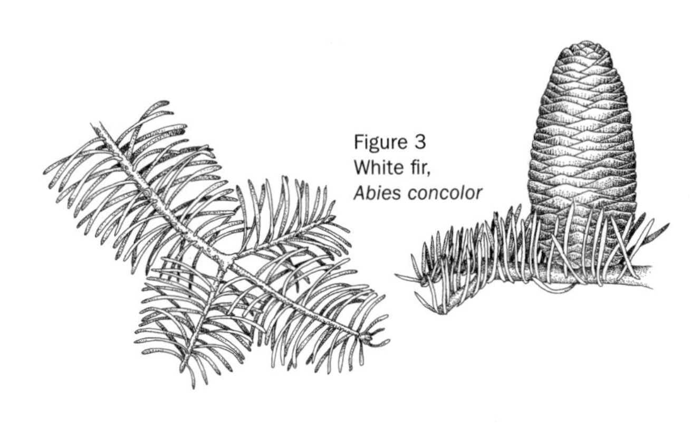
Figure 3 White fir, *Abies concolor*

National Park and is 66 m (217 ft) tall and 2.2 m (88 in.) in diameter. Young trees have elongated crowns, while older trees have rounded crowns with long, branchless stems. Trees can live more than 300 years. **LEAVES** are linear, 4 cm (1.5 in.) to 7 cm (2.75 in.) long, and arranged in either flat or V-shaped sprays. Tips are generally round. Leaf bases are twisted. Stomatal bloom occurs on both sides. **CONES** are barrel shaped, green turning olive brown to purple at maturity, and 7.5 cm (3 in.) to 12.5 cm (5 in.) long. Bracts are shorter than scales. Scales and bracts are deciduous. **BARK** on mature trees is gray and deeply fissured. The inner bark has "baconlike" alternating bands of reddish brown and pinkish yellow. Young trees have smooth, light gray bark with numerous resin blisters.

HABITAT AND RANGE: An important tree in montane coniferous forests in mountain ranges throughout California. It is often found growing at elevations between 900 m (3,000 ft) and 3,000 m (10,000 ft). Outside of California it grows in the Rocky Mountains. The lower elevation limit is roughly coincident with the lower limit of the winter snowpack.

REMARKS: White fir is renowned as being shade tolerant and somewhat fire resistant when mature. It has increased in abundance because of fire suppression. Old trees are often infected with a variety of rots and insects. Two varieties of white fir are recognized: *A. c.* var. *concolor* is found in the Rocky Mountains and the desert mountains of southern California, and *A. c.* var. *lowiana* is found in northern California and the Great Basin. *A. c.* var. *concolor* may occur in the eastern Klamath and Cascade Mountains as well. *A. c.* var. *lowiana* leaves have green upper surfaces with a few white lines of stomatal bloom. Leaves of *A. c.* var. *concolor,* in contrast, are more uniformly colored, with evident stomatal bloom on both surfaces. White fir readily hybridizes with grand fir in the western Klamath Mountains of California and Oregon.

GRAND FIR (Fig. 4) ***Abies grandis***

DESCRIPTION: A tall, erect, single-stemmed tree. Mature trees are typically 43 m (140 ft) to 48 m (160 ft) tall and 60 cm (24 in.) to 120 cm (48 in.) in diameter. The largest tree is 78 m (257 ft) tall and 2 m (78 in.) in diameter and grows in Redwood National Park. Trees can live more than 300 years. Mature trees have long, clear stems and domelike crowns. Tree stems are nearly cylindrical. Internodes are evident for many years on mature trees and are seen as more or less regularly spaced bands on the trunk. **LEAVES** are linear, 2.5 cm (1 in.) to 4 cm (1.5 in.) long, and arranged in flat sprays. They have stomatal bloom only underneath. Leaves vary in length and are flat, and upper surfaces are grooved. The apex is usually notched and leaf bases are twisted. **CONES** are barrel shaped, green, and 6 cm (2.5 in.) to 10 cm (4 in.) long. Bracts are shorter than scales. Scales and bracts are deciduous. **BARK** on mature trees is gray, with deep, irregular furrows. The inner bark is purplish red. Young trees have smooth, light gray bark with numerous resin blisters.

HABITAT AND RANGE: In California the species grows in coastal coniferous forests at elevations from sea level to 700 m (2,300 ft). In the Pacific Northwest, it is common at lower elevations throughout western Oregon and Washington, and at middle elevations in interior mountains. The southern limit is Willow Creek, just south of the Russian River. The most interior stand occurs on Asbill Creek in northeastern Mendocino County, at an elevation of 600 m (2,000 ft).

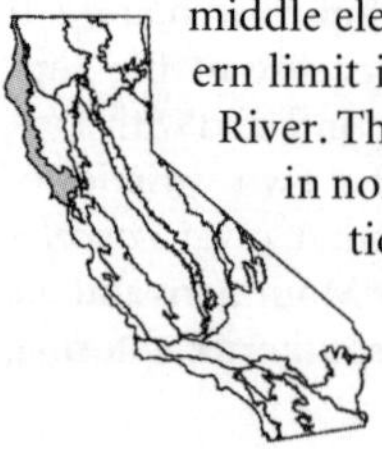

REMARKS: Grand fir is tolerant of shade and readily killed by fire. Like most true firs, the species is susceptible to a va-

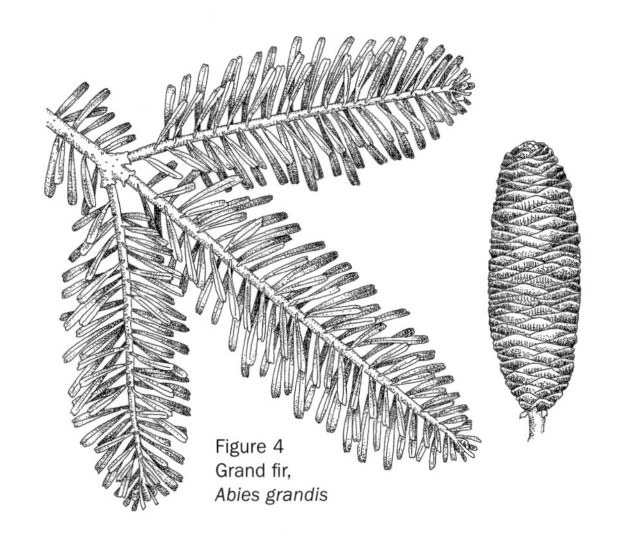
Figure 4
Grand fir,
Abies grandis

riety of heart and root rots. Grand fir readily hybridizes with white fir in the western Klamath Mountains of California and Oregon. It can be difficult to distinguish between the two species in those areas where they occur together. Fortunately, this is usually not a problem in California, since their ranges are relatively distinct. Grand fir grows at low elevations, and white fir generally grows in the snow zone. Another common name for this species is lowland white fir.

SUBALPINE FIR (Fig. 5) ***Abies lasiocarpa***

DESCRIPTION: An erect, single- or multistemmed, medium-sized tree. Mature trees are typically 18 m (60 ft) to 30 m (100 ft) tall and 45 cm (18 in.) to 60 cm (24 in.) in diameter. The largest grows in Olympic National Park and is 39 m (129 ft) tall and 2 m (80 in.) in diameter. Trees may live more than 250 years. The species has adapted to heavy winter snow conditions by having a conical or spirelike crown. When growing near the tree line, the trees often have a dwarfed shrub form. **LEAVES** are linear and 2.5 cm (1 in.) to 4 cm (1.5 in.) long. They have sprays arranged bottlebrush style. Most leaves, however, are clustered on the upper sides of twigs. Stomatal bloom can be found on both surfaces. Leaves are mostly flat, with tips that are either

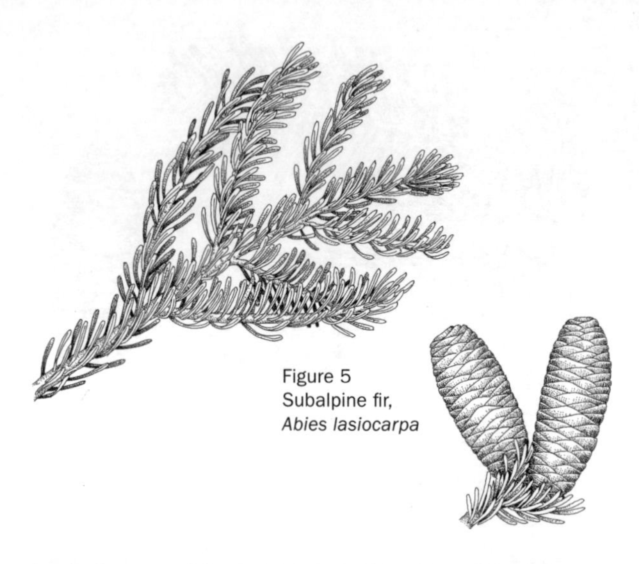

Figure 5
Subalpine fir,
Abies lasiocarpa

notched or round. **CONES** are cylindrical, 6 cm (2.5 in.) to 10 cm (4 in.) long, and deep purple. Bracts are shorter than scales. Scales and bracts are deciduous. **BARK** on mature trees is gray and has numerous resin blisters.

HABITAT AND RANGE: Grows in high-elevation coniferous forests from California to Alaska. In California it is rare and grows in isolated stands in the Russian Wilderness Area, Marble Mountains, Trinity Alps, and Siskiyou Mountains, at elevations from 1,700 m (5,500 ft) to 2,100 m (7,000 ft).

REMARKS: Subalpine fir is treated as 1 or 2 species. As 1 species, it has 2 varieties: *A. l.* var. *lasiocarpa* and *A. l.* var. *arizonica* (corkbark fir). When considered as 2 species, subalpine fir is split into *A. lasiocarpa* and *A. bifolia*. *A. lasiocarpa* is found in California and the Cascades and *A. bifolia* is found in the Rocky Mountains. Populations in the southern Rocky Mountains have thick, corky bark. Subalpine fir is shade tolerant and easily killed by fire. It often reproduces vegetatively by layering, and a mother tree can develop "skirts" of saplings around it. In the Rocky Mountains the species is used as lumber, but on the Pacific Slope most subalpine fir trees are in parks or wilderness areas. Several horticultural and ornamental types have been recognized.

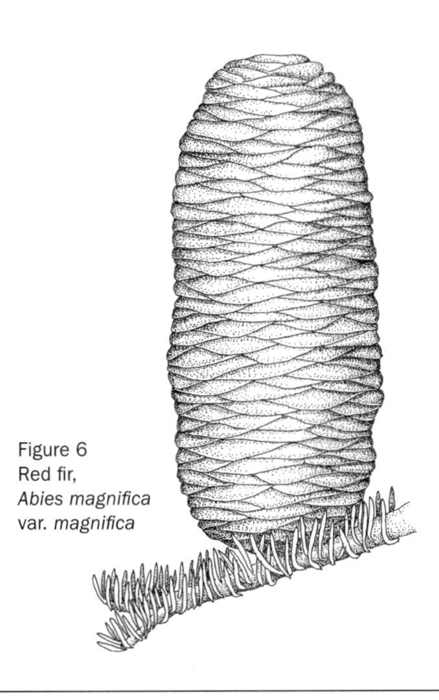

Figure 6
Red fir,
Abies magnifica
var. *magnifica*

RED FIR (Fig. 6) ***Abies magnifica* var. *magnifica***

DESCRIPTION: A tall, erect, single-stemmed tree. It is one of the largest true firs in the world, achieving typical heights of 45 m (150 ft) to 55 m (180 ft) and diameters of 1.2 m (4 ft) to 1.5 m (5 ft). The largest specimen grows in Sierra National Forest and is 55 m (180 ft) tall and 2.6 m (102 in.) in diameter. Trees are relatively long-lived, with individuals exceeding 300 years in age. Young trees have elongated crowns and older trees have domelike crowns. **LEAVES** are linear and 2 cm (.75 in.) to 4 cm (1.5 in.) long. The leaves have bases that tend to parallel the twigs, making them resemble hockey sticks, and are clustered on upper sides of twigs. Stomatal bloom occurs on both sides of leaves. Leaf tips are generally round. Leaves are usually ridged on upper surfaces, and on higher branches are 4-angled in cross section. **CONES** are barrel shaped and 15 cm (6 in.) to 23 cm (9 in.) long. Bracts are shorter than scales, are dark purple to purplish brown, and have a tail-like appendage. Scales and bracts are deciduous. **BARK** on mature trees is reddish within, and outside it is reddish brown and furrowed. Young trees have smooth gray bark with resin blisters.

HABITAT AND RANGE: Grows in montane and subalpine coniferous forests throughout the Sierra Nevada, southern Cascades, Klamath Mountains, and northern Coast Ranges, at elevations between 1,200 m (4,000 ft) and 2,800 m (9,200 ft). Cool, moist conditions characterize its habitat.

REMARKS: Red fir is moderately shade tolerant and is susceptible to a variety of rots and insects. The species is used as lumber, plywood, and Christmas trees (sold as "silvertip fir") and greens. Considerable confusion revolves around the taxonomic status of red fir, Shasta red fir, and noble fir. Traditionally, the length and shape of the cone bracts have been used to distinguish these taxa, but hybrids between noble fir and red fir are distinguished by chemical analyses. Exserted bracts on red fir cones in the southern Sierra Nevada cloud the issue even more.

SHASTA RED FIR (Fig. 7) ***Abies magnifica* var. *shastensis***

DESCRIPTION: A tall, erect, single-stemmed tree. It achieves a height and diameter similar to that of *A. m.* var. *magnifica*. The largest Shasta red fir is in Rogue River National Forest in Oregon and is 70 m (228 ft) tall and 2 m (78 in.) in diameter. The species can live more than 300 years. It is indistinguishable in shape from *A. m.* var. *magnifica*. **LEAVES** are linear and 2 cm (.75 in.) to 4 cm (1.5 in.) long. The leaves have bases that tend to parallel the twigs, making them resemble hockey sticks, and are clustered on upper sides of twigs. Stomatal bloom occurs on both sides of leaves. The leaf tips are generally round. Leaves are usually ridged on upper surfaces, and on higher branches they are 4-angled in cross section. **CONES** are barrel shaped and 15 cm (6 in.) to 23 cm (9 in.) long. Bracts are longer than the scales and cover about 25% to 50% of the cone surface. Scales and bracts are deciduous. **BARK** is reddish brown to purplish black on the outside; the inner bark is reddish brown.

HABITAT AND RANGE: Grows in the northern range of *A. magnifica* and in the southern Sierra Nevada. It is found north of Mount Lassen into Oregon, and south of the Klamath River

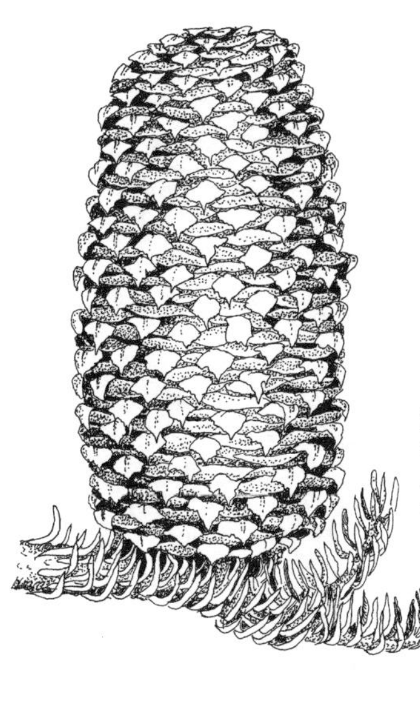

Figure 7
Shasta red fir,
Abies magnifica
var. *shastensis*

in the northern Coast Ranges. In the Klamath Mountains it can be confused with noble fir. Noble fir, however, is considered to grow north and west of the Klamath River in the Siskiyou Mountains and on South Fork Mountain.

REMARKS: The cone bracts of Shasta red fir and noble fir differ in shape. Noble fir's bracts taper to a point and have spinelike "tails" about as long as the exposed bracts. Shasta red fir's bract tips are blunt and have "tails" shorter than the bracts. Shasta red fir hybridizes with noble fir. Like red fir, it is used as lumber, plywood, Christmas trees (sold as "silvertip fir") and greens. Considerable confusion revolves around the taxonomic status of red fir, Shasta red fir, and noble fir (see remarks under red fir).

NOBLE FIR (Fig. 8) ***Abies procera***

DESCRIPTION: A tall, erect, single-stemmed tree, it is the largest of the world's firs. On productive growing sites mature trees are 43 m (140 ft) to 67 m (220 ft) tall and 90 cm (36 in.) to 150 cm (60 in.) in diameter. The largest grows in Mount Saint

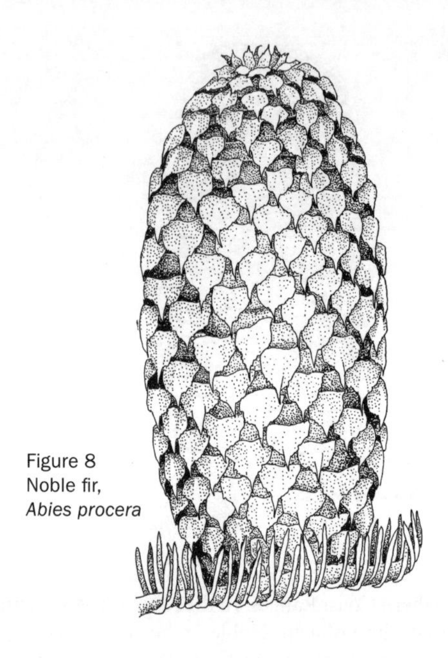

Figure 8
Noble fir,
Abies procera

Helens National Volcanic Monument in Washington and is 83 m (272 ft) feet tall and 2.4 m (95 in.) in diameter. Mature trees have long, columnar trunks with short, rounded crowns. Trees can exceed 400 years in age. **LEAVES** are linear and about 2.5 cm (1 in.) to 4 cm (1.5 in.) long. As with red fir, noble fir leaves tend to parallel the twigs, making them resemble hockey sticks, and are clustered on upper sides of twigs. Noble fir leaves have brilliant, silvery stomatal bloom on both surfaces of the most recent year's growth. Older leaves have faded stomatal bloom. The tip is generally round and the leaves are generally grooved on their upper surfaces. Leaves are 4-angled in cross section. **CONES** are barrel shaped and 10 cm (4 in.) to 15 cm (6 in.) long. Straw-colored bracts are exserted, nearly covering the scales. Bract tips are spinelike and point down. Cones are olive brown to purple. Scales and bracts are deciduous. **BARK** on mature trees is purplish within, and outside it is grayish purple and furrowed. Young trees have smooth gray bark with resin blisters.

HABITAT AND RANGE: Grows in upper montane to subalpine coniferous forests, at elevations from 1,500 m (5,000 ft) to 2,000 m (6,500 ft). In California, intermediate forms of noble fir and Shasta red fir (*A. magnifica* var. *shastensis*) are found in the Klamath Mountains and northern Coast Ranges (see Shasta red fir).

REMARKS: Unlike other true firs, noble fir is intolerant of shade. Young trees are easily killed by fire, but the thick bark of old trees makes them somewhat fire resistant. Noble fir is used as lumber, as Christmas trees and greens, and as an ornamental. Considerable confusion revolves around the taxonomic status of red fir, Shasta red fir, and noble fir (see remarks under red fir).

CALOCEDRUS (INCENSE-CEDAR)

The genus *Calocedrus* has 3 species, which grow in eastern Asia and western North America. One of these species is native to California.

INCENSE-CEDAR (Fig. 9) ***Calocedrus decurrens***

DESCRIPTION: A tall, erect, single-stemmed tree with long, branchless trunks and rounded, full crowns. Old-growth trees often have fluted bases. Mature trees are typically 24 m (80 ft) to 46 m (150 ft) tall and 90 cm (3 ft) to 180 cm (6 ft) in diameter. The largest grows in the Marble Mountains Wilderness and is 46 m (152 ft) tall and 3.6 m (12 ft) in diameter. Trees can live to be over 500 years old. **LEAVES** are scalelike, blunt tipped, yellowish green, and 6 mm (.25 in.) to 20 mm (.75 in.) long. They are set in opposite, alternating pairs. Lateral scales overlap the facial scales, circumscribing a "pilsner beer glass" or "wine glass" outline. Foliage sprays are flat. **CONES** are leathery, pendant, 20 mm (.75 in.) to 31 mm (1.25 in.) long, and cinnamon brown. Cones have 3 pairs of fused scales. When green, cones resemble duck bills. On mature cones the 2 outside pairs of fused scales curve outward and the central fused pair is straight. **BARK** of mature trees can be up to 15 cm (6 in.) thick and is fibrous, furrowed, and cinnamon brown.

HABITAT AND RANGE: Grows in montane coniferous forests and woodlands. Its range extends from northern Oregon into Baja California. In California it is commonly found growing in mixed coniferous forests and on serpentine soils, at ele-

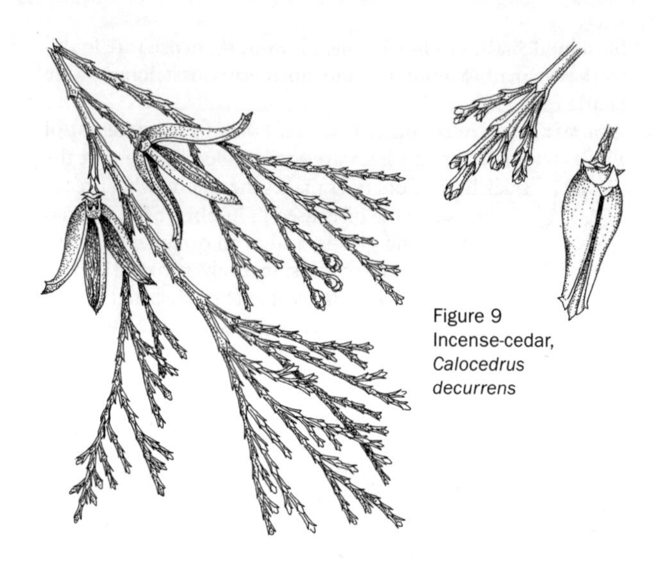

Figure 9
Incense-cedar,
Calocedrus decurrens

vations from 45 m (150 ft) to 3,000 m (9,700 ft). The species competes well on hot, dry sites as well as wet sites.

REMARKS: Incense-cedar is more shade tolerant than most of its potential canopy competitors. It is easily killed or scarred by fire because of its thin, flammable bark. Incense-cedar regenerates well in the understory and has been increasing in abundance in many mixed conifer forests since fire suppression began. As old-growth pine and Douglas-fir trees die, incense-cedar may become a prominent feature in many of California's mixed coniferous forests. Incense-cedar wood is decay resistant and is good for exterior use. Most wooden pencils are made of incense-cedar wood because it can be sharpened at any angle without splintering.

CHAMAECYPARIS (WHITECEDAR)

The genus *Chamaecyparis* has 7 species, all found in the Northern Hemisphere: 4 are native to China, Japan, or Taiwan; 1 is native to the eastern United States; and 2 are native to California. Whitecedars are evergreen, tall, and long-lived. Crowns are usually conical, with droopy branches and leaders. Old-growth trees have swelled, fluted butts. Whitecedars generally grow in cool, moist environments dominated by conifers.

Leaves are sharp-pointed evergreen scales that occur in opposite, alternating pairs. Lateral scales are shaped like the keel of a boat, and facial scales are flat. There may be evident stomatal bloom on the leaves. The leaves form flat sprays that fall as intact units. After 2 or 3 years the leaves turn yellow or brown. Cones are erect, spherical, leathery when green, and woody once the scales have opened. They resemble little soccer balls prior to opening. Each cone has 6 to 8 umbrella-like scales with central prickles. Cones most commonly mature in 1 season, although those of *C. nootkatensis* mature in 2 seasons. Pollen cones and seed cones can be found at the ends of twigs, but not on the same branch. Cone scales have 1 to 5 seeds. Twigs are markedly flat, becoming 4-angled when older. Whitecedar bark can be either thin and scaly or thick and deeply furrowed.

Chamaecyparis species are highly prized as ornamentals. For example, over 250 cultivars of *C. lawsoniana* have been developed in less than 100 years. These cultivars are noted for exceptional diversity in growth habit and foliage color. The wood is highly regarded for its rot resistance and durability. Woodworkers cherish the white wood, fine grain, and workability of *Chamaecyparis*.

The Jepson Manual recognizes *Chamaecyparis* as being in the genus *Cupressus*.

1. Stomatal bloom on the undersides of leaves is in an X pattern. Cone-scale prickles are minute to small . . . . **Port Orford–cedar** (***C. lawsoniana***)
1. There is no stomatal bloom on the leaves. Cone-scale prickles are large and prominent . . . . **Alaska yellow-cedar** (***C. nootkatensis***)

PORT ORFORD–CEDAR (Fig. 10) — *Chamaecyparis lawsoniana*

DESCRIPTION: A tall, erect, single-stemmed tree. Mature trees are typically 38 m (125 ft) to 55 m (180 ft) tall and 90 cm (3 ft) to 180 cm (6 ft) in diameter. The largest grows in Siskiyou National Forest and is 67 m (219 ft) tall and 3.6 m (12 ft) in diameter. Trees can live to be over 500 years old. Mature trees commonly have swollen butts and long, clear stems leading to

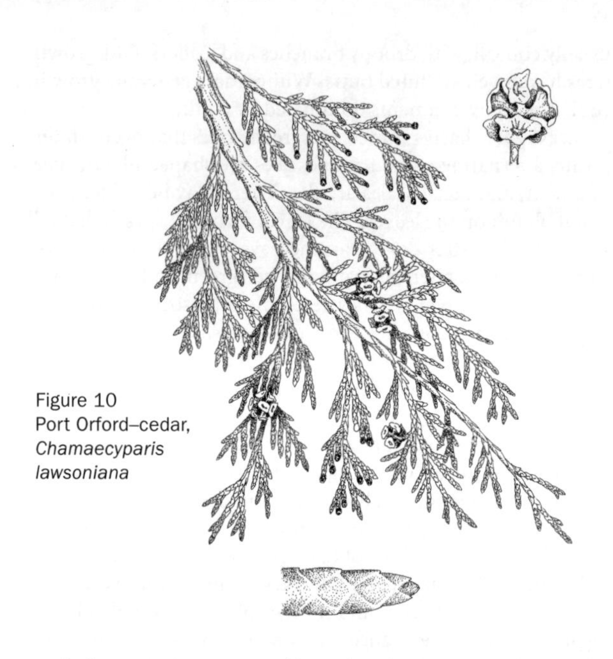

Figure 10
Port Orford–cedar, *Chamaecyparis lawsoniana*

conical crowns. Treetops and branches droop. **LEAVES** are scale-like, flat, somewhat blunt tipped, yellowish green to bluish green, and about 3 mm (.12 in.) long. Conspicuous glands can be found on facial scales. Stomatal bloom appears in an X pattern on the undersides of lacy, flat foliage sprays. Leaves are set in opposite, alternating pairs. **CONES** are woody, spherical, about 6 mm (.25 in.) in diameter, and green, with a whitish bloom before the cone scales open. Immature cones resemble little soccer balls. Mature cones are brown and have small prickles on peltate scales. Cones mature in 1 season. **BARK** on mature trees is up to 25 cm (10 in.) thick, grayish brown, fibrous, and deeply furrowed.

HABITAT AND RANGE: The species is uncommon and grows in coastal and montane coniferous forests in southwestern Oregon and northwestern California, at elevations from sea level to 1,500 m (5,000 ft). Coastal stands are often found on moist, lower slopes, while inland populations are more likely to occur on serpentine soils. Disjunct populations lie along the upper Trinity and Sacramento Rivers.

REMARKS: Port Orford–cedar is shade tolerant. Owing to its thick bark, it is fire-resistant. Over 250 cultivars have been developed in the last 100 years, attesting to Port Orford–cedar's great morphological variation. Boat builders and woodworkers prize its wood. Casket makers in Asia use Port Orford–cedar wood because of its pungent, gingerlike aroma and its resistance to rotting. The species' greatest threat comes from root rot (*Phytophthora lateralis*). Fungal spores transported by water can affect trees downstream. Great care must be taken so spores are not inadvertently dispersed by infected mud on vehicles or boots.

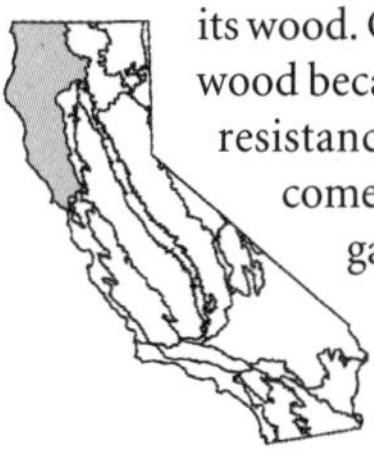

ALASKA YELLOW-CEDAR (Fig. 11) — *Chamaecyparis nootkatensis*

DESCRIPTION: A medium-sized, erect, single-stemmed tree. Shrub forms occur near the tree line in the Northwest and at lower elevations in California. Mature trees are typically 20 m (65 ft) to 30 m (100 ft) tall and 60 cm (2 ft) to 120 cm (4 ft) in diameter. The largest grows in Olympic National Park and is 38 m (124 ft) tall and 3.6 m (12 ft) in diameter. Trees can live to be over 1,000 years old, and one tree is reputed to be 3,500 years old. The species is renowned for droopy branches with loosely hanging, flat leaf sprays. Tree trunks have buttressed bases and conical crowns. **LEAVES** are scalelike, somewhat sharp pointed, bluish green when young and yellowish after 2 or 3 years. They measure about 3 mm (.12 in.) long. Leaves occur in opposite, alternating pairs and have no evident stomatal bloom. Foliage sprays are flat and lacy. **CONES** are woody, spherical, 6 mm (.25 in.) to 12 mm (.5 in.) in diameter, and green, with a whitish bloom before the cone scales open. The immature cones resemble little soccer balls. Mature cones are brown and have peltate scales. Cone scales have stout prickles. **BARK** is thin and grayish brown, with shallow furrows.

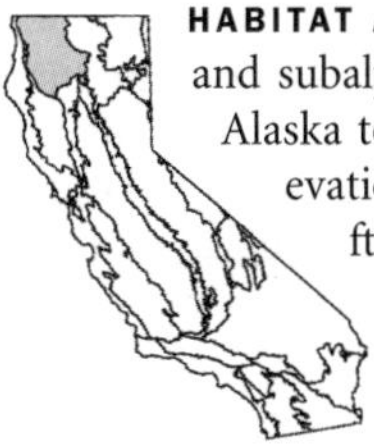

HABITAT AND RANGE: Grows in coastal, montane, and subalpine coniferous forests from south-central Alaska to northwestern California. It ranges in elevation from sea level in Alaska to 2,300 m (7,500 ft) in the Cascades. In California the species is rare and grows in disjunct stands in the Siskiyou Mountains at elevations of about 1,500 m (5,000 ft). It grows best in cold,

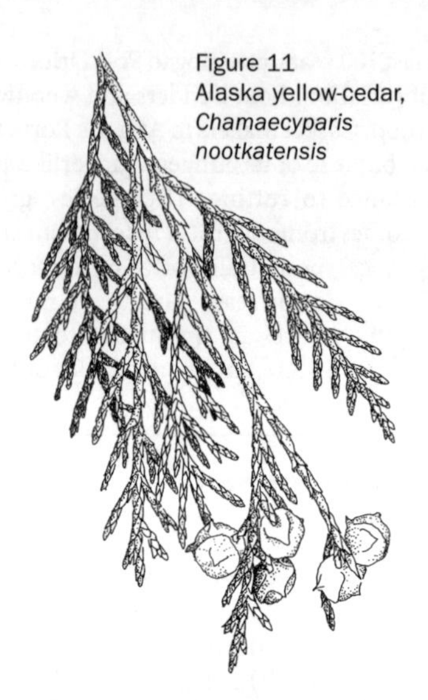

Figure 11
Alaska yellow-cedar, *Chamaecyparis nootkatensis*

wet habitats that receive abundant winter snow. It can be found growing on rocks with no apparent soil and on better-drained, moist soils.

REMARKS: Alaska yellow-cedar is shade tolerant and very susceptible to fire. Maintaining groves of Alaska yellow-cedar in California will take careful fire management. Alaska yellow-cedar can reproduce vegetatively by layering under snowpacks. Its wood is rot resistant and easily worked, and it is used in boats, saunas, and moldings. Another common name for this species is Alaska-cedar.

CUPRESSUS (CYPRESS)

The genus *Cupressus* has about 20 species; of the 15 found in North America, 8 are native to California. Species found outside North America occur in the Mediterranean region and Asia. Cypresses are small- to medium-sized evergreen trees and shrubs. Many species grow no taller than 12 m (40 ft), while others can reach 30 m (100 ft). Six of the 8 California species are classified as endangered, rare, or uncommon because of their restricted ranges.

Leaves are evergreen scales, about 3 mm (.12 in.) long, and 3-

or 4-angled in cross section. Some species have leaves that are pitted with resinous glands. Leaves occur in opposite, alternating pairs. Foliage sprigs are 3- or 4-angled in cross section. Mature cones are woody and spherical and have central prickles on their peltate scales. Unopened cones resemble soccer balls. Cones mature in 2 seasons. Pollen cones and seed cones are found on separate branches on the same tree. Seed cones are usually clustered at the branch tips, and they persist for many years. Most species have cone scales that do not open readily at maturity. Cone scales generally have from 6 to many seeds. The bark is generally of 2 kinds: exfoliating, which peels away to reveal a reddish brown or cherry red inner bark; or fibrous and grayish brown to brown.

In California, cypresses generally occur on soils with significant fertility problems. Several species grow on serpentine soils and one grows on coastal podzolized soils. They grow in chaparrals, chaparrals intermixed with oak woodlands, and coniferous forests and woodlands. Their distributions are highly localized and you must seek them out to find them. McNab cypress and Sargent cypress are the more common species.

Cypresses are widely used as ornamentals and for windbreaks. Monterey cypress is used as timber in the Southern Hemisphere.

1. Leaves form flat spray . . . . **McNab cypress** (*C. macnabiana*)
1. Leaves form round or 3-dimensional sprays 2
   2. Bark exfoliates to reveal reddish brown to cherry red inner bark 3
   2. Bark is fibrous and gray, grayish brown, or brown . . . 5
3. New leaves are light green and do not have pits with resinous glands . . . . **Tecate cypress** (*C. forbesii*)
3. New leaves are grayish green and have conspicuous pits with resinous glands 4
   4. Cones have conspicuous pointed projections. Trees grow in northern California . . . . **Baker cypress** (*C. bakeri*)
   4. Cones have inconspicuous pointed projections. Trees grow in central and southern California . . . . **Arizona cypress** (*C. arizonica*)
5. Leaves are grayish green and have some pits with resinous glands. Trees grow on serpentine soils . . . . **Sargent cypress** (*C. sargentii*)
5. Leaves are light green to dark green and have no pits with resinous glands. Trees do not grow on serpentine soils . . . 6

6. Cones are 2.5 cm (1 in.) to 4 cm (1.5 in.) long. Leaf sprays are thick and coarse. Mature trees are 20 m (65 ft) to 24 m (80 ft) tall .... **Monterey cypress** (*C. macrocarpa*)
6. Cones are mostly less than 2.5 cm (1 in.) long. Leaf sprays are slender and delicate. Mature trees in their native habitat are less than 7.6 m (25 ft) tall 7

7. Cones are usually more than 16 mm (.62 in.) long. Trees grow in the Santa Cruz Mountains .... **Santa Cruz cypress** (*C. abramsiana*)
7. Cones are usually less than 16 mm (.62 in.) long. Trees grow near the coast on the Monterey Peninsula and near Fort Bragg .... **Gowen cypress** (*C. goveniana*)

SANTA CRUZ CYPRESS *Cupressus abramsiana*

DESCRIPTION: A short, erect, evergreen tree with a symmetrical, pyramidal crown. Open-grown trees have branches reaching nearly to the ground. Trees grow to be about 9 m (30 ft) tall. **LEAVES** are scalelike, 3- or 4-angled in cross section, less than 3 mm (.12 in.) long, and light bright green. Leaves occur in opposite, alternating pairs and lack pits with resinous glands. Leaf sprays are slender and delicate. **CONES** are woody, brown, serotinous, spherical, and 16 mm (.62 in.) to 31 mm (1.25 in.) long. Scales are umbrella-like, with small humplike projections on their centers. **BARK** is gray, thin, fibrous, and broken into plates.

HABITAT AND RANGE: Grows in chaparrals and coniferous forests. It is classified as an endangered species and is restricted to 4 populations in the Santa Cruz Mountains: Bonny Doon, Eagle Rock, Butano Ridge, and Brachenbrae groves. Populations occur on sandy, infertile granitic soils, at elevations from 480 m (1,600 ft) to 760 m (2,500 ft).

REMARKS: Some taxonomists have included *C. abramsiana* in the species *C. goveniana*. Others have considered it to be a form of *C. sargentii* that grows on nonserpentine soils.

ARIZONA CYPRESS *Cupressus arizonica*

DESCRIPTION: A small- to medium-sized, erect to spreading, evergreen tree. Crowns can be slender or flat. Mature trees are typically 6 m (20 ft) to 15 m (50 ft) tall. **LEAVES** are scale-like, 3- or 4-angled in cross section, about 3 mm (.12 in.) long, and grayish green. They occur in opposite, alternating pairs and have pits with resinous glands. **CONES** are woody, spherical, serotinous, grayish brown, and 10 mm (.38 in.) to 35 mm (1.4 in.) long. Scales have inconspicuous projections and appear warty. **BARK** exfoliates, revealing smooth, cherry red inner bark. Older bark becomes brown to gray.

HABITAT AND RANGE: Grows in chaparrals, oak woodlands, and pinyon-juniper woodlands. It is a wide-ranging species in Arizona, New Mexico, and Mexico. In California it ranges in elevation from 760 m (2,500 ft) to 1,800 m (6,000 ft).

REMARKS: Two rare subspecies are recognized in California: *C. a.* ssp. *arizonica* (Cuyamaca cypress) and *C. a.* ssp. *nevadensis* (Piute cypress). The Cuyamaca cypress, with warty cones, is found in and near Cuyamaca State Park in San Diego County. The Piute cypress, with cones lacking warts, grows in the Sierra Nevada along the Kern River. Arizona cypress is fire adapted but can be extirpated by frequent fires.

BAKER CYPRESS *Cupressus bakeri*

DESCRIPTION: A small- to medium-sized, erect, evergreen tree with a symmetrical, narrow, pyramidal crown. Open-grown trees have branches reaching nearly to the ground. Trees typically grow to be about 12 m (40 ft) tall in Sierra Nevada–Cascades populations and up to 27 m (90 ft) tall in the Siskiyou Mountains. **LEAVES** are scalelike, 3- or 4-angled in cross section, less than 3 mm (.12 in.) long, and grayish green. They occur in opposite, alternating pairs and have pits with resinous glands. Leaf sprays are graceful. **CONES** are woody, serotinous, gray to silvery, spherical, and about 12 mm (.5 in.) to 25 mm (1 in.) long. Scales are somewhat warty, are umbrella-like, and have prominent pointed projections on their midpoints. **BARK** exfoliates, revealing reddish brown to cherry red inner bark. Older bark becomes gray and breaks into plates.

HABITAT AND RANGE: Grows in montane coniferous woodlands and mixed forests. Siskiyou Mountains populations often grow on serpentine soils, and Sierra Nevada–Cascades populations often grow on dry volcanic soils or lava beds. The species ranges in elevation from 1,060 m (3,500 ft) to 2,100 m (6,900 ft).

REMARKS: Baker cypress is considered uncommon and is a fire-dependent species, although the thin-barked trees most often die when exposed to fire. Fire helps to break cone serotiny, permits seeds to be dispersed, and prepares the seedbed.

TECATE CYPRESS *Cupressus forbesii*

DESCRIPTION: A short, shrubby, multistemmed, evergreen tree. Crowns are spreading and irregular. Branches are ascending on young trees and are more horizontal on older individuals. Trees are typically less than 9 m (30 ft) tall and may only reach 90 cm (3 ft) to 180 cm (6 ft) in high-density stands. **LEAVES** are scalelike, 3- or 4-angled in cross section, about 3 mm (.12 in.) long, and light green to dull green. They occur in opposite, alternating pairs and do not have conspicuous pits with resinous glands. **CONES** are woody, spherical, serotinous, brown to gray, and 2 cm (.75 in.) to 3 cm (1.25 in.) long. Scales have inconspicuous projections. **BARK** exfoliates, revealing a mottled inner bark that can be cherry red, reddish brown, or light green. The mottled bark is reminiscent of sycamore or eucalyptus bark. Older bark becomes brown.

HABITAT AND RANGE: This rare cypress grows in chaparrals from the Santa Ana Mountains in Orange County south into Baja California. The most extensive California population is on Otay Mountain. In California it ranges in elevation from 300 m (1,000 ft) to 2,500 m (8,000 ft).

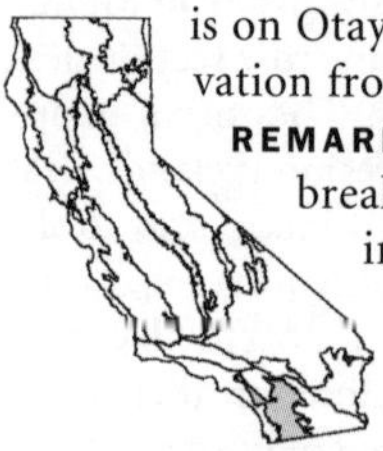

REMARKS: Tecate cypresses are used as windbreaks and hedges in hot, dry interior locations in southern California. The species is fire adapted but can be extirpated by frequent fires. It is also threatened by development.

GOWEN CYPRESS *Cupressus goveniana*

DESCRIPTION: Most commonly a short, single-stemmed, shrubby, evergreen tree. Crowns are slender and branches scarce. Trees near Monterey grow to be between 6 m (20 ft) and 7.5 m (25 ft) tall. Most trees near Fort Bragg are 1 m (3.2 ft) to 2 m (6.5 ft) tall on barren soils and up to 46 m (150 ft) tall on fertile soils. **LEAVES** are scalelike, 3- or 4-angled in cross section, less than 3 mm (.12 in.) long, and light yellow-green (near Monterey) to dark green (near Fort Bragg). They occur in opposite, alternating pairs and lack pits with resinous glands. Leaf sprays are slender and delicate. **CONES** are woody, brown to gray, serotinous, spherical, and 10 mm (.38 in.) to 25 mm (1 in.) long. Scales are umbrella-like, with inconspicuous projections on their centers. **BARK** is brown to gray and fibrous.

HABITAT AND RANGE: Grows in coastal coniferous forests. Two subspecies are recognized: the light yellow-green form, *C. g.* ssp. *goveniana* (Gowen cypress), and the dark green form, *C. g.* ssp. *pigmaea* (pygmy cypress). The Gowen cypress is found in 2 populations near Monterey, and the pygmy cypress occurs in Mendocino and Sonoma Counties. Both are found at elevations of less than 150 m (500 ft).

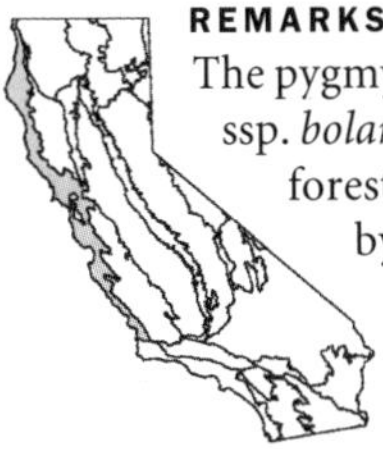

REMARKS: Both subspecies are classified as rare. The pygmy cypress is often found with *Pinus contorta* ssp. *bolanderi* (Bolander pine) and grows in "pygmy forests" on highly leached, white soils underlain by hardpan. The Gowen cypress grows with *P. muricata* (Bishop pine) and a variety of endemic shrubs on the Monterey Peninsula.

MCNAB CYPRESS (Fig. 12) ***Cupressus macnabiana***

DESCRIPTION: A small, shrublike tree with a bushy, spreading crown and multiple stems. Often, its crown is broader than tall. Stems are very short. Mature plants typically range in height from 2 m (6.5 ft) to 9 m (30 ft). The largest grows in Amador County and is 16.7 m (55 ft) tall and 124 cm (49 in.) in diameter. **LEAVES** are scalelike, 3- or 4-angled in cross section, less than 3 mm (.12 in.) long (up to 10 mm [.38 in.] on fast-growing sprays), and grayish green. They are aromatic and resinous, occur in opposite, alternating pairs, and have conspicuous pits with resinous glands. Unlike other California cypresses, McNab cypress leaves form flat sprays. **CONES** are woody, spherical, strongly serotinous, brown to gray, and 16 mm

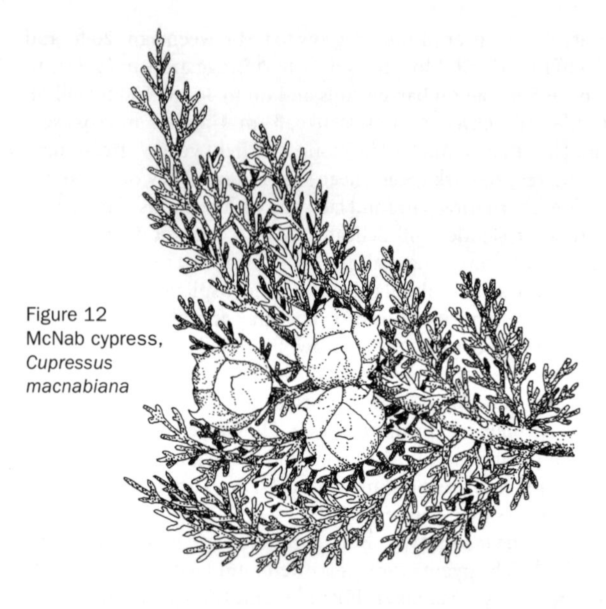

Figure 12
McNab cypress,
Cupressus macnabiana

(.62 in.) to 25 mm (1 in.) long. Scales are peltate and have conspicuous pointed projections on their centers. **BARK** is gray, fibrous, and furrowed. Bark on old trees has ridges separated by deep, diamond-shaped furrows.

HABITAT AND RANGE: More widely distributed than any of the other California cypresses. It grows in chaparrals, oak woodlands, and coniferous woodlands along the inner northern Coast Ranges and in the foothills of the northern Sierra Nevada, at elevations from 275 m (900 ft) to 840 m (2,750 ft). The species lives on serpentine and other dry, infertile soils. Its range overlaps that of Sargent cypress, and when they occur together McNab cypress is often found on the more exposed sites.

REMARKS: McNab cypress is shade intolerant and fire dependent. Its cones are strongly serotinous, and regeneration is in large part a function of renewing fires. Ranchers and farmers use McNab cypress for fence posts and firewood, and it serves to protect watersheds from erosion and landslides.

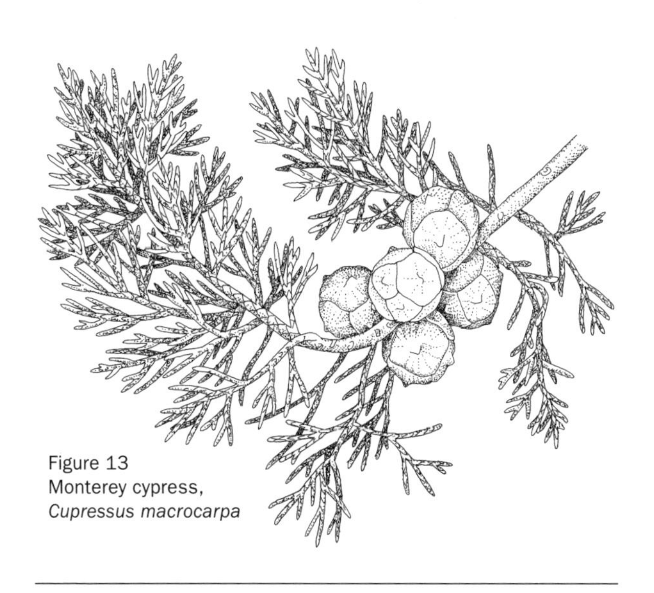

Figure 13
Monterey cypress,
Cupressus macrocarpa

MONTEREY CYPRESS *Cupressus macrocarpa*

(Fig. 13)

DESCRIPTION: A small- to medium-sized, erect, single-stemmed, evergreen tree. Trees growing next to the ocean often have twisted, contorted forms with broad, spreading crowns. Mature trees usually range in height from 20 m (65 ft) to 24 m (80 ft). The largest is in Pescadero and is 32 m (106 ft) tall and 4 m (13 ft) in diameter. Crowns are rounded and have large, irregularly spaced branches. **LEAVES** are scalelike, 3- or 4-angled in cross section, less than 3 mm (.12 in.) long (up to 10 mm [.38 in.] on fast-growing sprays), and bright green. Leaves occur in opposite, alternating pairs and lack conspicuous pits with resinous glands. Leaf sprays are thick and coarse. **CONES** are woody, spherical, weakly serotinous, brown, and 2.5 cm (1 in.) to 4 cm (1.5 in.) long. Scales are peltate and have inconspicuous pointed projections. **BARK** is brown to ashy gray, thick, and fibrous. Bark on large trees is fluted at the base.

HABITAT AND RANGE: As a native

tree, this cypress is rare; it is found in only 2 small populations, at Cypress Point and Point Lobos State Park, both near Carmel. In its native habitat, it is restricted to rocky, granitic soils near the ocean. There are only a few thousand individuals in each population, with the Cypress Point population being the largest.

REMARKS: Photographers and tourists flock to the Monterey Peninsula to see the gnarled, rugged silhouettes of Monterey cypresses. Strong ocean winds, salt spray, and skeletal soils combine to create stunted, distorted growth forms. When cultivated outside its native range, Monterey cypress is a relatively fast-growing, erect tree. Horticulturists and farmers have planted it as an ornamental and made windbreaks out of it along the Pacific Coast. It is also planted as a timber species in the Southern Hemisphere because of its durable heartwood. Monterey cypress is shade intolerant and on more productive soils is out-competed by other species. It is adapted to fire but not dependent on it.

SARGENT CYPRESS *Cupressus sargentii*

DESCRIPTION: A small- to medium-sized, single-stemmed, evergreen tree. It can be shrublike on exposed sites. Crowns are slender when young, and spreading and flat on older trees. Trees grow to be from 9 m (30 ft) to 23 m (75 ft) tall. **LEAVES** are scalelike, 3- or 4-angled in cross section, less than 3 mm (.12 in.) long (up to 10 mm [.38 in.] on fast-growing sprigs), and green to grayish green. They occur in opposite, alternating pairs and have pits with resinous glands. **CONES** are woody, dark brown, serotinous, spherical, and 2 cm (.75 in.) to 2.5 cm (1 in.) long. Scales are peltate and have inconspicuous projections on their centers. **BARK** is brown to gray, thick, and fibrous.

HABITAT AND RANGE: Grows in chaparrals and coniferous forests. It ranges from Mendocino County to Santa Barbara County, at elevations between 200 m (650 ft) and 1,000 m (3,200 ft), and is characteristically found in the interior Coast Ranges growing on serpentine soils (see McNab cypress).

REMARKS: Sargent cypress is more widely distributed than all but McNab cypress. It is adapted to fire and is fire dependent. Fire helps to break cone serotiny, permits seeds to be dispersed, and prepares the seedbed.

EPHEDRA (MORMON-TEA)

Ephedra is a monotypic genus that grows in many warm deserts worldwide. Six of the 42 species in the genus are found in California.

The genus is thought of as a gymnosperm, but its cones are so distinctive that it is placed in the division Gnetophyta rather than Pinophyta, the conifer division. The shrubs are broomlike and have grooved, photosynthetic, whorled stems that may be thornlike. The leaves are small and scalelike. Pollen and seeds are borne in cones found on different plants. Stem color and the number of scalelike leaves at the nodes are important characteristics in identifying *Ephedra* species.

A tea has long been made from Mormon-tea (*ma huang*) for its stimulating effects. The use of its active ingredient, ephedrine, in many herbal products is controversial, since deaths have been attributed to its consumption. Other common names for this species are desert-tea, ephedra, jointfir, and Mexican-tea.

1. There are 3 scalelike leaves at each stem node .... **California ephedra** (***E. californica***)
1. There are 2 scalelike leaves at each stem node .... **green ephedra** (***E. viridis***)

CALIFORNIA EPHEDRA *Ephedra californica*

DESCRIPTION: A seemingly leafless shrub that can grow to be 1 m (3.3 ft) tall. **LEAVES** are represented by 3 scales at the swollen nodes. Scales are deciduous; persistent brown bases remain on the stems. **POLLEN CONES** number 1 to 3 at a node; each is spherical and has 2 to 5 stamenlike structures. **SEED CONES** consist of single, 4-sided seeds that protrude from a set of bracts. **TWIGS** are erect, rigid, and pale green.

HABITAT AND RANGE: An associate of chaparrals and woodlands below 900 m (3,000 ft), it is found from the southern Sierra Nevada and interior Coast Ranges south through the Mojave and Colorado Deserts.

REMARKS: California ephedra's 3 scalelike leaves distinguish it from the other common species, green ephedra. Deer and livestock eat California ephedra during the winter.

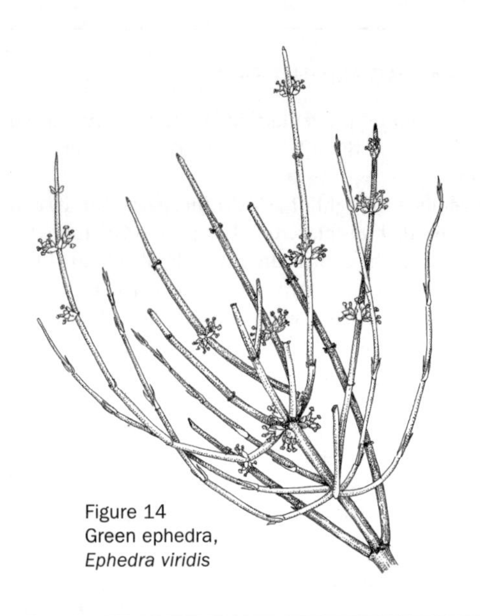

Figure 14
Green ephedra,
Ephedra viridis

GREEN EPHEDRA (Fig. 14; Pl. 2) *Ephedra viridis*

DESCRIPTION: A broomlike, apparently leafless shrub that may grow to be 1.5 m (5 ft) tall. **LEAVES** are a pair of deciduous scales at the swollen nodes; persistent brown bases remain on the stems. **POLLEN CONES** number 2 to 5 per node; each is spherical and has 2 to 5 stamenlike structures. **SEED CONES** consist of a pair of 3-sided seeds that protrude from a set of bracts. **TWIGS** are slender, erect, and dark or bright yellowish green.
HABITAT AND RANGE: Common in woodlands east of the Cascades and Sierra Nevada, and west through the Transverse Ranges to the coast, at elevations from 900 m (3,000 ft) to 2,300 m (7,500 ft). It also grows in the intermountain West.
REMARKS: Green ephedra is easily identified by habit and location. Its dark green stems distinguish it from its generally grayish woodland associates. When the cones are shedding pollen, the shrubs become golden orbs, looking more like a flowering plant than a gymnosperm. Deer and livestock eat green ephedra in the winter.

JUNIPERUS (JUNIPER)

The genus *Juniperus* has about 60 species; of the 13 native to the United States, 4 are native to California. All, except for one African species, are restricted to the Northern Hemisphere. Junipers are evergreen and they vary in height from prostrate shrubs to erect trees.

Leaves are evergreen and aromatic and have 2 forms: scales and awls. Leaves on young plants are always awl-like. Depending on the species, older plants can have leaves that are only scalelike, only awl-like, or both scalelike and awl-like. Leaves are either set in 3s or are opposite. Pits with resinous glands or resin dots are noticeable on some species. Stomatal bloom may be present. A tree may bear both pollen cones and seed cones, depending upon the species. Mature seed cones are spherical, berrylike, and composed of 3 to 6 enlarged, fleshy scales. Cones are red, reddish brown, or bluish black, with an ephemeral whitish bloom. Bark is thin, brown to reddish brown, and fibrous.

Junipers grow in a wide variety of habitats, from the cold environs of the Arctic and the high Himalaya to the hot, arid American Southwest and North Africa to temperate, humid China and eastern North America. In California, junipers grow in coastal, montane, and subalpine coniferous forests. Two California species (*J. occidentalis* and *J. osteosperma*) are predominantly treelike, 1 is shrublike (*J. communis*), and 1 is usually shrubby but can also occur as a tree (*J. californica*).

Junipers are widely used as ornamentals, and many cultivars have been developed. Perfumes and medicines have been derived from juniper wood and leaf extracts. The wood of some species is especially aromatic and is used to make "cedar" chests. Some cooks use the berries as a flavoring, and the berries are a favorite of birds and small mammals.

1. Leaves are awl-like and have stomatal bloom on their upper surfaces. Plants are prostrate shrubs .... **common juniper (*J. communis*)**
1. Leaves are mostly scalelike (young leaves can be awl-like) and lack stomatal bloom. Plants are trees 2
    2. Mature berrylike cones are bluish black under an ephemeral whitish bloom 3
    2. Mature berrylike cones are reddish brown under an ephemeral whitish bloom 4

3. Bark is grayish brown. Trees grow in juniper woodlands in the Cascades and on the Modoc Plateau . . . . **western juniper (*J. occidentalis* ssp. *occidentalis*)**
3. Bark is reddish brown. Trees grow on exposed, rocky slopes in montane to subalpine habitats, mostly in the Sierra Nevada and southern California mountains . . . . **mountain juniper (*J. occidentalis* ssp. *australis*)**
    4. Leaves are generally set in whorls of 3. Trees are usually multistemmed . . . . . **California juniper (*J. californica*)**
    4. Leaves are set opposite in pairs. Trees are usually single stemmed . . . . **Utah juniper (*J. osteosperma*)**

CALIFORNIA JUNIPER *Juniperus californica*

(Fig. 15)

DESCRIPTION: A shrubby, crooked, multitrunked tree. Crowns are conical to round, and branches stiff and irregular. Mature plants are usually 1 m (3.2 ft) to 4.5 m (15 ft) tall and 30 cm (1 ft) to 60 cm (2 ft) in diameter. The largest lives in Colusa County and is 10 m (33 ft) tall and 75 cm (30 in.) in diameter. The trees can live approximately 250 years. **LEAVES** are scalelike, yellowish green, and set in whorls of 3. They have pronounced pits with resinous glands. Juvenile leaves are awl-like. **CONES** are spherical, berrylike, reddish brown at maturity beneath an ephemeral whitish bloom, and about 6 mm (.25 in.) to 12 mm (.5 in.) in diameter. **BARK** is thin, fibrous, and ashy gray. Stems are conspicuously fluted.

HABITAT AND RANGE: Grows in low-elevation to montane coniferous woodlands, desert scrubs, and chaparrals. It may occur with singleleaf pinyon or Mojave yucca and is found at elevations lower than those of Utah juniper. California juniper occurs on very shallow, coarse-textured, rocky, infertile soils, at elevations from 45 m (150 ft) to 1,500 m (5,000 ft).

REMARKS: California juniper seedlings need shade to become established, while mature trees are shade intolerant. It has thin bark and is easily killed by fire. California junipers are used as fuelwood and Christmas trees and to make fence posts. The species provides important cover and forage for wildlife. Birds and small mammals are beneficiaries of the plenti-

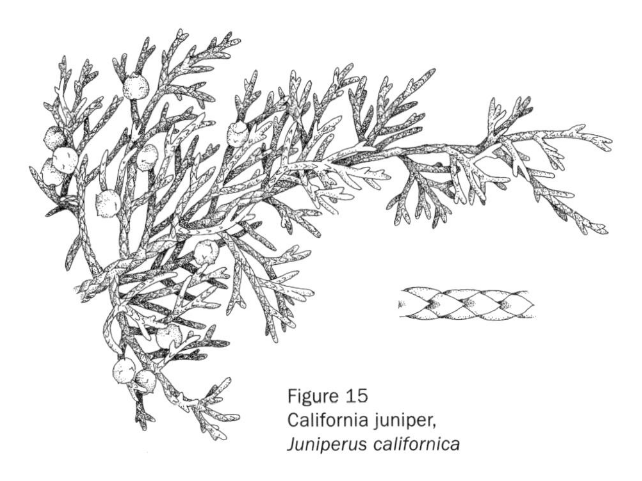

Figure 15
California juniper,
Juniperus californica

ful, sweet berries. Native Americans used California juniper to make bows and juniper berry cakes.

COMMON JUNIPER (Fig. 16) *Juniperus communis*

DESCRIPTION: Can be either an erect tree, a waist-high shrub, or a prostrate shrub. In California it is a prostrate shrub less than 1 m (3.3 ft) tall, with a spreading, matlike crown measuring 1 m (3.3 ft) to 2 m (6.5 ft) across. The largest grows in Washtenaw County, Michigan, and is 14 m (46 ft) tall and 30 cm (1 ft) in diameter. **LEAVES** are awl-like and set in whorls of 3. They are stiff and spine tipped and have a wide stripe of stomatal bloom on their upper surfaces. The undersides of the leaves are dark green, shiny, and convex. Leaves are pungent when crushed. **CONES** are spherical, berrylike, bluish black beneath an ephemeral whitish bloom, and about 6 mm (.25 in.) in diameter. Cones are fleshy and sweet and take 3 years to mature. **BARK** has shallow furrows separating thin, papery, reddish brown scales.

HABITAT AND RANGE: Widely distributed throughout the Northern Hemisphere. In California it grows in montane and subalpine coniferous forests. In the Sierra Nevada and Cascades it is found in disjunct populations north of Mono Pass on high

Figure 16
Common juniper,
Juniperus communis

slopes and ridges, at elevations from 2,000 m (6,400 ft) to 3,300 m (11,000 ft). In Del Norte and Siskiyou Counties it occurs more extensively in Jeffrey pine woodlands and chaparrals on serpentine soils, at elevations from 90 m (300 ft) to 1,500 m (5,000 ft).

REMARKS: Several varieties and subspecies have been proposed for common juniper. Those that may apply to California are *J. c.* var. *montana* and *J. c.* var. *saxatilis,* although recently published manuals do not recognize them. The species is a popular ornamental, and dozens of cultivars have been developed. Juniper berries and foliage have been used as incense, perfume, medicine, and liniment and for beverages. Juniper berries are supposedly helpful in alleviating urinary tract problems and gout, as well as numerous other ailments. Cooks use the berries as a flavoring and in marinades for various kinds of meat and wild game. In the 1500s, juniper berries were used to create an inexpensive diuretic called gin. Wildlife species eat the foliage sparingly; the berries, however, are devoured by many bird and mammal species.

MOUNTAIN JUNIPER — *Juniperus occidentalis* var. *australis*

(Pl. 3)

DESCRIPTION: A short, erect tree with stout trunks (usually multistemmed) and large, rough branches. Crowns are rounded and full. Mature trees are typically 4.5 m (15 ft) to

15 m (50 ft) tall and 30 cm (1 ft) to 90 cm (3 ft) in diameter. The largest is in Stanislaus National Forest and is 26 m (86 ft) tall and 3.9 m (12.75 ft) in diameter. Some individuals live to be 2,000 to 3,000 years old. On exposed, harsh sites the species has a shrubby growth form. **LEAVES** are scalelike, grayish green to bluish green, and set in whorls of 3 (sometimes opposite). They have pits with resinous glands and/or resin deposits. Juvenile leaves are awl-like. **CONES** are spherical, berrylike, bluish black beneath an ephemeral whitish bloom, and about 6 mm (.25 in.) to 10 mm (.38 in.) in diameter. **BARK** is thin, fibrous, and reddish brown to cinnamon brown.

HABITAT AND RANGE: Grows in montane and subalpine coniferous woodlands in California and western Nevada, at elevations from 2,000 m (6,500 ft) to 3,000 m (10,000 ft). The species is best represented among granite slabs on higher-elevation Sierran slopes and ridges, where Jeffrey pine is a common associate. Mountain juniper also grows in lower-elevation pinyon-juniper woodlands and in disjunct juniper stands in the northern Coast Ranges and the San Gabriel and San Bernardino Mountains.

REMARKS: Mountain juniper is intolerant of shade and is readily killed by fire. It is perhaps the most picturesque tree in the Sierra Nevada when growing on exposed, windy granite slopes. There, the mountain juniper develops a tortured, twisted form with a sparse, wind-whipped crown. The short stature and rough, spreading branches of such a tree make it appear nearly shrublike, but its massive, squat trunk reveals it to be a rugged, stalwart tree that has weathered hundreds if not thousands of winters. The bark of the mountain juniper varies in color from reddish brown to a tawny, cinnamon brown when found on more exposed sites. The berrylike cones are eaten by birds and rodents and are an important source of food during the winter.

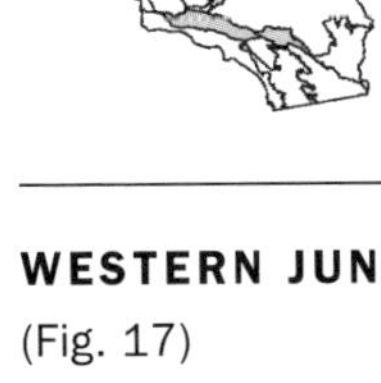

WESTERN JUNIPER — *Juniperus occidentalis* var. *occidentalis*

(Fig. 17)

DESCRIPTION: A short, erect tree with a short trunk and large, rough branches. Crowns are rounded and full. Mature trees are usually 4.5 m (15 ft) to 10 m (33 ft) tall and 30 cm (1 ft) to 90 cm (3 ft) in diameter. The species is long-lived, with some

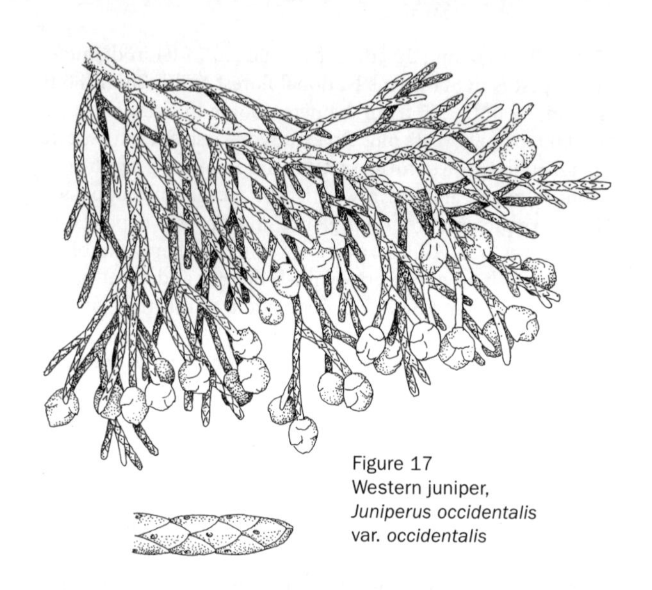

Figure 17
Western juniper,
Juniperus occidentalis
var. *occidentalis*

individuals surviving for hundreds of years. On exposed, harsh sites trees have a shrubby growth habit. **LEAVES** are scalelike, grayish green to bluish green, and usually opposite or set in whorls of 3. They have pits with resinous glands and/or resin deposits. Juvenile leaves are awl-like. **CONES** are spherical, berrylike, bluish black beneath an ephemeral whitish bloom, and about 6 mm (.25 in.) to 12 mm (.5 in.) in diameter. **BARK** is thin, fibrous, and brown.

HABITAT AND RANGE: Grows in interior coniferous woodlands in eastern Washington, eastern Oregon, and California north of Plumas County, at elevations from 700 m (2,300 ft) to 2,300 m (7,500 ft). It is usually found growing between sagebrush flats and pine forests and is a quintessential rimrock species. The species dominates open woodlands, with western juniper trees scattered through big sagebrush and/or bitterbrush. In much of its range, it grows on soils derived from volcanic rock.

REMARKS: Western juniper is intolerant of shade and readily killed by fire. Fire suppression, livestock grazing, and perhaps climate change in northeastern California have allowed western juniper to expand

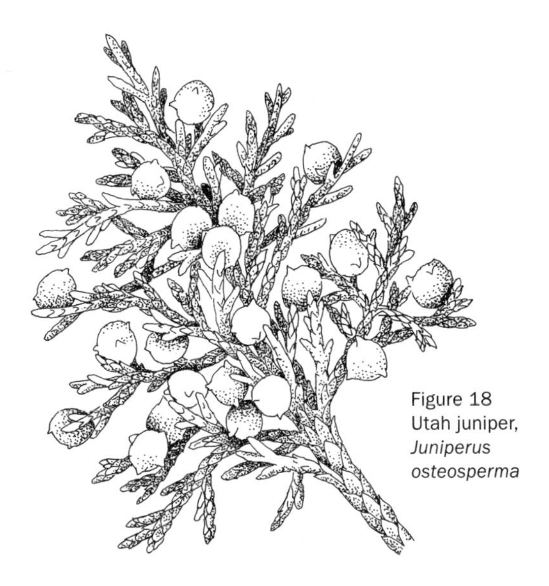

Figure 18
Utah juniper,
Juniperus osteosperma

into sagebrush-dominated flats. The species is used for firewood and to make fence posts and various novelty products. Its wood is aromatic and similar to the wood of eastern redcedar (*J. virginiana*), which is used in cedar chests. The berrylike cones are eaten by birds and rodents and are an important source of food during the winter. The foliage is palatable to elk, deer, and some small mammals. Western juniper hybridizes with Utah juniper east of the Warner Mountains.

UTAH JUNIPER (Fig. 18) *Juniperus osteosperma*

DESCRIPTION: A short, single-stemmed (occasionally multistemmed), erect tree with a short trunk and large, rough branches. Crowns are rounded and full. Mature trees are typically 4.5 m (15 ft) to 9 m (30 ft) tall and 15 cm (6 in.) to 30 cm (12 in.) in diameter. The largest grows at Crowley Lake, California, and is 11.5 m (38 ft) tall and 2.1 m (85 in.) in diameter. The species is long-lived, with some individuals thought to be more than 1,000 years old. **LEAVES** are scalelike and yellowish green. Most are set opposite (or sometimes in 3s) and lack pits with resinous glands. Juvenile leaves are awl-like. **CONES** are spherical, berrylike, reddish brown beneath an ephemeral whitish bloom, and about 6 mm (.25 in.) to 20 mm (.75 in.) in diameter. **BARK** is thin, fibrous, and grayish brown.

HABITAT AND RANGE: Grows in montane coniferous woodlands throughout the Great Basin. In California it is best represented in mountains in southeastern California and, farther north, along the California-Nevada border. The species grows on dry, shallow, rocky soils at elevations from 1,000 m (3,200 ft) to 2,500 m (8,400 ft).

REMARKS: Utah juniper is tolerant of drought and intolerant of fire. Along with singleleaf pinyon it dominates most of the Great Basin's pinyon-juniper woodlands. Utah juniper has long been used for firewood and to make fence posts, mine timbers, and charcoal. Juniper berries and foliage are eaten by many species of birds and mammals. The berries are especially palatable to small desert rodents. Utah juniper hybridizes with western juniper east of the Warner Mountains.

PICEA (SPRUCE)

The genus *Picea* has about 35 tree species; of the 7 native to North America, 3 are native to California. Spruces generally grow in cool, moist environments throughout the Northern Hemisphere.

Leaves are evergreen, linear or needlelike, and often sharp pointed. They are spirally arranged, usually in bottlebrush fashion. Leaves arise from stubby, woody pegs on twigs that differ in color from the attached leaves. They can be either flat or 4-angled in cross section. Stomatal bloom can be seen on either the lower or upper surfaces. When bruised the leaves emit a pungent aroma. Seed cones are pendant and attached near the ends of twigs. Scales are much longer than bracts and usually have ragged edges. Mature trees have scaly bark that resembles potato chips or corn flakes.

Spruces are evergreen, and on fertile sites they can achieve large sizes. Sitka spruce, for example, reaches dimensions surpassed only by giant sequoia, redwood, and western redcedar. In California, Sitka spruce grows close to the coast. Engelmann and Brewer spruce grow inland. Spruces are widely used as ornamentals, and there are many cultivars.

1. Leaf tips are blunt to round. Leaves are not 4-angled in cross section. Twigs and branches are very droopy. Cones are 7.5 cm (3 in.) to 15 cm (6 in.) long. Cone-scale margins are smooth. This tree is endemic to the Klamath Mountains . . . . **Brewer spruce (*P. breweriana*)**
1. Leaf tips are sharp. Leaves are 4-angled or diamond shaped in cross section. Twigs and branches droop slightly. Cones are less than 9 cm (3.5 in.) long. Cone-scale margins are ragged 2
   2. Leaves point forward and are 4-angled in cross section. Cones are less than 6 cm (2.5 in.) long. This tree grows in the Klamath Mountains and Cascades . . . . **Engelmann spruce (*P. engelmannii*)**
   2. Leaves are more or less perpendicular to the twig and are diamond shaped in cross section. Cones are less than 5 cm (2 in.) long. This tree grows close to the Pacific Ocean north of Fort Bragg . . . **Sitka spruce (*P. sitchensis*)**

BREWER SPRUCE *Picea breweriana*

(Fig. 19; Pl. 4)

DESCRIPTION: A tall, erect, single-stemmed tree. It typically attains heights of 15 m (50 ft) to 30 m (100 ft) and diameters of 45 cm (18 in.) to 75 cm (30 in.). The largest grows in Klamath National Forest and is 54 m (176 ft) tall and 1.6 m (64 in.) in diameter. The oldest is around 460 years old. The species has a narrow crown with many droopy branchlets that may reach nearly to the ground. Its droopy habit has caused some to refer to it as the weeping spruce. **LEAVES** are linear, flat, and 20 mm (.75 in.) to 30 mm (1.25 in.) long. They have round tips and white stomatal bloom below. **CONES** are cylindrical, pendant, leathery, and 7.5 cm (3 in.) to 15 cm (6 in.) long. Cone scales are thin and leathery, with smooth, round edges. **BARK** is thin and reddish brown, with long, thin, exfoliating scales that give it the appearance of being studded with corn flakes. **TWIGS** are long, thin, droopy, and covered with fine hairs.

HABITAT AND RANGE: Found only in montane and subalpine coniferous forests throughout the Klamath Mountains of California and Oregon. It may be locally common where it occurs on the upper slopes of high ridges and in hollows at the heads of small, north-facing watersheds. It thrives in cool, moist

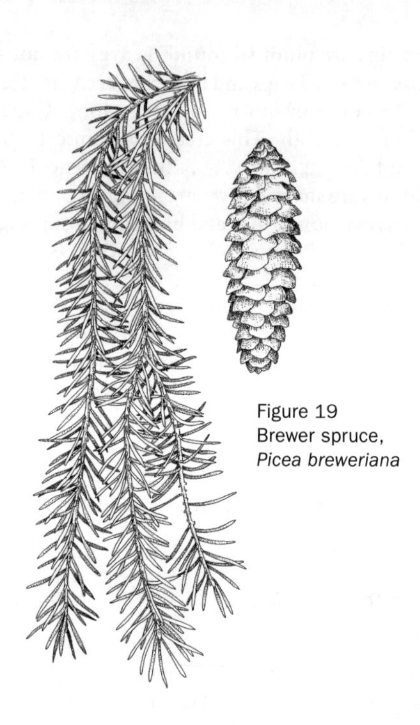

Figure 19
Brewer spruce,
Picea breweriana

environments that receive considerable winter snowfall. In California the species ranges in elevation from 1,000 m (3,300 ft) to 2,300 m (7,500 ft). The most northerly stand occurs on Iron Mountain, Oregon, and the most southerly stand is at East Weaver Lake in Trinity County. The species is one of several rare conifers in the Russian Peak Wilderness.

REMARKS: Brewer spruce is shade tolerant, readily killed by fire, and susceptible to being uprooted and overthrown by wind. It has no special uses and is not sought after as a timber species. In Europe, it is a popular ornamental.

ENGELMANN SPRUCE (Fig. 20) *Picea engelmannii*

DESCRIPTION: A tall, erect, single-stemmed tree. It is one of the largest of the montane and subalpine species in Califor-

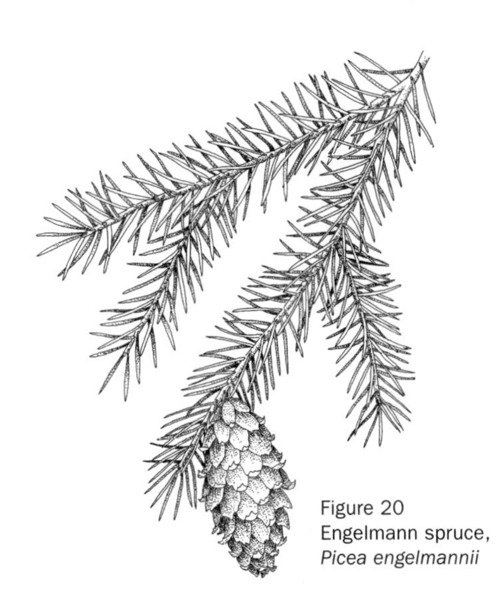

Figure 20
Engelmann spruce,
Picea engelmannii

nia. Mature trees are typically 30 m (100 ft) to 36 m (120 ft) tall and 45 cm (18 in.) to 75 cm (30 in.) in diameter. The largest grows near Payette Lake, Idaho, and is 55 m (179 ft) tall and 2.3 m (92 in.) in diameter. The species is long-lived, with some trees surviving more than 500 years. It has a long-pointed, slender, sometimes spirelike crown. **LEAVES** are linear, 4-angled in cross section, and 20 mm (.75 in.) to 30 mm (1.25 in.) long, with sharp-pointed tips. When crushed, the leaves give off a foul odor. Leaves tend to point toward the ends of the twigs. They are more flexible than those of Sitka spruce and have stomatal bloom on all surfaces. **CONES** are pendant, papery, and 4 cm (1.5 in.) to 6 cm (2.5 in.) long. Cone scales have ragged edges and are tan. **BARK** is very thin and purplish to reddish brown, with loosely attached scales.

HABITAT AND RANGE: Grows in montane and subalpine coniferous forests in only 2 areas in California. Stands can be found near Russian Peak in Siskiyou County and along Clark Creek in Shasta County, at elevations between 1,000 m (3,300 ft) and 2,000 m (6,500 ft). Outside of California it grows in the Cascades and Rocky Mountains.

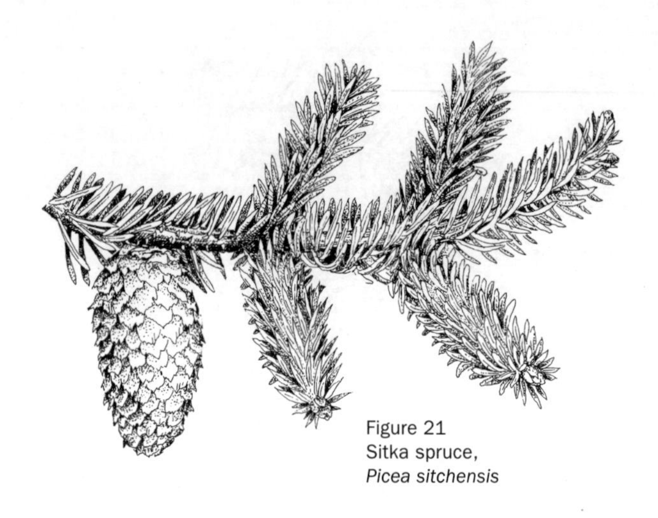

Figure 21
Sitka spruce,
Picea sitchensis

REMARKS: Engelmann spruce is rare in California. Elsewhere in its range, it is used for lumber and to make pulp, aircraft parts, and musical instruments. It is quite shade tolerant and, because of its thin bark, easily killed by fire. Engelmann spruce hybridizes with white spruce (*P. glauca*) in the northern Rocky Mountains.

SITKA SPRUCE (Fig. 21) ***Picea sitchensis***

DESCRIPTION: A tall, erect, single-stemmed tree. It is the largest of the world's spruces. Mature trees are typically 40 m (130 ft) to 55 m (180 ft) tall and 1 m (40 in.) to 1.5 m (60 in.) in diameter. The largest is found in Olympic National Park and is 58 m (191 ft) tall and 5.7 m (18.75 ft) in diameter. Trees can live more than 800 years. The species has long, cylindrical stems with either horizontal or ascending branches and pendant branchlets. Trees established on old logs form colonnades and have buttressed roots. When growing adjacent to the ocean in California, the species is often contorted and prostrate. **LEAVES** are linear and 12 mm (.5 in.) to 38 mm (1.5 in.) long. The leaves form a bottlebrush arrangement on the twigs. They are 4-angled in cross section, stiff, very sharp pointed, and generally set at right angles to twigs. Leaves arise from small woody pegs attached to twigs. Newer leaves have conspicuous white stomatal bloom. **CONES** are oblong and 5 cm (2 in.) to 9 cm (3.5 in.) long. Scales are papery, have ragged tips, and are light brown to tan. **BARK** is thin and gray to silvery gray, with large exfoliating scales that give it the appearance of being studded with potato chips.

HABITAT AND RANGE: In northwestern California, the species grows in coastal coniferous forests at elevations from sea level to 200 m (650 ft). It is known as a tidewater species, seldom occurring more than a few miles from the Pacific Ocean. The southern end of its range is near Fort Bragg. Sitka spruce forests are more extensive in coastal Oregon, Washington, British Columbia, and Alaska.

REMARKS: Sitka spruce is shade tolerant and readily killed by fire. It has thc highest strength-to-weight ratio of the world's conifers. During World War II, some British bombers were made from Sitka spruce. It is also valued as a pulp and lumber species and is used in the making of musical instruments.

PINUS (PINE)

The genus *Pinus* has about 95 species; 42 are native to the United States, and 18 of these are native to California. All but 1 are restricted to the Northern Hemisphere. No other genus in California has as many tree species as *Pinus*. Pines are evergreen and vary in height from short shrubs to ponderosa and sugar pines that reach 61 m (200 ft) tall. Plants growing near the tree line can have a prostrate habit due to the shaping effects of desiccation and sand- and ice-blasting.

Leaves are evergreen and have 2 forms: primary scales and secondary needles. Needles are set in bundles of 1 to 5 and persist for at least several years in healthy trees. Bristlecone and foxtail pines, for example, have needles that persist for decades. In cross section, needles can be triangular, semicircular, or round. Lines of stomata are evident. Primary scales have a papery texture, are inconspicuous, and may be deciduous. The primary scales are not useful identification characteristics. Papery sheaths surround bundles of needles. The bundle sheaths are deciduous for most 5-needled pines as well as pinyons (which have 1 to 5 needles per bundle). Most 2- and 3-needled pines have persistent bundle sheaths. Both pollen cones and seed cones are found on the same tree, but rarely on the same branch. Seed cones are often located near the tops of trees. Pollen cones are catkinlike, clustered, many scaled, and often colorful (yellow, orange, blue, or red). They are deciduous following pollen dispersal in the spring, and for this reason are not usually used in species identification. Mature seed cones originate from nodes on stems and branches. Individual cones have spirally arranged, persistent, woody scales that may have prickles. Cones mature in 2 years. Bark is scaly, thin to thick, and furrowed. In some species the scales look like puzzle pieces.

In California, pines grow in a wide variety of habitats, from the cool, moist environs along the sand dunes of the north coast to the cold, windswept ridges (up to 3,300 m [11,000 ft] high) of California's highest mountains. They often grow in relatively dry, sandy soils or in areas subject to high evaporation. Pines are generally intolerant of shade.

Pines are widely used as lumber and to make paper pulp, resins, turpentine, and pharmaceuticals, as well as in landscaping. Pine "nuts" have been important food sources for wildlife and Native Americans and are sold commercially throughout California.

1. Needles occur in bundles of 1 .... **singleleaf pinyon (*P. monophylla*)**
1. Needles occur in bundles of more than 1 2
    2. Needles occur in bundles of 2 3
    2. Needles occur in bundles of more than 2 6
3. Seed-cone scales have no prickles. The papery sheaths at the bases of leaf bundles are deciduous .... **singleleaf pinyon (*P. monophylla*)**
3. Seed-cone scales have prickles. The papery sheaths at the bases of leaf bundles are persistent 4
    4. Seed cones are whorled and are found along branches and trunks. Cones are serotinous and usually occur in clusters of 3 to 5. Cone-scale prickles resemble flat, thick spines. Leaves are 10 cm (4 in.) to 15 cm (6 in.) long .... **Bishop pine (*P. muricata*)**
    4. Seed cones are solitary, pendant, and found near the tips of branches. Cones may be persistent. Cone-scale prickles are conspicuous and recurved. Leaves are 3 cm (1.25 in.) to 7.5 cm (3 in.) long 5

5. Seed cones are persistent on trees. Bark is furrowed. Trees grow in coastal habitats . . . . **beach pine (*P. contorta* ssp. *contorta*)**
5. Seed cones are not persistent on trees. Bark is scaly. Trees grow in inland habitats . . . . **lodgepole pine (*P. contorta* ssp. *murrayana*)**
    6. Needles occur in bundles of 3 7
    6. Needles occur in bundles of more than 3 14
7. Mature seed cones persist on trees for more than 2 years . . . 8
7. Mature seed cones persist on trees for 1 year 11
    8. Leaves are 7.5 cm (3 in.) to 15 cm (6 in.) long. Seed cones are asymmetrical, have prickles on cone scales, and persist on trees for many years 9
    8. Leaves are 15 cm (6 in.) to 43 cm (17 in.) long. Seed cones are symmetrical, have massive clawlike cone-scale tips, and persist on trees for only 2 or 3 years 10
9. The upper cone surfaces have enlarged, round scales. Scales have a minute prickle . . . . **Monterey pine (*P. radiata*)**
9. Upper cone surfaces have enlarged, knobby scales. Scales have a short, stout prickle . . . . **knobcone pine (*P. attenuata*)**
    10. Seed cones are 25 cm (10 in.) to 35 cm (14 in.) long and yellowish brown. Leaves are 15 cm (6 in.) to 30 cm (12 in.) long, dark green, and stiff . . . . **Coulter pine (*P. coulteri*)**
    10. Seed cones are 15 cm (6 in.) to 25 cm (10 in.) long and brown. Leaves are 18 cm (7 in.) to 43 cm (17 in.) long, grayish green, and flexible . . . . **ghost pine (*P. sabiniana*)**
11. Seed-cone scales have no prickles or claws. The papery sheaths at the bases of leaf bundles are deciduous . . . . **Parry pinyon (*P. quadrifolia*)**
11. Seed-cone scales have prickles or claws. The papery sheaths at the bases of leaf bundles are persistent 12
    12. Leaves are 7.5 cm (3 in.) to 12.5 cm (5 in.) long. Seed cones are 5 cm (2 in.) to 9 cm (3.5 in.) long . . . . **Washoe pine (*P. washoensis*)**
    12. Leaves are generally more than 15 cm (6 in.) long. Seed cones are more than 9 cm (3.5 in.) long 13

13. Seed cones are 12.5 cm (5 in.) to 25 cm (10 in.) long. Dry cone-scales have prickles that point downward. Leaves are bluish green. Tips of new twigs are bluish green. Bark smells like vanilla, pineapple, or butterscotch . . . . **Jeffrey pine (*P. jeffreyi*)**

13. Seed cones are 9 cm (3.5 in.) to 15 cm (6 in.) long. Dry cone scales have prickles that curve outward. Leaves are green. Tips of new twigs are yellowish green. Bark does not smell like vanilla, pineapple, or butterscotch . . . . **ponderosa pine (*P. ponderosa*)**

14. Needles occur in bundles of 4 . . . . **Parry pinyon (*P. quadrifolia*)**

14. Needles occur in bundles of 5. 15

15. Needle bundle sheaths are persistent . . . . **Torrey pine (*P. torreyana*)**

15. Needle bundle sheaths are deciduous 16

16. Seed-cone scales are thickest at their tips. Leaves are less than 6 cm (2.5 in.) long 17

16. Seed-cone scales are thinnest at their tips. Leaves are 5 cm (2 in.) to 10 cm (4 in.) long 20

17. Seed-cone scales lack prickles. Plants grow in the Peninsular Ranges . . . . **Parry pinyon (*P. quadrifolia*)**

17. Seed-cone scales have small prickles. Plants do not grow in the Peninsular Ranges 18

18. Seed cones are weakly serotinous. Cones are typically torn apart by animals. Twigs are very flexible . . . . **whitebark pine (*P. albicaulis*)**

18. Seed cones are not serotinous. Cone scales are not commonly torn apart by animals. Twigs are not very flexible 19

19. Prickles on seed-cone scales are slender, persistent, and up to 6 mm (.25 in.) long. Tree grows in the White, Inyo, and Panamint Mountains and is often multistemmed . . . . **Great Basin bristlecone pine (*P. longaeva*)**

19. Prickles on seed-cone scales are minute and deciduous. Tree grows in either the Klamath Mountains or the southern Sierra Nevada and is usually single stemmed . . . . **foxtail pine (*P. balfouriana*)**

20. Seed-cone stalks are less than 25 mm (1 in.) long. Twigs are very flexible . . . . **limber pine (*P. flexilis*)**

20. Seed-cone stalks are more than 25 mm (1 in.) long. Twigs are not very flexible 21

21. Seed cones are 28 cm (11 in.) to 46 cm (18 in.) long. Mature bark is reddish brown to purplish and platelike, with deciduous puzzle-piece-like flakes. Stomatal bloom can be seen on all needle surfaces .... **sugar pine (*P. lambertiana*)**
21. Seed cones are 12.5 cm (5 in.) to 28 cm (11 in.) long. Mature bark has dark gray to purplish gray, block-patterned bark that resembles an alligator's hide. Stomatal bloom can be seen only on inner needle surfaces .... **western white pine (*P. monticola*)**

WHITEBARK PINE (Fig. 22) ***Pinus albicaulis***

DESCRIPTION: A short, usually multistemmed tree whose multiple trunks are often partially fused at the base. The trunks may originate from different seeds in the same seed cache. In California a typical mature tree ranges in height from 7.5 m (25 ft) to 15 m (50 ft) and in diameter from 30 cm (12 in.) to 60 cm (24 in.). The largest grows in Sawtooth National Recreation Area in Idaho and is 21 m (69 ft) tall and 2.6 m (105 in.) in diameter. Trees can live to be 400 to 700 years old. At the tree line, the species often has a dwarfed, shrub form and grows in isolated clumps that measure 30 cm (1 ft) to 90 cm (3 ft) high. **LEAVES** are set in bundles of 5 and measure 3 cm (1.25 in.) to 7 cm (2.75 in.) long. Needles are dark green, stiff, and somewhat curved. Stomatal bloom is evident on all surfaces. Bundle sheaths are deciduous. **CONES** have a broad egg shape to globe shape. They are often erect, are purplish brown to tan, and measure 4 cm (1.5 in.) to 9 cm (3.5 in.) long. Stalks are very short. Cone scales are thick and have sharp-pointed tips. Scales do not open fully to release seed until they disintegrate on the ground or animals tear them apart. **TWIGS** are thick and very flexible. **BARK** is usually less than 12 mm (.5 in.) thick and is grayish white to white. The bark of mature trees has scaly plates separated by furrows.

HABITAT AND RANGE: Grows in subalpine coniferous forests and woodlands at or near the tree line. It is found in the higher mountains of the Pacific Northwest and in the northern Rocky Mountains. In California the species is widespread in the Sierra Nevada, Cascades, and Warner Mountains and is localized in the Klamath Mountains. Its lowest elevation

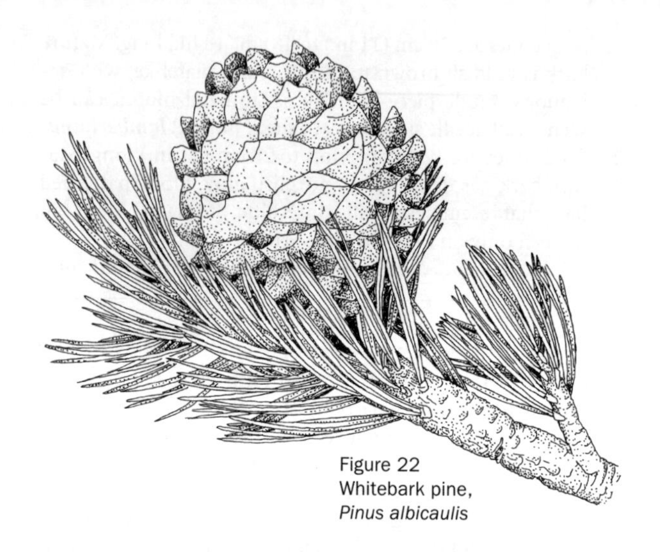

Figure 22
Whitebark pine,
Pinus albicaulis

limit is around 1,800 m (6,000 ft), and its highest limit, in the southern Sierra Nevada, is around 3,600 m (12,000 ft).

REMARKS: Whitebark pine is intolerant of shade and is easily killed by fire. Over 90% of its seeds are dispersed and stored in seed caches by birds (especially the Clark's nutcracker), squirrels, and chipmunks. Whitebark pine is threatened by white pine blister rust, the mountain pine beetle, limber pine dwarf mistletoe, and replacement by other subalpine tree species.

KNOBCONE PINE (Fig. 23) *Pinus attenuata*

DESCRIPTION: A small- to medium-sized, single-stemmed, erect tree typically ranging in height from 6 m (20 ft) to 12 m (40 ft) and in diameter from 30 cm (12 in.) to 60 cm (24 in.). Trees often have multiple, erect, crooked trunks. The largest tree grows in Shasta County and is 36 m (117 ft) tall and 109 cm (43 in.) in diameter. The species is short-lived, with a usual life span of only about 60 years. The oldest trees are probably less than 100 years old. Mature trees have an irregular, open-branching pattern. Crowns are sparse; they have a narrow conical shape on younger trees and are rounded on older ones. **LEAVES** are set in bundles of 3, are yellowish green, and measure 7.5 cm (3 in.) to 15 cm (6 in.) long. Bundle sheaths are persistent. **CONES** are asymmetrical, elongated, conical, and set in whorls of 3 to

Figure 23
Knobcone pine,
Pinus attenuata

5. They are recurved, serotinous, persistent, and 7.5 cm (3 in.) to 15 cm (6 in.) long. Scales have prickles and are massive on the upper sides of cones, especially near their bases, where they are pronouncedly knobby. Elsewhere they are thick and flat. **BARK** on mature trees is grayish brown and has shallow furrows separated by superficially scaly ridges.

HABITAT AND RANGE: Widely distributed in chaparrals and coniferous forests and woodlands from western Oregon to Baja California, at elevations below 1,900 m (6,200 ft). In California there are extensive stands in the Klamath Mountains and northern Coast Ranges. Elsewhere in California, the species has a patchy distribution. It usually grows on shallow, rocky, infertile soils, especially serpentine soils.

REMARKS: Knobcone pine has a strongly serotinous cone. In most circumstances, it requires fire to crack the resin seal and allow the scales to open. Following a fire, seed is rapidly dispersed. Individual trees are quite susceptible to fire. Knobcone pine is shade intolerant. The species forms a hybrid with *P. radiata* in the Santa Cruz Mountains.

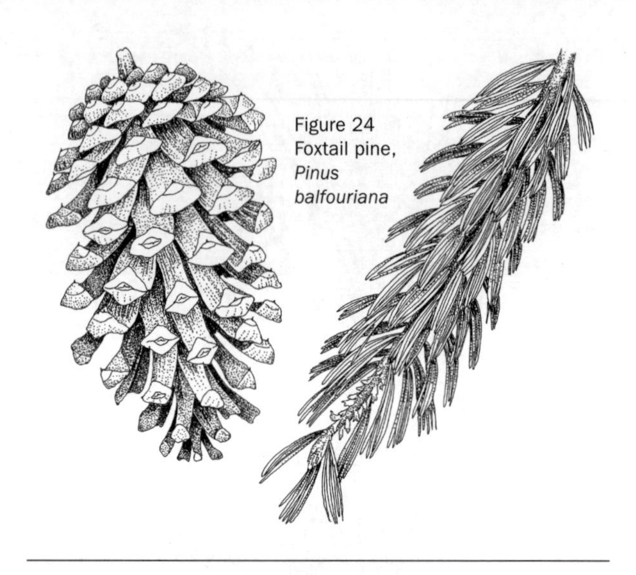

Figure 24
Foxtail pine,
Pinus balfouriana

FOXTAIL PINE (Fig. 24; Pl. 5) ***Pinus balfouriana***

DESCRIPTION: A small- to medium-sized, erect, single-stemmed tree, typically with an irregular crown and spreading branches. Some tree-line plants exposed to excessive sand- and ice-blasting develop contorted, shrubby forms. Thin strips of cambium on their leeward sides support sparse canopies. Mature trees often attain heights of 9 m (30 ft) to 18 m (60 ft). Typical diameters range from 30 cm (1 ft) to 60 cm (2 ft). The largest tree grows in Shasta-Trinity National Forest and is 23 m (76 ft) tall and 2.5 m (100 in.) in diameter. Some trees are estimated to be from 2,500 to 3,000 years old. **LEAVES** are set in bundles of 5 (sometimes 4) and are 2 cm (.75 in.) to 4 cm (1.5 in.) long. They can persist on branches for decades and are tightly arranged bottlebrush style, producing a foxtail-like appearance. Needles are dark green, stiff, and somewhat curved. They have scattered resin ducts on their surfaces and white stomatal lines on their inner surfaces. Bundle sheaths are deciduous. **CONES** are egg shaped to barrel shaped, pendant, reddish brown, and 9 cm (3.5 in.) to 12.5 cm (5 in.) long. Cone stalks are less than 2 cm (.75 in.) long. Cone-scale tips are somewhat thick and have minute or absent prickles. **BARK** is gray to reddish brown and deeply fissured on older trees. Young bark is grayish white.

HABITAT AND RANGE: Grows in upper montane and subalpine coniferous forests and woodlands. It is endemic to California and is further classified as 2 disjunct subspecies: the blue-gray-leaved *P. b.* ssp. *balfouriana* in the Klamath Mountains, and the yellow-green-leaved *P. b.* ssp. *austrina* in the southern Sierra Nevada. Both subspecies grow near the tree line, at elevations from 2,000 m (6,700 ft) to 2,700 m (9,000 ft) in the north and 2,600 m (8,500 ft) to 3,600 m (12,000 ft) in the south.

REMARKS: Foxtail pine resembles Great Basin bristlecone pine. In the southern Sierra Nevada the two species are separated by as few as 20 air miles. Winds blowing pollen from the Sierra Nevada to the White Mountains may have helped to produce natural hybrids.

BEACH PINE (Fig. 25) ***Pinus contorta* ssp. *contorta***

DESCRIPTION: A small- to medium-sized, erect, single-stemmed tree. Mature trees are 6 m (20 ft) to 15 m (50 ft) tall and 15 cm (6 in.) to 45 cm (18 in.) in diameter. The largest tree grows in Snohomish County, Washington, and is 31 m (101 ft) tall and 111 cm (44 in.) in diameter. The species is short-lived, and trees probably do not live more than 200 years. Mature trees have many branches and irregularly shaped crowns. **LEAVES** are set in bundles of 2; they are twisted, green, and 2.5 cm (1 in.) to 7.5 cm (3 in.) long. Bundle sheaths are persistent. **CONES** are asymmetrically egg shaped, pendant, yellowish brown to brown, and 4 cm (1.5 in.) to 5 cm (2 in.) long. The reflexed cones persist for many years on branches and are not serotinous. Closely related races are either serotinous or variably serotinous. Scales are thick and have long, recurved prickles. **BARK** on mature trees is thick and fissured.

HABITAT AND RANGE: Grows in coastal and low-elevation montane coniferous forests from Alaska to Mendocino County, at elevations between near sea level and 150 m (500 ft). In California it is restricted to a narrow coastal strip, with the exception of a closely related race growing in the Siskiyou Mountains. Along the coast, the species is usually found growing on sandy soils or sand dunes.

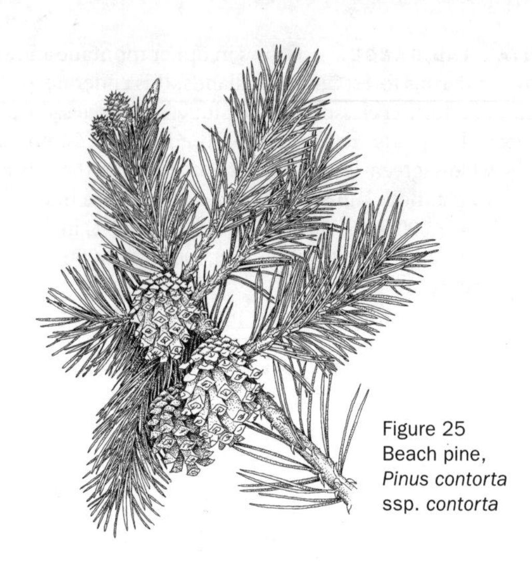

Figure 25
Beach pine,
Pinus contorta
ssp. *contorta*

REMARKS: Two forms of lodgepole pine are offshoots of beach pine: *P. c.* ssp. *bolanderi* and an unnamed form growing in Del Norte County. The rare subspecies *P. c.* ssp. *bolanderi* is known as the "pygmy" race (or Mendocino White Plains race). It differs from *P. c.* var. *contorta* in that it is mostly serotinous, has knobby cones, is less than 2.1 m (7 ft) tall at maturity, and grows on highly acidic soils underlain by hardpan on coastal terraces in Mendocino County. The Del Norte form grows on serpentine or other ultramafic soils and is variably serotinous. Beach pine is shade intolerant and has little resistance to fire. The coastal forms either are fire-free or have very long intervals between fires. Wind may be an important disturbance agent with this species. Beach pine has been used to stabilize sand dunes. Native Americans used the pitch to treat sores and relieve the discomfort of sore throats. Another common name for the species is shore pine.

LODGEPOLE PINE *Pinus contorta* ssp. *murrayana* (Fig. 26)

DESCRIPTION: A medium-sized, erect, single-stemmed tree that typically ranges from 18 m (60 ft) to 33 m (110 ft) in height and from 25 cm (10 in.) to 45 cm (18 in.) in diameter. The largest

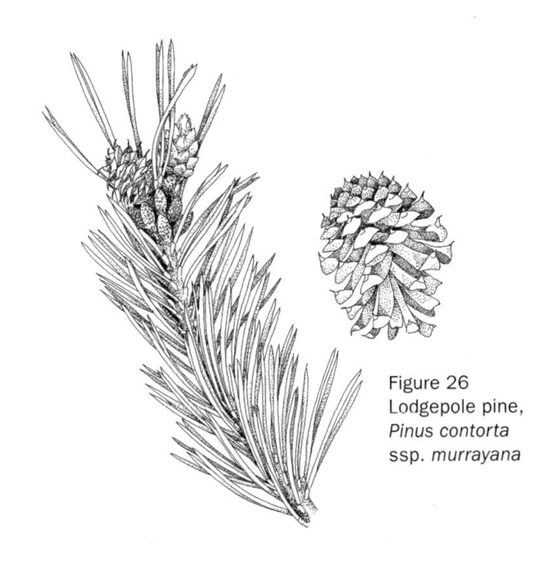

Figure 26
Lodgepole pine,
Pinus contorta
ssp. *murrayana*

tree grows in Stanislaus National Forest and is 38 m (124 ft) tall and 1.9 m (73 in.) in diameter. Relative to other lodgepole pine subspecies, *P. c.* ssp. *murrayana* is long-lived; it can reach 500 to 600 years in age. Crowns are short and narrow and branches are short. Trees near the tree line can have a dwarfed, shrubby form. **LEAVES** are set in bundles of 2; they are twisted, yellowish green, and 2.5 cm (1 in.) to 7.5 cm (3 in.) long. Bundle sheaths are persistent. **CONES** are egg shaped to spherical, pendant, yellowish brown to brown, and 4 cm (1.5 in.) to 5 cm (2 in.) long. The cones are nonserotinous and are deciduous after a year or two. Cone scales have thick tips and long, recurved prickles. **BARK** on mature trees is thin, scaly, and tan to gray. **HABITAT AND RANGE:** Grows in montane and subalpine coniferous forests in mountains from Washington to Baja California. Its lowest elevation limit is around 1,500 m (5,000 ft) in the north and 3,500 m (11,600 ft) in the south. The species has a wide ecological amplitude, being able to thrive in dry, rocky sites as well as in wet meadows and riparian areas. On more fertile sites it is usually out-competed by other trees.

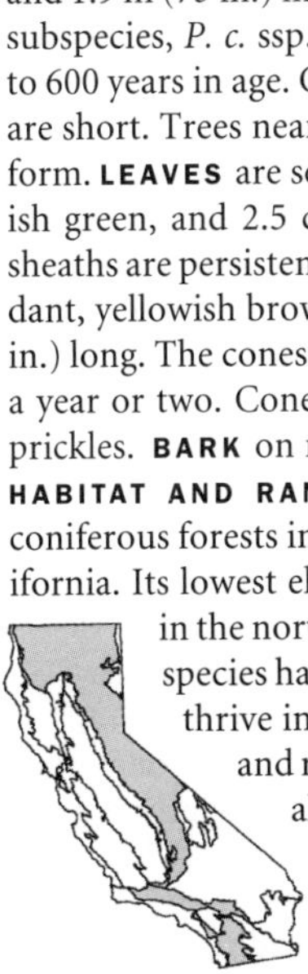

REMARKS: Lodgepole pine is shade and fire intolerant. In Washington, Oregon, and northern California, the species experiences a cycle of mountain pine beetle outbreak, wildfire, and regeneration. Different dynamics occur to the south. In central and southern California, lodgepole pine is susceptible to the defoliating lodgepole needleminer and less so to fire. Occasionally, the species develops persistent, very dense "dog hair thickets" of trees. Lodgepole pine growing in the Sierra Nevada is inaccurately termed "tamarack"—which is actually the common name for eastern larch, a conifer of the Great Lakes states and the Northeast. In the Rocky Mountains, *P. c.* ssp. *latifolia* has variably serotinous cones; stands often develop after widespread, stand-replacing disturbances such as wildfires and insect outbreaks.

COULTER PINE (Fig. 27) *Pinus coulteri*

DESCRIPTION: A medium-sized, erect, single-stemmed tree typically ranging in height from 9 m (30 ft) to 26 m (85 ft) and in diameter from 30 cm (12 in.) to 75 cm (30 in.). The largest tree grows in San Diego County and is 24 m (80 ft) tall and 1.7 m (66 in.) in diameter. Stems are single, erect, and crooked. Crowns are pyramidal among trees grown in relatively dense stands; among open-grown trees, crowns are open and spreading, with branches extending nearly to the ground. The species is short-lived, with an average life span of around 100 years. **LEAVES** are set in bundles of 3; they are stout, dark green to bluish green, and 15 cm (6 in.) to 30 cm (12 in.) long. Bundle sheaths are persistent. **CONES** are symmetrical, oblong to egg shaped, stiff, yellowish brown, variably serotinous, persistent, and 25 cm (10 in.) to 35 cm (14 in.) long. They often occur in whorls of 4. A green cone weighs between 1.8 kg (4 lb) and 2.7 kg (6 lb), making it the most massive pinecone. Cone stalks are up to 10 cm (4 in.) long. Cone scales have massive, sharp, flat "claws." **BARK** on mature trees is thick and fissured and has yellowish plates.

HABITAT AND RANGE: Grows in chaparrals and lower montane woodlands and forests. Its range extends from Contra Costa County in the north to Baja California in the south, at elevations from 150 m (500 ft) to 2,100 m (7,000 ft). The species typically grows on dry, harsh, south-facing slopes and ridges and is tolerant of serpentine soils.

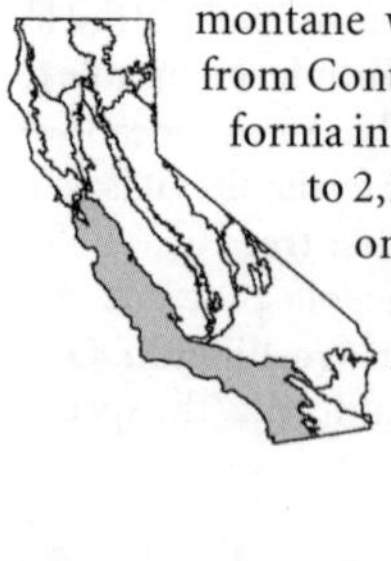

REMARKS: As a mature tree, Coulter pine is intolerant of shade; younger trees

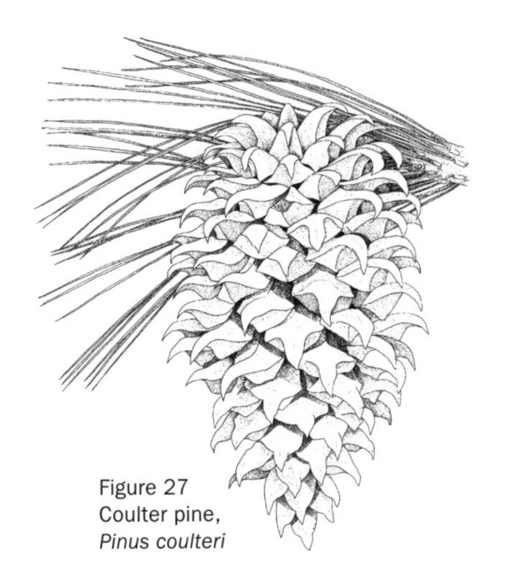

Figure 27
Coulter pine,
Pinus coulteri

can be found growing in some shade. The species is fire and drought tolerant. Coulter pines growing among chaparral shrubs, canyon live oak, or Sargent cypress have serotinous cones. The heat of a fire is necessary to crack the cones' resin seal and allow cone scales to open. Fire suppression has hampered regeneration of Coulter pine in these vegetation types. Coulter pines growing in association with coast live oak, however, are non-serotinous. Native Americans ate Coulter pine seeds. The species is used as firewood and lumber.

LIMBER PINE (Fig. 28) *Pinus flexilis*

DESCRIPTION: A small- to medium-sized, usually erect and single-stemmed tree. Individuals are often multistemmed and crooked. At the tree line, the species can have a dwarfed, shrub form that occurs in isolated clumps 30 cm (1 ft) to 90 cm (3 ft) high. Crowns are broadly rounded. Trees are often open-grown, with branches extending nearly to the ground. A typical mature tree is 7.5 m (25 ft) to 15 m (50 ft) tall and 30 cm (12 in.) to 75 cm (30 in.) in diameter. The largest grows in Uinta National Forest in Utah and is 17.7 m (58 ft) tall and 2.2 m (87 in.) in diameter. Trees can live up to 2,500 years. **LEAVES** are set in bundles of 5, measure 5 cm (2 in.) to 9 cm (3.5 in.) long, and are tufted at the ends of branches. Needles are dark green, curved,

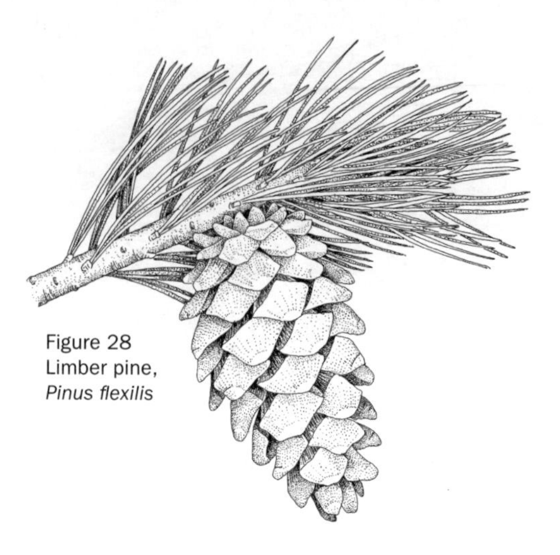

Figure 28
Limber pine,
Pinus flexilis

and somewhat stiff. Stomatal bloom is evident on all surfaces. Bundle sheaths are deciduous. **CONES** are egg shaped to cylindrical, pendant, yellowish brown, and 5 cm (2 in.) to 18 cm (7 in.) long. Stalks are less than 2.5 cm (1 in.) long. Cone scales are thick, but thin at their tips. Scale tips are slightly reflexed and lack prickles. **TWIGS** are thick and very flexible. **BARK** on mature trees is 4 cm (1.5 in.) to 5 cm (2 in.) thick, dark brown, deeply furrowed, and broken into scaly, rectangular plates.

HABITAT AND RANGE: Outside of California the species grows in montane and subalpine coniferous forests and woodlands on dry sites throughout the Rocky Mountains, the Great Basin, and northeastern Oregon. In California it is common east of the central and southern Sierra Nevada crest and atop a few dry mountaintops in southern California, as well as in desert mountains. It is found growing at 900 m (3,000 ft) in the northern part of its range and above 3,300 m (11,000 ft) in the southern part.

REMARKS: Limber pine is intolerant of shade and moderately tolerant of fire. Its seed provides critical nutrition for wildlife, especially the Clark's nutcracker. Limber pine and whitebark pine resemble each other in growth form, although their cones are quite different. Limber pine forms hybrids with *P. albicaulis*.

Figure 29
Jeffrey pine,
Pinus jeffreyi

JEFFREY PINE (Fig. 29) ***Pinus jeffreyi***

DESCRIPTION: A tall, erect, single-stemmed tree. Mature trees are typically 24 m (80 ft) to 43 m (140 ft) tall and 90 cm (3 ft) to 120 cm (4 ft) in diameter. The largest grows in Stanislaus National Forest and is 60 m (197 ft) tall and 2.5 m (98 in.) in diameter. Trees can live up to 500 years. The species typically has long, clear stems with regularly spaced branches. Crowns of mature trees are rounded and somewhat open. **LEAVES** are set in bundles of 3; they are bluish green and 19 cm (7.5 in.) to 28 cm (11 in.) long. Bundle sheaths are persistent. **CONES** are egg shaped to cylindrical, brown to reddish brown, deciduous, and 12.5 cm (5 in.) to 25 cm (10 in.) long. Cone stalks are about 2.5 cm (1 in.) long. Cone scales are brown and have slender, incurved prickles that point downward. **TWIGS** have bluish green tips. **BARK** on mature trees is up to 10 cm (4 in.) thick, with broad plates composed of puzzle-piece–like scales separated by irregular, deep furrows. The bark can smell like vanilla, pineapple, or butterscotch.

HABITAT AND RANGE: Grows in montane and subalpine coniferous forests in southwestern Oregon, throughout much of California and western Nevada, and into Baja California. It primarily grows on harsh sites, such as serpentine soils, and on high-elevation rocky sites, and it is found in pure stands on dry sites east of the Sierra Nevada and Cascade Mountains. It ranges in elevation from 60 m (200 ft) to 3,000 m (10,000 ft).

REMARKS: Jeffrey pine is shade intolerant, resistant to fire, and more tolerant of climatic extremes than ponderosa pine. It is very susceptible to the damaging effects of ozone pollution and has suffered significant damage in the mountains of southern California. Jeffrey pine is used as lumber and for making moldings. Birds (e.g., Clark's nutcracker) eat its seeds, and a variety of mammals eat its stems and roots. Jeffrey pine provides important hiding cover for birds and thermal cover for deer and elk. It hybridizes with ponderosa and Coulter pines.

SUGAR PINE (Fig. 30) *Pinus lambertiana*

DESCRIPTION: An erect, single-stemmed tree. It is the largest and tallest of the world's pines. Mature trees are typically 53 m (175 ft) to 61 m (200 ft) tall and 90 cm (36 in.) to 150 cm (60 in.) in diameter. The largest is in Yosemite National Park and is 82 m (270 ft) tall and 2.8 m (111 in.) in diameter. Trees can live more than 500 years. These trees characteristically have long, clear stems with long, horizontal branches high in the crown. Branches are asymmetrical and often have tufts of leaves extending along the entire branch. In late summer and fall, long, pendant cones hang from the branch ends. The species' growth form is so distinctive that it is possible to identify its silhouette on ridges at distances of up to 8 km (5 mi). **LEAVES** are set in bundles of 5 and measure 5 cm (2 in.) to 10 cm (4 in.) long. Needles are dark green to bluish green, with lines of white stomatal bloom on all surfaces. Bundle sheaths are deciduous. **CONES** are cylindrical, 25 cm (10 in.) to 60 cm (24 in.) long, 10 cm (4 in.) to 12.5 cm (5 in.) in diameter, and pendant. Cone stalks range in length from 5 cm (2 in.) to 15 cm (6 in.). Cone scales are round, thin tipped, and usually pitchy; they lack prickles. **BARK** on mature trees can be up to 10 cm (4 in.) thick and has broad plates composed of reddish brown to purplish puzzle-piece-like scales separated by irregular, deep furrows. The bark on younger trees is grayish green.

Figure 30
Sugar pine,
Pinus lambertiana

HABITAT AND RANGE: Grows in montane and subalpine coniferous forests, woodlands, and chaparrals throughout much of California. Disjunct populations are found in the western Great Basin. The species' northernmost populations occur in north-central Oregon, and its southernmost populations in Baja California. In California it generally grows above 1,200 m (4,000 ft). Its lowest elevation limit is around 300 m (1,000 ft), and its highest limit, in the southern Sierra Nevada, is around 3,000 m (10,000 ft).

REMARKS: Sugar pine is intermediate in its tolerance of shade. Mature trees are resistant to surface fires, and trees of all ages are susceptible to the introduced disease white pine blister rust, which has reduced the species' populations significantly. The nonresinous wood of sugar pine is used as lumber and to make moldings. Native Americans used sugar pine pitch to repair canoes and to fasten feathers to shafts. Native Americans and settlers ate the sugary, mildly cathartic resin.

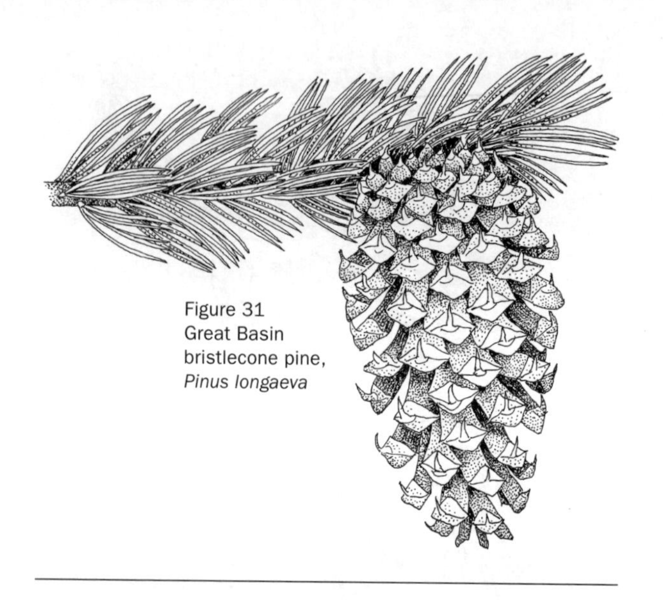

Figure 31
Great Basin
bristlecone pine,
Pinus longaeva

GREAT BASIN BRISTLECONE PINE (Fig. 31) ***Pinus longaeva***

DESCRIPTION: A short, stocky, contorted tree that often has multiple stems. Crowns are sparse to absent on individual stems. Older trees may have thin strips of cambium on their leeward sides supporting exceedingly spare canopies. In California, mature trees are typically 9 m (30 ft) to 12 m (40 ft) tall and 30 cm (1 ft) to 1.8 m (6 ft) in diameter. The largest grows in the White Mountains and is 14 m (47 ft) tall and 3.8 m (12.5 ft) in diameter. Trees can live to be 3,000 to 4,000 years old. A specimen from Wheeler Peak in eastern Nevada, now dead, was determined to have lived more than 5,000 years and is considered to be the oldest tree. **LEAVES** are set in bundles of 5 and measure 12 mm (.5 in.) to 40 mm (1.5 in.) long. They can persist on branches for decades and are tightly arranged bottlebrush style, producing a foxtail-like appearance. Needles are dark green, stiff, and somewhat curved. Scattered resin droplets dot the needles, and white stomatal lines appear on their inner surfaces. Bundle sheaths are deciduous. **CONES** are egg shaped to barrel shaped, pendant, reddish brown, and 5 cm (2 in.) to 12.5 cm (5 in.) long. Cone stalks are less than 2 cm (.75 in.) long. Cone-scale tips are somewhat thick and have slender, persistent, bristlelike prickles that measure up to 6 mm (.25 in.) long. **BARK** is reddish brown.

HABITAT AND RANGE: Grows in subalpine coniferous woodlands on desert mountains in the Great Basin regions of California, Nevada, and western Utah, at elevations from 2,300 m (7,500 ft) to 3,600 m (12,000 ft). In California it is most extensive on dolomitic soils in the White Mountains. Other populations are found in the Inyo, Last Chance, and Panamint Mountains. It is an uncommon species.

REMARKS: Great Basin bristlecone pine is exceedingly slow growing. One tree growing in the White Mountains was determined to be 90 cm (3 ft) tall, 7.5 cm (3 in.) in diameter, and 700 years old. Dendrochronologists have created a chronology extending thousands of years beyond the ages of the oldest living trees by cross dating tree-ring patterns of standing live trees with standing dead trees, and standing dead trees with logs. The closely related Rocky Mountain bristlecone pine (*P. aristata*) and Great Basin bristlecone pine were once considered a single species.

SINGLELEAF PINYON (Fig. 32) *Pinus monophylla*

DESCRIPTION: A low, spreading tree, usually with multiple trunks. Mature trees are usually 4.5 m (15 ft) to 15 m (50 ft) tall and 15 cm (6 in.) to 38 cm (15 in.) in diameter. The largest tree grows in Inyo County and is 13.7 m (45 ft) tall and 1.3 m (52 in.) in diameter. Trees are often more than 400 years old. Crowns are conical on young trees, becoming rounded on older trees. **LEAVES** are set in bundles of 1 (sometimes 2); they are 2.5 cm (1 in.) to 6 cm (2.5 in.) long and often incurved. Needles are grayish green, circular in cross section, thick, stiff, and sharp pointed. Bundle sheaths are deciduous. **CONES** are egg shaped, very broad at the base, 4 cm (1.5 in.) to 11 cm (4.5 in.) long, and often erect. Scales are thick, with deep recesses housing 2 large, egg-shaped seeds. **BARK** on mature trees is dark brown and furrowed and has narrow, scaly ridges.

HABITAT AND RANGE: Grows in inland coniferous woodlands and is the most common tree in the Great Basin. In Cal-

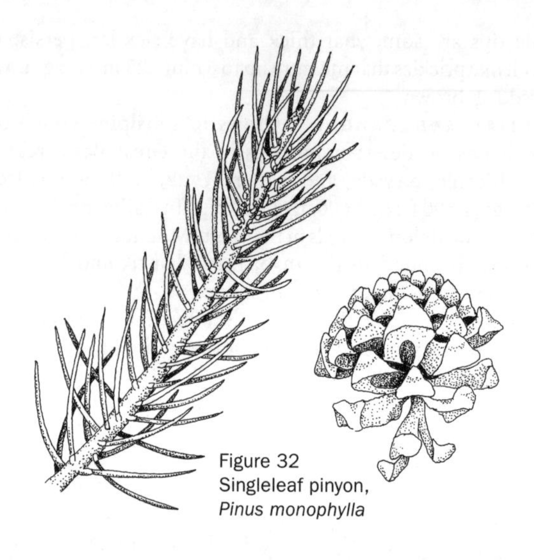
Figure 32
Singleleaf pinyon,
Pinus monophylla

ifornia it is mostly found growing in the Desert Province above valley floors. Significant populations are found in the Transverse Ranges and southern Sierra Nevada. Scattered populations can be found west of the Sierra Nevada crest, some of which are thought to have been sown by Native Americans after they returned from trading with tribes east of the Sierra Nevada. The species can occur in pure stands but is more commonly found growing in mixed woodlands. Utah juniper is its most common associate. Singleleaf pinyon ranges in elevation from 900 m (3,000 ft) to 3,000 m (10,000 ft).

REMARKS: Two-leaved forms of singleleaf pinyon in southeastern and southern California have been suggested to be the result of hybridization with the 2-needled *P. edulis* or the 5-needled *P. juarezensis* that grows in Mexico. Two-needled pinyon populations in the Mid Hills and New York Mountains may be *P. edulis*. Singleleaf pinyon is intolerant of shade and is very susceptible to fire. The seeds of singleleaf pinyons are edible and have been an important staple for Native Americans and wildlife. The species is also used as Christmas trees and for making fence posts.

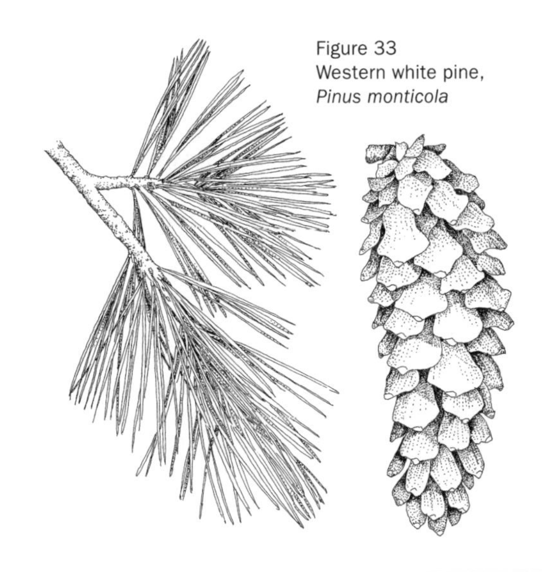

Figure 33
Western white pine,
Pinus monticola

WESTERN WHITE PINE *Pinus monticola*

(Fig. 33; Pl. 6)

DESCRIPTION: A tall, erect, single-stemmed tree. Mature trees are typically 37 m (120 ft) to 55 m (180 ft) tall and 60 cm (24 in.) to 120 cm (48 in.) in diameter. The largest grows in El Dorado National Forest and is 46 m (151 ft) tall and 2.7 m (105 in.) in diameter. The trees can live up to 600 years. They typically have long, clear stems with regularly spaced branches. Crowns are usually narrow, short, and somewhat open. **LEAVES** are set in bundles of 5 and are 5 cm (2 in.) to 10 cm (4 in.) long. Needles are slender and bluish green, with lines of white stomatal bloom on their inner surfaces only. Bundle sheaths are deciduous. **CONES** are cylindrical, pendant, yellowish brown, curved, 12.5 cm (5 in.) to 25 cm (10 in.) long, and 4 cm (1.5 in.) to 5 cm (2 in.) wide. They are often found in clusters of 1 to 7 at the ends of branches. Cone stalks range in length from 2.5 cm (1 in.) to 5 cm (2 in.). Cone scales are round, thin lipped, and usually pitchy; they lack prickles. **BARK** on mature trees is thin, rarely more than 4 cm (1.5 in.) thick. It is grayish brown and split into rectangular plates, much like the skin of an alligator. The bark on young trees is light brown to grayish green.

HABITAT AND RANGE: Grows in montane and subalpine coniferous forests from the mountains of California north into the Pacific Northwest and the northern Rocky Mountains. In California it primarily grows in the upper montane zone or higher, from 1,500 m (5,000 ft) to 3,300 m (11,000 ft). In Del Norte County it grows as low as 150 m (500 ft) on serpentine soils.

REMARKS: Western white pine is intermediate in its shade tolerance and resistance to fire. Trees of all ages are susceptible to the introduced disease white pine blister rust, which has reduced the species' populations significantly. The nonresinous wood is used as lumber and to make moldings. Western white pine trees provide important cover and food for squirrels, mice, and various birds.

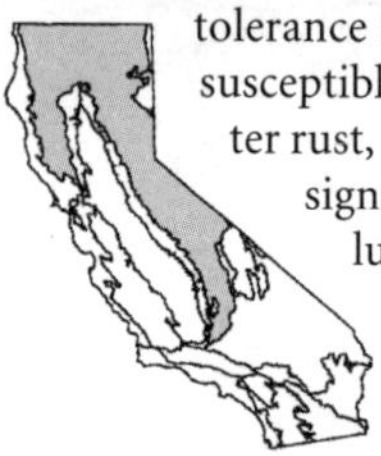

BISHOP PINE (Fig. 34) *Pinus muricata*

DESCRIPTION: A medium-sized tree typically ranging in height from 12 m (40 ft) to 24 m (80 ft) and in diameter from 60 cm (2 ft) to 90 cm (3 ft). Stems are single, erect, and somewhat crooked and irregular. The largest tree grows in Mendocino County and is 34 m (112 ft) tall and 1.4 m (54 in.) in diameter. The species is short-lived, and trees often succumb to disease by the time they are 80 to 100 years old. Old trees probably do not exceed 200 years in age. Mature trees have open, irregularly branched, rounded crowns. **LEAVES** are set in bundles of 2; they are twisted, thick, and 10 cm (4 in.) to 15 cm (6 in.) long. Northern populations of Bishop pine have bluish green needles, while southern populations have green to yellowish green needles. **CONES** are asymmetrically egg shaped and serotinous. They are set in whorls of 3 to 5, persist for many years, and are 5 cm (2 in.) to 10 cm (4 in.) long. New cones are brown and weather to gray while attached to the tree. Scales are thick and have stout, flat or curved prickles at their tips. **BARK** on mature trees is thick, brown, and irregularly furrowed.

HABITAT AND RANGE: In California the species grows in disjunct coastal and island coniferous forests, from Humboldt

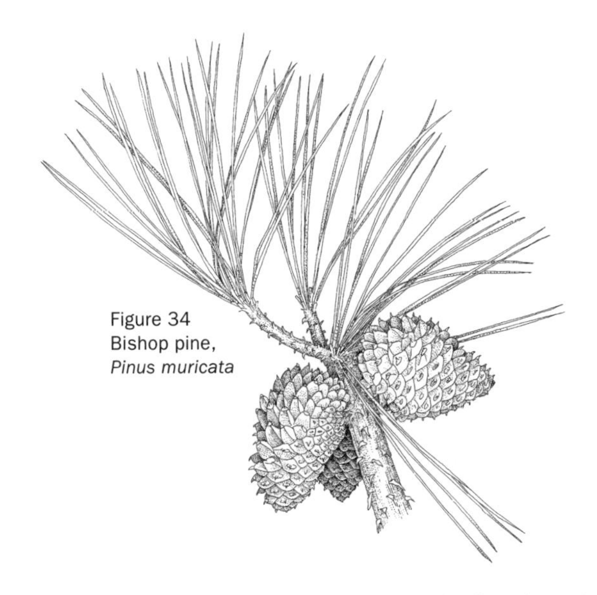

Figure 34
Bishop pine,
Pinus muricata

County to Santa Barbara County. Stands are also found in Baja California and on Cedros Island. The species occurs from near sea level to 400 m (1,300 ft).

REMARKS: Bishop pine leaf and cone characteristics are highly variable within populations and make determination of status difficult. Two forms are generally recognized, based on leaf color: the green leaf form of southern California and the bluish green form found in northern California. Most Bishop pine stands are even-aged, having originated following a fire. The heat of a fire or from a hot, low-humidity day may break the resin seal of cones and allow the scales to open.

PONDEROSA PINE (Fig. 35) ***Pinus ponderosa***

DESCRIPTION: A tall, erect, single-stemmed tree. Mature trees are typically 30 m (100 ft) to 55 m (180 ft) tall and 90 cm (3 ft) to 120 cm (4 ft) in diameter. The largest grows in the Trinity Alps Wilderness and is 68 m (223 ft) tall and 2.4 m (94 in.) in diameter. Trees are long-lived and may reach 600 years in age.

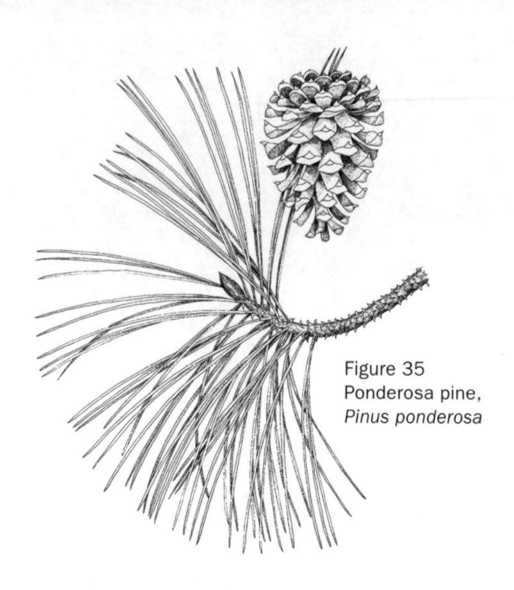

Figure 35
Ponderosa pine,
Pinus ponderosa

Trees typically have long, clear stems with regularly spaced branches. Crowns are conical and somewhat open. **LEAVES** are set in bundles of 3; they are green to yellowish green and 10 cm (4 in.) to 28 cm (11 in.) long. Bundle sheaths are persistent. **CONES** are more or less symmetrical, egg shaped to conical, brown, deciduous, and 7.5 cm (3 in.) to 16 cm (6.5 in.) long. Cone stalks are less than 2.5 cm (1 in.) long. Cone scales are dark reddish brown and have slender prickles that point outward. **BARK** on mature trees can be up to 10 cm (4 in.) thick, with broad plates composed of puzzle-piece-like scales separated by deep furrows. Young trees have reddish brown plates and blackened furrows. Old trees often develop yellowish brown plates and scales. The undersides of these scales can be yellow.

HABITAT AND RANGE: Grows in montane coniferous forests, woodlands, and chaparrals throughout much of western North America. In California it is widely distributed and generally grows at low and mid elevations, from 150 m (500 ft) in the Sacramento Valley to over 2,900 m (9,500 ft) in the Sierra Nevada. The species occurs along creeks east of the Sierra Nevada and the White Mountains.

REMARKS: Three varieties of ponderosa pine are recognized: *P. p.* var. *ponderosa* (in the Pacific states) with 3 needles per bundle; *P. p.* var. *arizonica* (in Arizona and New Mexico) with 5 needles per bundle; and *P. p.* var. *scopulorum* (in the Rocky Mountains) with 2 or 3 needles per bundle. In addition, 3 races within *P. p.* var. *ponderosa* have been recognized. They are the Pacific Race, with its larger needles, cones, and seeds, found north of the Transverse Ranges and west of the Sierra Nevada into Oregon; the North Plateau Race, with its thick hypodermal layers and sunken stomata, found east of the Sierra Nevada into the eastern Pacific Northwest; and a Southern California Race with distinctive monoterpenes. Ponderosa pine hybridizes with Jeffrey and Washoe pines. It is shade intolerant and fire resistant. Ponderosa pine is a valuable lumber species and provides significant cover and food for wildlife. Bark beetle epidemics can occur in ponderosa pine stands following droughts. Native Americans used the species' resin for a variety of medicinal purposes.

PARRY PINYON (Fig. 36) *Pinus quadrifolia*

DESCRIPTION: A short-trunked, spreading tree. Mature trees are usually 4.5 m (15 ft) to 11 m (35 ft) tall and 30 cm (12 in.) to 60 cm (24 in.) in diameter. The largest grows in Riverside County and is 16 m (53 ft) tall and 68 cm (27 in.) in diameter. The trees can live to be 200 to 500 years old. Younger trees have a pyramidal crown, while older ones have a rounder crown. **LEAVES** are set in bundles of 4 (sometimes 3 or 5), measure 2.5 cm (1 in.) to 5 cm (2 in.) long, and are incurved. Needles have white stomatal bloom on their inner surfaces and are bluish green on the outside. They are stiff and thick. Bundle sheaths are deciduous. **CONES** are egg shaped to spherical, often erect, 3 cm (1.25 in.) to 7.5 cm (3 in.) long, and light brown. Scales are thick at their tips, with deep recesses housing the seeds. Stalks are less than 12 mm (.5 in.) long. **BARK** on mature trees is thin, reddish brown, and furrowed, with narrow, scaly ridges.

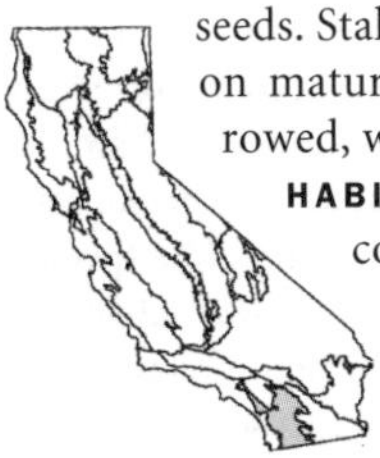

HABITAT AND RANGE: Grows in montane coniferous woodlands and chaparrals. It is widespread in northern Baja California. In California, populations are found in Riverside County growing in chaparrals,

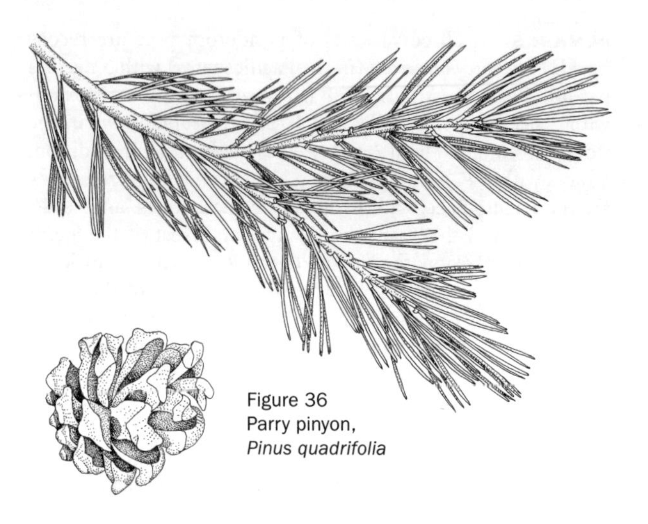

Figure 36
Parry pinyon,
Pinus quadrifolia

and in southeastern San Diego County. The species occurs in association with California juniper as well as other species in the mixed woodlands of southern California and northern Baja California. It inhabits rocky, thin soils at elevations from 1,200 m (4,000 ft) to 2,600 m (8,500 ft).

REMARKS: Parry pinyon is intolerant of shade but tolerant of drought. It is easily killed by fire because of its thin bark and low branches. Some people believe that Parry pinyon is not a separate species but a hybrid between *P. monophylla* and the recently described 5-needled pinyon, *P. juarezensis,* found in Mexico. As with other pinyons, the seeds of Parry pinyon are edible and have been an important staple of Native Americans and a variety of wildlife species. It is also used for firewood and fence posts.

MONTEREY PINE (Fig. 37; Pl. 7) *Pinus radiata*

DESCRIPTION: A medium-sized tree usually ranging in height from 15 m (50 ft) to 38 m (125 ft) and in diameter from 60 cm (2 ft) to 90 cm (3 ft). Stems are single, erect, and somewhat crooked. The largest tree grows in Napa, California, and is 37 m (120 ft) tall and 1.9 m (74 in.) in diameter. The species is short-lived and often succumbs to disease or insects at age 80 to 100. Mature trees have an irregular, rough branching pattern, which, in special situations, can be quite picturesque. Crowns

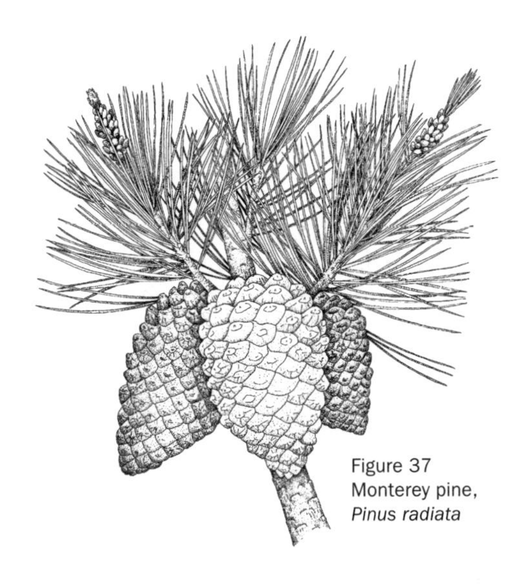

Figure 37
Monterey pine,
Pinus radiata

are rounded. **LEAVES** are set in bundles of 3 (sometimes 2); they are dark green and 10 cm (4 in.) to 15 cm (6 in.) long. Bundle sheaths are persistent. **CONES** are asymmetrical and have a broad egg shape. They are set in whorls of 3 to 5; are recurved, serotinous, and persistent; and measure 7.5 cm (3 in.) to 14 cm (5.5 in.) long. Scales have minute prickles and are massive on the upper side of the cone, especially near the base, where they have pronounced, round knobs. Elsewhere they are thick and flat. **BARK** on mature trees has deep fissures separating narrow dark brown to blackened ridges.

HABITAT AND RANGE: In California, native Monterey pine is considered rare and grows in 3 disjunct coastal areas: Año Nuevo–Swanton, Monterey-Carmel, and Pico Creek–Cambria. It also occurs in Mexico on Cedros and Guadalupe Islands. The species is found at elevations below 400 m (1,300 ft). Dry, sandy loam soils, a coastal climate, and summer fog are site characteristics for this species.

REMARKS: Millions of acres of Monterey pine have been planted worldwide, especially in the Southern Hemisphere. It is known for rapid growth and its potential for genetic improvement. The Cedros and Guadalupe Island populations have 2 leaves per bundle and are recognized as a separate variety (*P. r.* var. *binata*). Monterey pine forms natural hybrids with *P. attenuata* and *P. muricata*. It is intermediate in its shade and fire tolerance. Monterey pine has been widely planted as an ornamental because of its rapid growth and beauty. Unfortunately, many trees that were planted 60 to 80 years ago are now succumbing to various native and introduced rots or insects. Native populations of Monterey pine are being threatened by urbanization, fire suppression, pathogens—notably pine pitch canker (*Fusarium subglutinans* forma *pini*)—and, on Guadalupe Island, goats.

GHOST PINE (Fig. 38; Pl. 8) ***Pinus sabiniana***

DESCRIPTION: A medium-sized, erect, single-stemmed tree. Mature trees are typically 12 m (40 ft) to 24 m (80 ft) tall and 30 cm (12 in.) to 90 cm (36 in.) in diameter. The largest tree grows in Redding and is 49 m (161 ft) tall and 1.5 m (59 in.) in diameter. Trees often have multiple, leaning, crooked trunks. Crowns of mature trees are open, with sparse foliage on large, irregularly spaced branches. The species self-prunes, resulting in branches high on the stems. Trees are thought to live more than 200 years. **LEAVES** are set in bundles of 3; they are grayish green, 18 cm (7 in.) to 43 cm (17 in.) long, and flexible. Bundle sheaths are persistent. **CONES** are more or less symmetrical, oblong to egg shaped, brown, variably serotinous, persistent, and 15 cm (6 in.) to 25 cm (10 in.) long. Green cones weigh between .33 kg (.75 lb) and 1 kg (2.2 lb). Cone stalks are as long as 7 cm (2.75 in.). Cone scales have massive, sharp, flat "claws." **BARK** on mature trees is dark gray and has irregular fissures.
HABITAT AND RANGE: Grows in foothill woodlands and chaparrals. It is commonly associated with blue oak around the Sacramento and San Joaquin Valleys. The species is endemic to California, and its range extends from the Salmon River in Siskiyou County south to Piru Creek in Ventura County. There is a notable 88-km (55-mi) gap in its range in the southern Sierra Nevada foothills. The species grows at elevations from 30 m (100 ft) to 2,100 m (7,000 ft). 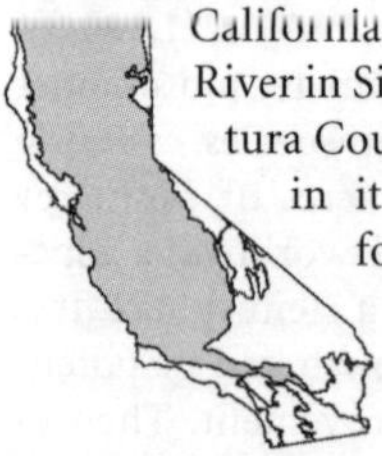
REMARKS: Ghost pine is intolerant of shade and is considered a fire-dependent

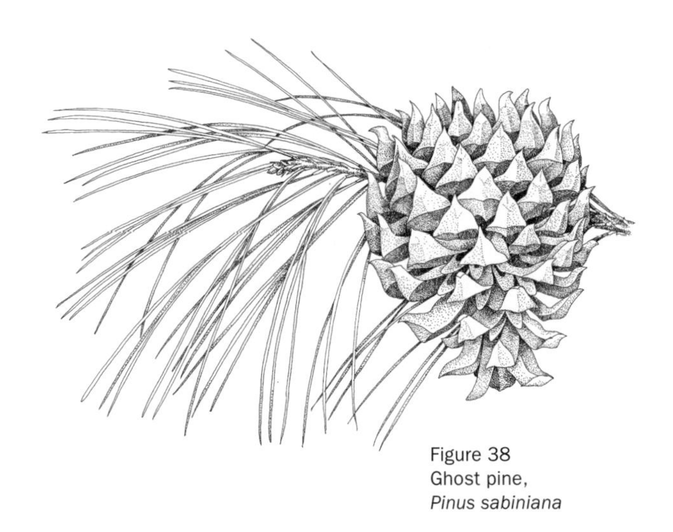

Figure 38
Ghost pine,
Pinus sabiniana

species. Mature trees are easily injured by fire, but fire is necessary to reduce competition and create suitable seedbeds. Ghost pine woodlands are valued as winter range for many wildlife species. Unfortunately, excessive fuelwood cutting, type conversion, and housing developments are reducing ghost pine populations. Native Americans ate the large-seeded ghost pine seeds and parts of its cones, bark, and buds. Its resin was used for medicinal purposes. Ghost pine is also known as foothill pine, bull pine, digger pine, and gray pine.

TORREY PINE (Fig. 39) *Pinus torreyana*

DESCRIPTION: A small- to medium-sized tree typically ranging in height from 7.5 m (25 ft) to 18 m (60 ft) and in diameter from 30 cm (12 in.) to 60 cm (24 in.). The largest tree grows in Carpenteria, outside of its native range, and is 38 m (126 ft) tall and 2 m (78 in.) in diameter. Stems are single and erect. Crowns are open and spreading, with many large branches. On exposed sites, the species can become shrubby. It is short-lived; the oldest individual is around 150 years old. **EAVES** are set in bundles of 5; they are grayish green to bluish green, thick, and 15 cm (6 in.) to 25 cm (10 in.) long. Bundle sheaths are persistent. **CONES** have a broad egg shape to spher-

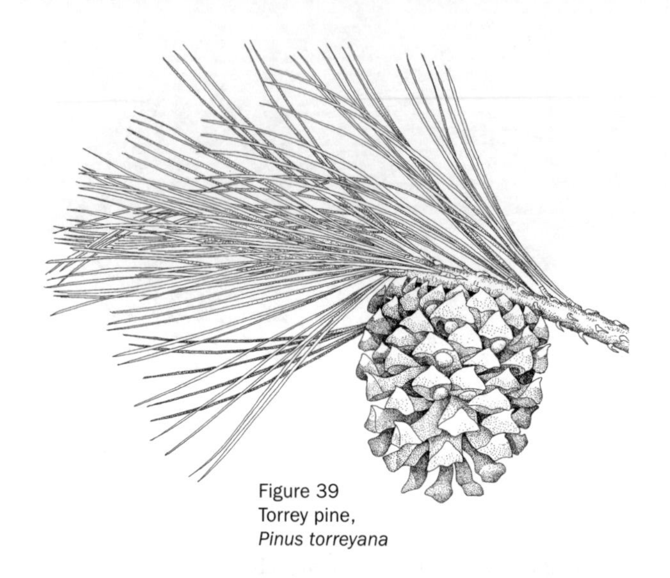

Figure 39
Torrey pine,
Pinus torreyana

ical shape. They are dark brown, weakly serotinous, persistent, and 10 cm (4 in.) to 15 cm (6 in.) long. Cone stalks are up to 4 cm (1.5 in.) long. Scales are thick, with pyramid-like tips and small prickles. **BARK** on mature trees is thick and fissured and has reddish brown scales on flat ridges.

HABITAT AND RANGE: A rare California endemic that grows in 2 areas: near San Diego and on Santa Rosa Island. There are fewer than 9,000 individuals of the species. It lives in coastal woodlands and chaparrals and is adapted to areas with high humidity, moderate temperatures, and infertile, dry soils. It grows at elevations between 60 m (200 ft) and 150 m (500 ft).

REMARKS: Torrey pine is shade tolerant and apparently resistant to surface fires. It has low genetic diversity, and maintaining this rare species will require careful management. Regeneration in the mainland population has been difficult in recent years, presumably because of lack of fire. Grazing pressure on Santa Rosa Island may have created enough growing space to account for higher regeneration rates there. Torrey pine is planted as an ornamental, and in the Southern Hemisphere it has been tested for its timber value.

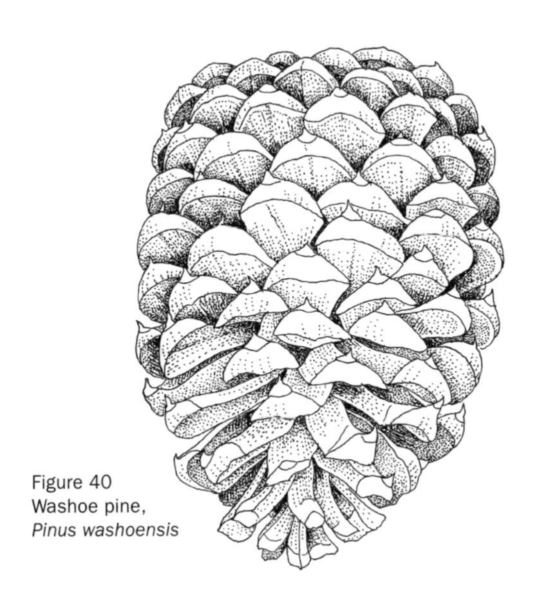

Figure 40
Washoe pine,
Pinus washoensis

WASHOE PINE (Fig. 40) ***Pinus washoensis***

DESCRIPTION: A medium-sized to tall, erect, single-stemmed tree. Mature trees are usually 12 m (40 ft) to 30 m (100 ft) tall and 30 cm (12 in.) to 60 cm (24 in.) in diameter. The largest grows in the Warner Mountains and is 49 m (161 ft) tall and 1.6 m (65 in.) in diameter. Trees can be more than 300 years old. Trees typically have long, clear stems with regularly spaced branches. Crowns are conical and somewhat open, or occasionally flat topped. **LEAVES** are set in bundles of 3; they are dark green and 7.5 cm (3 in.) to 12.5 cm (5 in.) long. Bundle sheaths are persistent. **CONES** are egg shaped, symmetrical, brown, deciduous, and 5 cm (2 in.) to 9 cm (3.5 in.) long. Cone stalks are more than 12 mm (.5 in.) long. Cone scales are brown and have slender, straight prickles that point downward. **BARK** on mature trees has shallow furrows and scaly, yellowish brown plates.

HABITAT AND RANGE: Grows in upper montane and subalpine coniferous forests. It is uncommon and is found in scattered stands in northwestern Nevada and northeastern California. In particular, sizable populations can be found on Mount Rose, Nevada; the southern Warner Mountains; and the Bald Mountain Range in northeastern California. The species has also been reported in Oregon and British Columbia. It grows at elevations between 1,450 m (4,800 ft) and 2,600 m (8,500 ft). It occurs in pure stands at the higher elevations and is associated with other conifer species elsewhere.

REMARKS: Washoe pine appears to be derived from the Rocky Mountain variety of ponderosa pine (*P. p.* var. *scopulorum*). It is known to hybridize with the Pacific form of ponderosa pine (*P. p.* var. *ponderosa*) and rarely with Jeffrey pine. White fir may dominate higher-elevation stands of Washoe pine in the absence of thinning fires. It appears to be shade intolerant and resistant to fire, although increased fuel loads because of fire suppression now present a fire hazard.

PSEUDOTSUGA (DOUGLAS-FIR)

The genus *Pseudotsuga* is generally recognized as having 5 species: 3 in eastern Asia and 2 in western North America, including California. Douglas-firs are evergreen and tall. *P. menziesii,* for example, can attain heights of over 91 m (300 ft). In overall appearance the trees are reminiscent of spruce trees.

Leaves are flat, linear, and grooved above, and they have tapered petioles. Stomatal bloom is on the undersides of leaves. Leaves arise from minute pegs. Bruised leaves emit a sweet aroma. Seed cones are pendant and are found near the ends of twigs or in the axils of upper leaves. The spirally arranged, leathery cone scales are subtended by 3-pronged bracts that are longer than the cone scales. Cones and seeds mature in 1 year. Buds are sharp pointed, with overlapping scales. They are broadest near the middle and taper at both ends.

1. Seed cones are less than 9 cm (3.5 in.) long. Three-pronged bracts are well exserted between cone scales. Leaf tips are round . . . . **Douglas-fir (*P. menziesii*)**
1. Seed cones are more than 9 cm (3.5 in.) long. Three-pronged bracts are barely exserted between cone scales. Leaf tips are pointed . . . . **bigcone Douglas-fir (*P. macrocarpa*)**

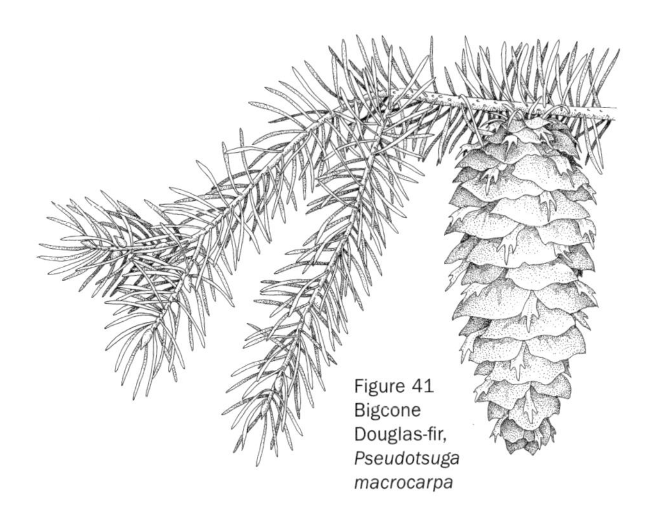

Figure 41
Bigcone
Douglas-fir,
Pseudotsuga macrocarpa

BIGCONE DOUGLAS-FIR *Pseudotsuga macrocarpa* (Fig. 41)

DESCRIPTION: A small- to medium-sized, erect, single-stemmed tree, averaging about 18 m (60 ft) tall and 75 cm (30 in.) in diameter. The largest specimen is in the Angeles National Forest and is 44 m (145 ft) tall and 2.3 m (91 in.) in diameter. Mature trees often have thick, tapering stems with long, sweeping branches. Leaves can be found along the entire length of most branches, a feature reminiscent of sugar pine. The oldest trees exceed 600 years old. **LEAVES** are linear, 2 cm (.75 in.) to 3 cm (1.25 in.) long, and blue green. The apex is pointed, and leaves arise from minute pegs. Stomatal bloom appears on the undersides of leaves. Leaves are arranged bottlebrush style. **CONES** are leathery, 10 cm (4 in.) to 15 cm (6 in.) long, and pendant. Three-lobed bracts are exserted between the cone scales and barely extend beyond the ends of the scales. **BARK** on mature trees is dark reddish brown and has deep furrows between large round ridges. **BUDS** are spindle shaped, sharp pointed, and encased with overlapping bud scales.

HABITAT AND RANGE: Grows in montane coniferous forests and is mostly confined to the Transverse and Peninsular Ranges of southern California. Its southern limit is in San Diego County and its northern limit is in Kern County. It primarily grows on dry slopes and canyons at elevations from 600 m (2,000 ft) to 1,800 m (6,000 ft). The species is often found in chaparrals or intermixed with Coulter pine, ponderosa pine, other conifers, and/or broadleaved trees.

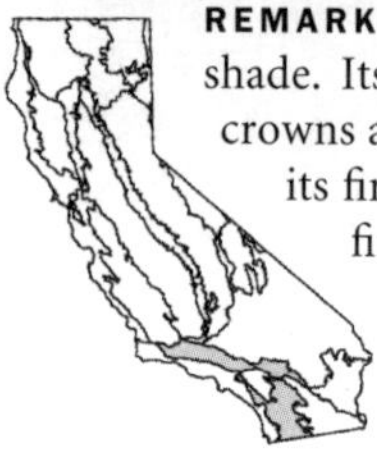

REMARKS: Bigcone Douglas-fir is intolerant of shade. Its thick bark and its ability to sprout new crowns and branches following fire are examples of its fire adaptations, although repeated or severe fire will kill the trees. The species' primary value is its association with maintaining watersheds, wildlife habitat, and scenic beauty.

DOUGLAS-FIR (Fig. 42) ***Pseudotsuga menziesii***

DESCRIPTION: An erect, tall, single-stemmed tree. Mature trees are typically 38 m (125 ft) to 61 m (200 ft) tall and 60 cm (24 in.) to 1.5 m (60 in.) in diameter. The largest is in Coos County, Oregon, and is 100 m (329 ft) tall and 3.5 m (139 in.) in diameter. The oldest trees live more than 750 years. The species has long, clear stems and rounded, spreading crowns. **LEAVES** are linear, flexible, 12 mm (.5 in.) to 4 cm (1.5 in.) long, and sweet smelling. Leaves are arranged bottlebrush style. Stomatal bloom can be found underneath. Leaf tips are round. The round leaf scars have raised lips. **CONES** are 5 cm (2 in.) to 10 cm (4 in.) long and pendant. Three-lobed bracts are exserted between the cone scales. The bracts resemble the hind end of a mouse entering a hole with its tail and rear legs protruding. **BARK** on mature trees is deeply furrowed and corky. The furrows often appear in an elongated diamond pattern. A cross section of the outer bark reveals alternating bands of reddish brown and yellow, similar to bacon. Bark on young trees has resin blisters. **BUDS** are spindle shaped, sharp pointed, and encased with overlapping bud scales. **HABITAT AND RANGE:** Found in coniferous and hardwood forests throughout much of northern California. Its southern limit in the Coast Ranges is in Santa Barbara County, and in the Sierra Nevada it is in Fresno County; the species' range extends well into British Columbia. The species is more widespread in the northern coastal counties. In California it can be found from sea

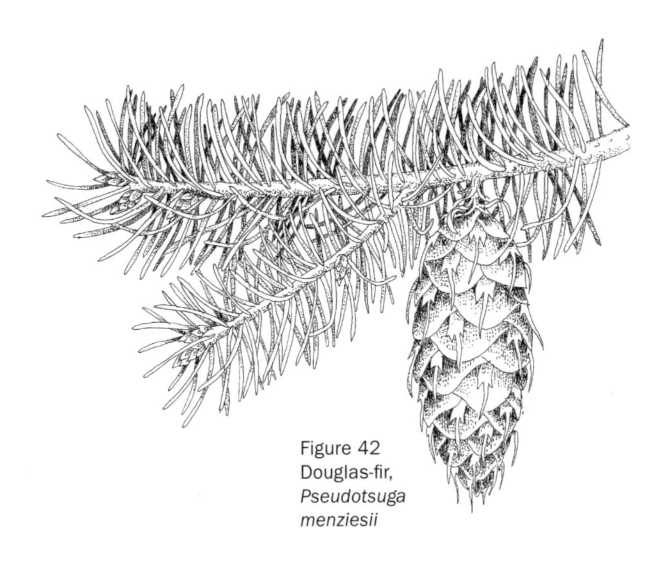

Figure 42
Douglas-fir,
Pseudotsuga menziesii

level to 2,200 m (7,200 ft). Plants growing in California are considered the typical variety (*P. menziesii* var. *menziesii*). A second variety, *P. menziesii* var. *glauca,* occurs in the Rocky Mountains.
REMARKS: Douglas-fir is intermediate in its shade tolerance. Its thick bark provides considerable protection from fire. The species is susceptible to numerous fungi and insects, as well as damage from mammals eating the inner bark. Douglas-fir is the major timber-producing species in western North America. It is used as lumber and to make pulp, plywood, and particleboard. Old-growth Douglas-fir forests are host to many rare or threatened species, such as the northern spotted owl and the marbled murrelet. Because of these and other species, forest management in the Pacific Northwest and California is adjusting to allow for wildlife recovery.

SEQUOIA (REDWOOD)

There is only 1 species of redwood. As many as 40 fossil forms are recognized as having been widely distributed in forests of the Northern Hemisphere.

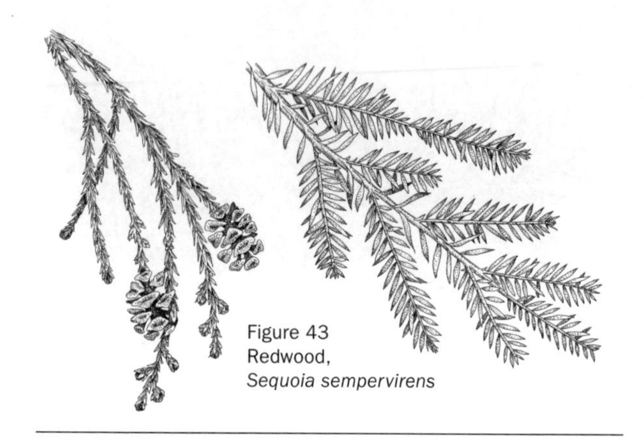

Figure 43
Redwood,
Sequoia sempervirens

REDWOOD (Fig. 43; Pl. 9) ***Sequoia sempervirens***

DESCRIPTION: This is the tallest tree in the world. Mature trees are typically 61 m (200 ft) to 91 m (300 ft) tall and 1.8 m (6 ft) to 4.5 m (15 ft) in diameter. The largest grows in Humboldt Redwoods State Park and is 111 m (363 ft) tall and 5 m (17 ft) in diameter. Trees more than 110 m (360 ft) tall can also be found in Mendocino County. One tree was determined to be 2,200 years old. Mature redwoods have long, clear stems and rounded, spreading crowns. Occasionally, rings of redwood sprouts can be found at the bases of injured trees or stumps. **LEAVES** are linear, 2 cm (.75 in.) to 2.5 cm (1 in.) long, dark green above, and arranged in flat sprays. Leaves have no petioles. Stomatal bloom appears only on the underneath surfaces. Leaf tips are pointed. Leaves from the upper canopy, especially on cone-bearing branches, can be awl-like, resembling giant sequoia foliage. Leaves fall as sprays. **CONES** are barrel shaped, brown, 2 cm (.75 in.) to 3 cm (1.25 in.) long, and persistent. Scales are umbrella-like. **BARK** is fibrous, deeply furrowed, reddish brown, and thick.

HABITAT AND RANGE: Grows in coastal coniferous forests from Curry County, Oregon, south to Monterey County. It ranges in elevation from sea level to over 900 m (3,000 ft). The species is intolerant of strong, salt-laden ocean breezes and is rarely found adjacent to the ocean; rather, forests dominated by coastal scrub or Sitka spruce separate redwood from the ocean. Redwood is thought of as a fog-belt species, but it can occur inland (e.g., in Napa and Del Norte Counties) on sites with sufficient summer soil moisture.

REMARKS: Redwood is shade tolerant, although it grows best in full sunlight. It regenerates best from seed on mineral soil and a variety of organic seedbeds, but not in its own litter. It can sprout prolifically on young stumps, but less so on older stumps or at the bases of mature trees following fire. Redwood is best known for its nearly pure stands on alluvial flats and lower slopes in southern Humboldt County. Only giant sequoia groves surpass the biomass of redwood stands. Old-growth redwood forests are home to the spotted owl and marbled murrelet as well as other rare or endangered species. Redwood is used as lumber and to make decking and souvenirs.

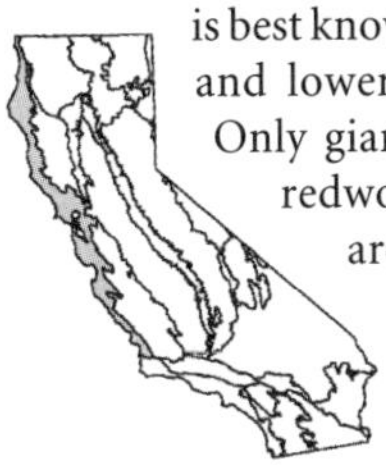

SEQUOIADENDRON (GIANT SEQUOIA)

There is only 1 species of giant sequoia. Like redwood, it has many recognized fossil forms.

GIANT SEQUOIA *Sequoiadendron giganteum*

(Fig. 44; Pl. 10)

DESCRIPTION: An erect, tall, single-stemmed tree. It is the largest tree in the world. Mature trees are usually 69 m (225 ft) to 84 m (275 ft) tall and 3 m (10 ft) to 6 m (20 ft) in diameter. The largest tree grows in Sequoia National Park and is 84 m (275 ft) tall and 9.8 m (32 ft) in diameter. The General Grant tree is over 12 m (40 ft) in diameter. Some trees are thought to live as long as 2,500 to 3,200 years. Mature trees have long, clear, stout stems and rounded, spreading crowns. **LEAVES** are sharp-pointed and awl-like, bluish green, and about 6 mm (.25 in.) long. Stomatal bloom appears on all surfaces. Leaves fall as sprays. **CONES** are barrel shaped, 4.5 cm (1.75 in.) to 7 cm (2.75 in.) long, serotinous, and brown. Scales are umbrella-like and wrinkled. **BARK** is fibrous, deeply furrowed, and cinnamon red. Bark on mature trees can be up to 60 cm (2 ft) thick.

HABITAT AND RANGE: Grows in montane coniferous forests in the western Sierra Nevada, at elevations from 800 m (2,700 ft) to 2,700 m (8,800 ft). There are about 75 scattered

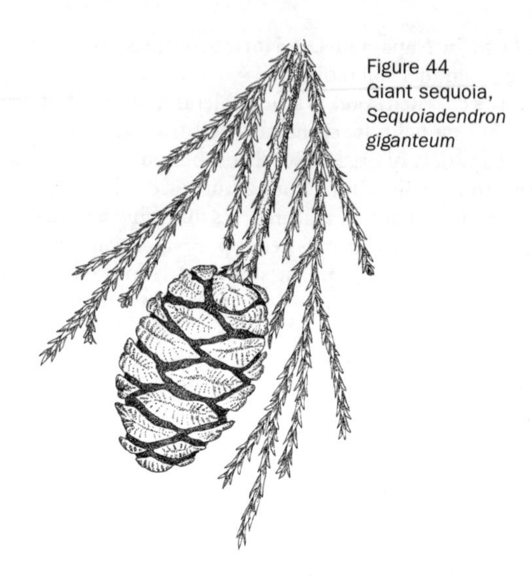

Figure 44
Giant sequoia, *Sequoiadendron giganteum*

groves of this species from Placer County to southern Tulare County. It is best represented in a 113-km-long (70 mi) section in the southern part of its range. Giant sequoia can grow on a variety of soils, but it needs stable, adequate soil moisture. In its native range, it often grows on lower slopes or on flat ridges where soil moisture is relatively plentiful. It is an uncommon species.

REMARKS: Giant sequoia is world renowned for its monumental size, statuesque form, and pleasing colors. More than 90% of giant sequoia groves are protected in parks. Giant sequoia is shade intolerant and does not reproduce well without canopy openings. Fire has been the primary agent for providing mineral soil and reduced competition. Fire suppression has reduced suitable seedbeds and has increased fuel loads. In recent years, land managers have improved the health of these groves with carefully planned prescribed fires that have increased areas of mineral soil and reduced the fuel loads. Giant sequoia is mostly regarded for its aesthetic and scientific value. It is also planted as an ornamental and for Christmas trees. It holds some promise as a timber tree because of its fast growth rate.

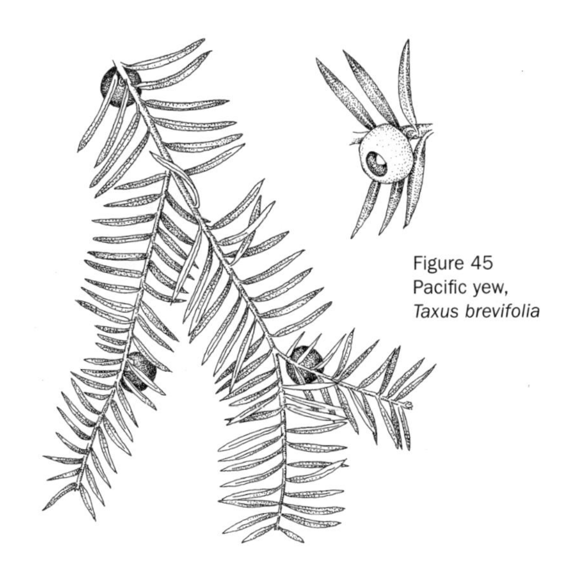

Figure 45
Pacific yew,
Taxus brevifolia

TAXUS (YEW)

The genus *Taxus* has about 10 species of trees and shrubs; 3 are native to the United States, and 1 of these is native to California. Yews are found in Asia, North Africa, Europe, and North America.

PACIFIC YEW (Fig. 45) ***Taxus brevifolia***

DESCRIPTION: An uncommon small- to medium-sized, erect, single- or multistemmed tree that grows in coastal and montane coniferous forests. In some cases it is a shrub. Mature trees usually range in height from 6 m (20 ft) to 18 m (60 ft) tall and are 30 cm (1 ft) to 90 cm (3 ft) in diameter. The largest grows in Lewis County, Washington, and is 16 m (54 ft) tall and 1.4 m (57 in.) in diameter. Trees can reach 250 to 350 years in age. The species typically has twisted, contorted stems and conical to spreading crowns. Crowns are usually sparse and have droopy twigs. **LEAVES** are linear, 12 mm (.5 in.) to 25 mm (1 in.) long, dark green above, light green below, and arranged in flat sprays.

Leaves have petioles. Stomatal bloom is absent. Leaf tips are pointed. **ARILS** are spherical, berrylike, pendant, coral colored, and from 6 mm (.25 in.) to 12 mm (.5 in.) in diameter. **TWIGS** droop. **BARK** is thin, with reddish brown to purplish scales over reddish to purplish inner bark.

HABITAT AND RANGE: Grows in coastal and coniferous forests from coastal Alaska southward to the northern Rocky Mountains and Cascades, at elevations from 10 m (33 ft) to 1,500 m (5,000 ft). In California it is scattered in the northern forests, usually in riparian zones or on lower slopes. The species lives in the cool, moist conditions often found in canyon bottoms and on north-facing lower slopes. It is typically an understory tree, often growing in old-growth forests.

REMARKS: Pacific yew is very shade tolerant and easily killed by fire. Native Americans used it to make bows, harpoons, spear handles, and paddles. Wild ungulates are known to browse Pacific yew's foliage, although some people believe it to be toxic to domestic livestock. The species' greatest commercial value lies in its pharmaceutical properties. Its bark and foliage contain taxol, a chemical that has cancer-fighting characteristics. Some botanists believe that Pacific yew is a variety of *T. baccata* (European yew).

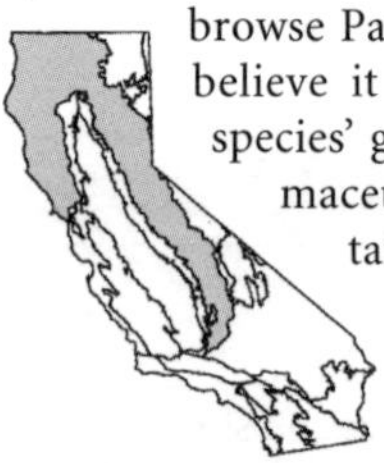

THUJA (ARBORVITAE)

The genus *Thuja* has 5 species of trees or large shrubs; 2 are native to the United States, and 1 of these is native to California. *Thuja* is found in Asia and North America.

WESTERN REDCEDAR (Fig. 46) ***Thuja plicata***

DESCRIPTION: A tall, erect, single-stemmed tree with long, clear trunks and rounded, full crowns. Old-growth trees often have fluted bases. Mature trees are typically 21 m (70 ft) to 40 m (130 ft) tall and 60 cm (2 ft) to 120 cm (4 ft) in diameter. The largest grows in Olympic National Park and is 48 m (159 ft) tall and 6 m (20 ft) in diameter. Trees are long-lived and may reach more than 1,000 years in age. **LEAVES** are scalelike, blunt tipped, dark green, and about 3 mm (.12 in.) long. The stomatal bloom pattern on the underside of a leaf resembles a butterfly or bow

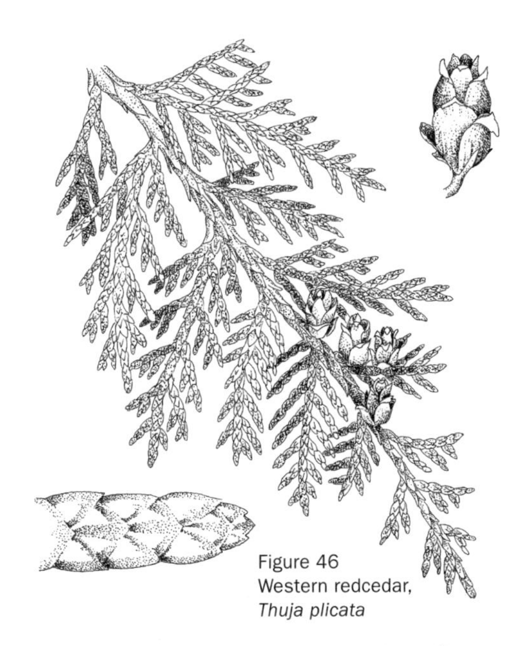

Figure 46
Western redcedar,
Thuja plicata

tie. Leaves are in opposite, alternating pairs. Foliage sprays are flat. **CONES** are leathery, erect, about 12 mm (.5 in.) long, and brown. Scales are in opposite, alternating pairs. **BARK** on mature trees is thin, fibrous, furrowed, and cinnamon brown.

HABITAT AND RANGE: Grows in coastal and montane coniferous forests from Alaska to northwestern California. Inland, it can be found in Idaho and western Montana. In California it is scattered on moist sites in Del Norte County and northern Humboldt County. It is abundant along the lower Mad River and in the hills south of Ferndale. The species grows only where there is abundant soil moisture, usually on lower slopes. It ranges in elevation from sea level to 2,100 m (7,000 ft).

REMARKS: Western redcedar is shade tolerant. It is very susceptible to fire because of its thin bark, shallow root system, and flammable foliage. The species is used as lumber, as well as for making shakes, shingles, and paneling. Native Americans fashioned totem poles, canoes, and houses out of western redcedar. Its bark was used to make clothing and fishing line. Its leaf oils have been used to manufacture perfume, insecticides, and medicines. Western redcedar's foliage makes good browse for deer and elk and habitat for many bird and small mammal species. The species is planted as an ornamental and, when properly pruned, makes excellent hedges. Western redcedar foliage is similar to that of Port Orford–cedar and Alaska yellow-cedar, but its cones are very different.

TORREYA (TORREYA)

The genus *Torreya* has 7 species of trees and shrubs; 2 are native to the United States, and 1 of these is native to California. *Torreya* species are found in Asia and North America.

CALIFORNIA-NUTMEG (Fig. 47) ***Torreya californica***

DESCRIPTION: An erect, single-stemmed, long-lived, small- to medium-sized tree. Mature trees are usually 4.5 m (15 ft) to 27 m (90 ft) tall and 20 cm (8 in.) to 50 cm (20 in.) in diameter. The largest grows in Swanton, California, and is 29 m (96 ft) tall and 2 m (80 in.) in diameter. Crowns of young trees are broadly conical, while mature trees often have long, clear stems and domed crowns. **LEAVES** are linear, stiff, spine tipped, and 2.5 cm (1 in.) to 7 cm (2.75 in.) long. Stomatal bloom is underneath. Leaves are generally arranged in flat sprays. Bruised leaves emit a foul odor. **ARILS** are olive shaped, 2.5 cm (1 in.) to 4 cm (1.5 in.) long, and green with purplish streaks. **BARK** is thin and has scaly, grayish brown plates.

HABITAT AND RANGE: Grows in coniferous forests, coniferous woodlands, and chaparrals. It is found in the Coast Ranges from southwestern Trinity County to Monterey County, and in the western Sierra Nevada from Shasta County to Tulare County. The species grows on diverse sites, from riparian zones to dry, hot chaparrals, at elevations ranging from 30 m (100 ft) to 2,100 m (7,000 ft). Best growth occurs on moist sites in the redwood region.

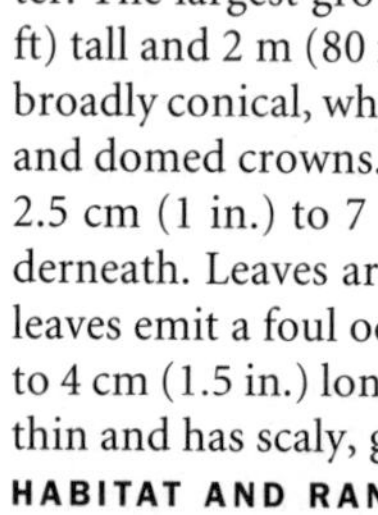

REMARKS: California-nutmeg is very shade tolerant. Its trunk is easily killed by fire, but the tree quickly resprouts. It is typically found growing in older stands and does not fare well following disturbance. Growth

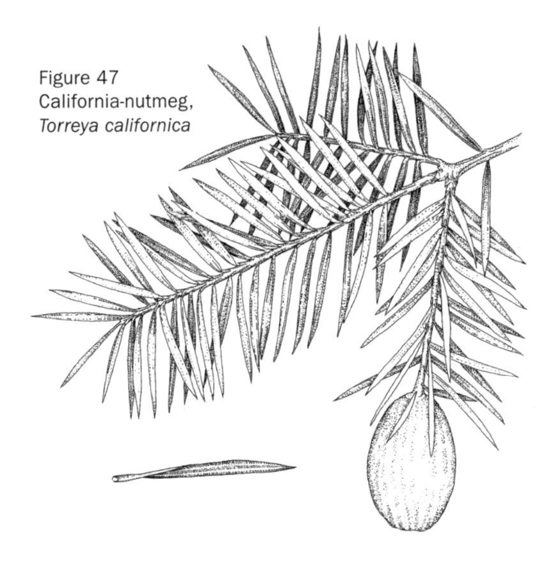
Figure 47
California-nutmeg,
Torreya californica

is slow. Understory trees measuring 10 cm (4 in.) to 20 cm (8 in.) in diameter, for example, may be over 100 years old. California-nutmeg's wood is very durable and has been used to make cabinets and novelties. Occasionally, it has been used for ornamental plantings, fuel, and fence posts. Native Americans ate the seeds (which supposedly taste like peanuts), wove its roots into baskets, and used its wood for bows. Stellar's and scrub jays eat the seeds.

TSUGA (HEMLOCK)

The genus *Tsuga* has 10 to 17 species, depending on botanical interpretation, found in North America and Asia. Four species are native to the United States, and 2 of these are native to California.

Leaves are linear and can be either flat or thick. They generally are flat or grooved on the upper sides and ridged beneath. They can have stomatal bloom only below or on both surfaces. Leaf tips are generally round to blunt and occasionally notched. Leaves have conspicuous petioles attached to short pegs that are nearly parallel with the twigs. Some species have flat sprays, while the leaves of others are arranged bottlebrush style. Seed cones are pendant and they mature in 1 year. They are egg shaped to oblong. Scales are leathery and bracts are shorter than the smooth-margined scales. Hemlock trees are usually conical and characteristically have droopy leaders and branches. Buds are small, nonresinous, and egg shaped. Bark is furrowed, brown to purplish brown in color, and high in tannin content.

Hemlocks typically grow in cool, moist environments in a wide variety of elevations. In California they grow at sea level as well as in montane and subalpine forests. Hemlocks are used as ornamentals and as windbreaks and screens.

1. Leaves are set more or less in flat sprays. Stomatal bloom can be found on the undersides of leaves. Seed cones are 12 mm (.5 in.) to 25 mm (1 in.) long . . . . **western hemlock** (*T. heterophylla*)
1. Leaves are set in 3-dimensional sprays. Stomatal bloom can be found on all surfaces. Seed cones are 2.5 cm (1 in.) to 7.5 cm (3 in.) long . . . . . . **mountain hemlock** (*T. mertensiana*)

WESTERN HEMLOCK (Fig. 48) *Tsuga heterophylla*

DESCRIPTION: An erect, tall, single-stemmed tree. Mature trees are typically 30 m (100 ft) to 45 m (150 ft) tall and 60 cm (24 in.) to 120 cm (48 in.) in diameter. The largest is in Olympic National Park and is 53 m (174 ft) tall and 2.7 m (108 in.) in diameter. Mature trees have narrow, short conical crowns with droopy leaders and branches. Trees can live more than 500 years. **LEAVES** form flat, droopy sprays. Individual leaves are linear, flat, petiolate, and grooved above. Leaf length is highly variable but uniformly short. Leaves typically range from 6 mm (.25 in.) to 20 mm (.75 in.) in length. Tips have a broad round shape. Stomatal bloom is only on the undersides of leaves. **CONES** are 20 mm (.75 in.) to 25 mm (1 in.) long and oblong to egg shaped. They have leathery scales, and scale margins are smooth to wavy and thin. **TWIGS** are slender, flexible, and pubescent. Small pegs lie parallel to twigs. **BARK** is thin and slightly furrowed. Outer bark is light brown to gray, and inner bark is dark red with purple streaks.

HABITAT AND RANGE: Grows in coastal and montane coniferous forests from northern California into Alaska. It has a continental range occurring from central Idaho north into central British Columbia. In California the species is locally com-

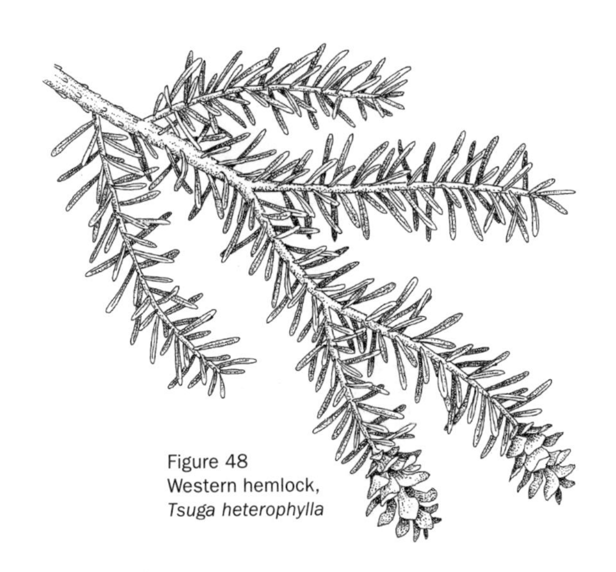

Figure 48
Western hemlock,
Tsuga heterophylla

mon from Humboldt County to the Oregon border and again in coastal Mendocino and Sonoma Counties. Within the state it rarely grows more than 32 km (20 mi) from the coast or at elevations less than 600 m (2,000 ft).

REMARKS: Western hemlock is very shade tolerant and is considered by some to be self-replacing where it grows. It is easily killed by fire, a variety of rots, and dwarf mistletoe. Snow and wind can knock over this shallow-rooted species. It is used as lumber and to make pulp, poles, pilings, and railway ties. The tannin in the outer bark has been used for tanning leather. Deer, elk, rabbits, mice, and beaver browse seedlings and saplings.

MOUNTAIN HEMLOCK *Tsuga mertensiana*

(Fig. 49; Pl. 11)

DESCRIPTION: An erect, tall, single-stemmed tree. It is one of the largest subalpine trees in western North America. On exposed ridges and near the tree line, however, it effects a prostrate, shrublike growth form. Mature trees are typically 30 m (100 ft) tall and 75 cm (2.5 ft) to 90 cm (3 ft) in diameter. The

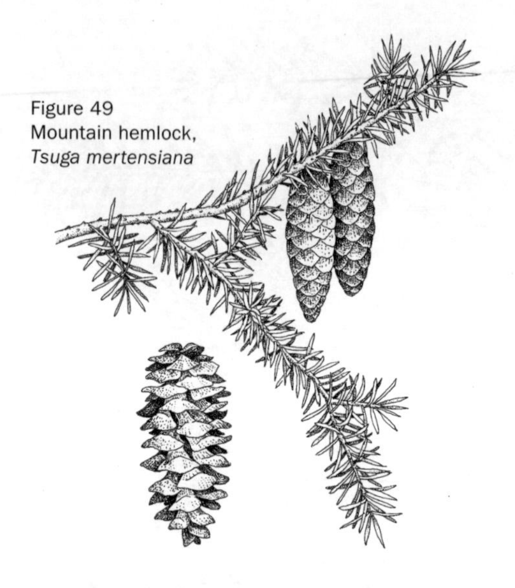

Figure 49
Mountain hemlock,
Tsuga mertensiana

largest is in Alpine County and is 34 m (113 ft) tall and 2.2 m (88 in.) in diameter. Open-grown trees have tapered stems and conical crowns with branches extending nearly to the ground. Like all hemlocks, this species has droopy leaders. Trees can live more than 800 years. **LEAVES** are linear and thick (3-angled) and spread in all directions. Stomatal bloom occurs on all leaf surfaces, creating blue-green leaves. Leaves are 6 mm (.25 in.) to 25 mm (1 in.) long and have blunt tips. **CONES** are 2.5 cm (1 in.) to 7.5 cm (3 in.) long, pendant, cylindrical, and initially purple, becoming brown at maturity. Scales are leathery, thin, and smooth margined, and they have prominent radial lines. **TWIGS** are slender, flexible, pubescent, and droopy. Small pegs or leaf cushions adhere to twigs. **BARK** is 2.5 cm (1 in.) to 4 cm (1.5 in.) thick, deeply furrowed, and reddish brown

HABITAT AND RANGE: Most often grows in subalpine coniferous forests and woodlands in the western United States. It has a continental distribution from central Idaho to southeastern British Columbia. Its coastal range is from southeastern Alaska to the central Sierra Nevada. In California it is locally common in the Klamath Mountains and Cascades. In the

Sierra Nevada its groves are sparsely scattered. The species grows at elevations from 1,200 m (4,000 ft) to 3,500 m (11,600 ft). The best stands are found on moist, northerly slopes.

REMARKS: Mountain hemlock is shade tolerant, shallow rooted, and easily killed by fire. Other destructive agents include wind and various fungal infections. A naturally occurring hybrid between *T. mertensiana* and *T. heterophylla* has been recognized. In California a form (*T. m.* ssp. *grandicona*) has been named based on the larger cones found on mountain hemlock in the Siskiyou Mountains.

BROADLEAVED TREES AND SHRUBS

KEY TO GROUPS

1. Plants appear to be leafless, or they have small scaly leaves . . . . **Group 1** (p. 112)
1. Plants have larger leaves 2
    2. Plants have rosettes of large swordlike or fanlike leaves . . . . **Group 2** (p. 112)
    2. Plants lack rosettes of large swordlike or fanlike leaves 3
3. Leaves are compound . . . . **Group 3** (p. 112)
3. Leaves are simple 4
    4. Leaves are opposite or whorled . . . . . **Group 4** (p. 116)
    4. Leaves are alternate 5
5. Branches and stems are armed with prickles, spines, or thorns . . . . **Group 5** (p. 118)
5. Branches and stems are not armed with prickles, spines, or thorns 6
    6. Leaves are lobed . . . . **Group 6** (p. 119)
    6. Leaves are unlobed 7
7. Plants are trees . . . . **Group 7** (p. 120)
7. Plants are shrubs 8
    8. Leaves are winter-deciduous . . . . **Group 8** (p. 122)
    8. Leaves are evergreen or drought-deciduous . . . . **Group 9** (p. 124)

GROUP 1: PLANTS APPEAR TO BE LEAFLESS, OR THEY HAVE SMALL SCALY LEAVES

1. Plants lose their leaves early, which makes them appear leafless 2
1. Plants have small scaly leaves 4
    2. Plants are broomlike shrubs less than 3 m (10 ft) tall. Stems do not have thorns . . . . **Spanish broom** (*Spartium junceum*)
    2. Plants are not broomlike shrubs and are more than 8 m (26 ft) tall at maturity. Stems have thorns 3
3. Stems are ashy gray. Leaves are simple . . . . **smoke tree** (*Psorothamnus spinosus*)
3. Stems are green. Leaves are compound . . . . **blue palo verde** (*Cercidium floridum*)
    4. Plants are shrubs or small trees from 3 m (10 ft) to 6 m (20 ft) tall . . . . **tamarisk** (*Tamarix*)
    4. Plants are shrubs less than 2 m (6.5 ft) tall 5
5. Plants are less than 30 cm (12 in.) tall . . . . **white heather** (*Cassiope mertensiana*)
5. Plants are more than 30 cm (12 in.) tall 6
    6. Stems resemble horse tails and are jointed . . . . **ephedra** (*Ephedra*)
    NOTE: A description of *Ephedra* is found in the conifer section.
    6. Stems do not resemble horse tails and are not jointed . . . . **scalebroom** (*Lepidospartum squamatum*)

GROUP 2: PLANTS HAVE ROSETTES OF LARGE SWORDLIKE OR FANLIKE LEAVES

1. Leaves are fanlike . . . . **California fan palm** (*Washingtonia filifera*)
1. Leaves are swordlike 2
    2. Leaves are spine tipped . . . . **yucca** (*Yucca*)
    2. Leaves are not spine tipped . . . . **nolina** (*Nolina*)

GROUP 3: LEAVES ARE COMPOUND

1. Leaves are opposite 2
1. Leaves are alternate 10

2. Leaves are palmately compound . . . . **California buckeye** (*Aesculus californica*)

2. Leaves are pinnately compound 3

3. Plants are trees 4

3. Plants are shrubs 7

4. Twigs have spongy piths. Fruits are fleshy, berrylike drupes . . . . **red elderberry** (*Sambucus racemosa*)

4. Twigs have solid piths. Fruits are dry 5

5. Twigs are flat at nodes . . . . **ash** (*Fraxinus*)

5. Twigs are round at nodes 6

6. Leaflets are 5 cm (2 in.) to 11 cm (4.5 in.) long and coarsely serrated. Fruits are paired samaras . . . . **box-elder** (*Acer negundo*)

6. Leaflets are 2.5 cm (1 in.) to 5 cm (2 in.) long and finely serrated. Fruits are inflated capsules . . . . **Sierra bladdernut** (*Staphylea bolanderi*)

7. Leaves are evergreen and have a strong smell; they have 2 leaflets per leaf. Leaflets are less than 10 mm (.4 in.) long . . . . **creosote bush** (*Larrea tridentata*)

7. Leaves are deciduous and lack a strong smell; they have 3 or more leaflets per leaf. Leaflets are more than 25 mm (1 in.) long 8

8. Leaves have 5 or more leaflets. Fruits are drupes . . . . **elderberry** (*Sambucus*)

8. Leaves have 3 leaflets. Fruits are samaras or capsules . . . 9

9. Leaflet margins are coarsely serrated. Fruits are paired samaras . . . . **mountain maple** (*Acer glabrum*)

9. Leaflet margins are finely serrated. Fruits are inflated capsules . . . . **Sierra bladdernut** (*Staphylea bolanderi*)

10. Leaves are palmately compound 11

10. Leaves are pinnately compound 12

11. Leaflets are 2.5 cm (1 in.) to 6 cm (2.5 in.) long. Flowers are usually yellow but may also be lilac, blue, or bicolored. Twigs lack spines . . . . **yellow bush lupine** (*Lupinus arboreus*)

11. Leaflets are 6 mm (.25 in.) to 13 mm (.5 in.) long. Flowers are purple or magenta. Twigs are spiny . . . . **chaparral pea** (*Pickeringia montana*)

12. Plants have fernlike leaves 13

12. Plants do not have fernlike leaves 14

13. Leaves are deciduous and 1 or 2 times pinnately compound . . . . **fern bush** (*Chamaebatiaria millefolium*)

13. Leaves are evergreen and 2 or 3 times pinnately compound . . . . **mountain misery** (*Chamaebatia foliolosa*)

14. Branches and stems are armed with prickles or spines 15
14. Branches and stems are unarmed, lacking prickles and spines 24
15. Plants are trees 16
15. Plants are shrubs 20
16. Leaves are once-pinnately compound 17
16. Leaves are twice-pinnately compound 18
17. Leaves lack a terminal leaflet . . . . **ironwood** (*Olneya tesota*)
17. Leaves have a terminal leaflet . . . . **black locust** (*Robinia pseudoacacia*)
18. Stout prickles are hooked and are found along the branches . . . . **catclaw acacia** (*Acacia greggii*)
18. Spines are not hooked and are found at the bases of the petioles 19
19. Plants have leaves throughout the growing season and stems are gray to brown. Primary leaflets are more than 5 cm (2 in.) long. Flowers are small and showy only in mass . . . . **honey mesquite** (*Prosopis glandulosa*)
19. Plants have leaves only during the early portion of the growing season. Stems are bright green. Primary leaflets are less than 2 cm (.75 in.) long. Individual flowers are large and showy . . . . **blue palo verde** (*Cercidium floridum*)
20. Leaves are densely covered with fine, grayish white hairs . . . . **Ironwood** (*Olneya tesota*)
20. Leaves are not densely covered with hairs 21
21. Leaves are once-pinnately compound or palmately compound 22
21. Leaves are twice-pinnately compound 23
22. Leaflets are less than 2 cm (.75 in.) long. Fruits are orange-red hips . . . . **rose** (*Rosa*)
22. Leaflets are more than 4 cm (1.5 in.) long. Fruits are "blackberry-like," i.e., sweet drupelets attached to an expanded receptacle . . . . **blackberry** (*Rubus*)
23. Stout prickles are hooked and are found along the branches . . . . **catclaw acacia** (*Acacia greggii*)
23. Spines are not hooked and are found at the base of the petiole . . . . **honey mesquite** (*Prosopis glandulosa*)
24. Plants are trees 25
24. Plants are shrubs 27
25. Leaves typically have 3 leaflets (sometimes 5) . . . . **California hop tree** (*Ptelea crenulata*)

25. Leaves have 7 to many leaflets 26

26. Leaf margins are entire except for 1 to 2 glandular teeth at their bases. Twigs have solid piths . . . . **tree-of-heaven** (*Ailanthus altissima*)

26. Leaf margins have fine serrations. Twigs have chambered piths . . . . **walnut** (*Juglans californica*)

27. Leaves generally have more than 3 leaflets 28

27. Leaves generally have 3 leaflets. WARNING: THIS CHOICE MAY LEAD YOU TO POISON-OAK 30

28. Leaves are evergreen . . . . **Oregon-grape** (*Berberis*)

28. Leaves are deciduous 29

29. Leaves are less than 25 mm (1 in.) long. Leaflets are oblong and have entire margins . . . . **shrub cinquefoil** (*Potentilla fruticosa*)

29. Leaves are more than 25 mm (1 in.) long. Leaflets have serrated margins . . . . **western mountain-ash** (*Sorbus scopulina*)

30. Leaves are evergreen . . . **bladderpod** (*Isomeris arborea*)

30. Leaves are deciduous 31

31. Leaflets are lobed in various ways 32

31. Leaflets are not lobed 33

32. Terminal leaflets are much larger than lateral leaflets and lack distinct petiolules. Flowers and fruits are located at the ends of short twigs . . . **skunkbush** (*Rhus trilobata*)

32. Terminal leaflets are about the same size as lateral leaflets and typically have distinct petiolules. Flowers and fruits are located in the axils of leaves. WARNING: THIS PLANT CAUSES DERMATITIS . . . . **poison-oak** (*Toxicodendron diversilobum*)

33. Leaflets are more than 2 cm (.75 in.) long. Fruits are round samaras . . . . **California hop tree** (*Ptelea crenulata*)

33. Leaflets are less than 15 mm (.6 in.) long. Fruits are legumes 34

34. Young twigs are hairy, and older ones are hairless. Foliage is sparse on stems. Inflorescences have 1 to 2 flowers. Seedpods have hairs only on their margins . . . . **Scotch broom** (*Cytisus scoparius*)

34. Mature twigs have silky, silvery hairs. Foliage is abundant on stems. Inflorescences have 4 to 10 flowers. Seedpods are covered with hairs . . . . **French broom** (*Genista monspessulana*)

GROUP 4: LEAVES ARE SIMPLE, OPPOSITE, OR WHORLED

1. Stems are armed with spines or thorns 2
1. Stems are unarmed 3
   2. Leaves are set in clusters. Blades are linear to club shaped and are 6 mm (.25 in.) to 12 mm (.5 in.) long. Blade surfaces are not silvery . . . . **black bush** (*Coleogyne ramosissima*)
   2. Leaves are not set in clusters. Blades are oblong and 2 cm (.75 in.) to 5 cm (2 in.) long. Blade surfaces are silvery . . . . **silver buffaloberry** (*Shepherdia argentea*)
3. Leaves are palmately lobed . . . . **maple** (*Acer*)
3. Leaves are not palmately lobed 4
   4. Leaves are leathery and evergreen 5
   4. Leaves are thin and deciduous 13
5. Leaves are scalelike and they nearly completely cover the branches . . . . **white heather** (*Cassiope mertensiana*)
5. Leaves are not scalelike 6
   6. Upper leaves do not have evident petioles 7
   6. Leaves have evident petioles 10
7. Leaves are needlelike and found on short side branches . . . . **heath** (*Erica lusitanica*)
7. Leaves are lance shaped, oblong, or egg shaped and not found on short side branches 8
   8. Flowers are set in clusters. Leaf margins are not curled under . . . . **keckiella** (*Keckiella*)
   8. Flowers are solitary. Leaf margins are curled under . . . 9
9. Smaller leaves are often bundled in the axils of larger leaves. Upper leaf surfaces are glandular and sticky. Plants grow in dry habitats . . . . **bush monkeyflower** (*Mimulus aurantiacus*)
9. Leaves are not bundled in the axils of larger leaves. Upper leaf surfaces are not glandular and sticky. Plants grow in or near fens and moist meadows . . . **laurel** (*Kalmia polifolia*)
   10. Leaves are entire or their margins are curled under . . . 11
   10. Leaves have serrated margins or are spiny tipped and not curled under 12
11. Leaves arise from spurlike or thornlike branches. Inflorescences are rounded clusters of colorful flowers . . . . **California-lilac** (*Ceanothus*)

11. Leaves do not arise from spurlike or thornlike branches. Inflorescences are catkinlike with drab flowers .... **silk tassel (*Garrya*)**

  12. Twigs are 4-angled in cross section. Flowers have 4 petals and are greenish to purplish .... **Oregon boxwood (*Paxistima myrsinites*)**

  12. Twigs are round to ridged. Flowers have 5 petals and are whitish, bluish, or pinkish .... **California-lilac (*Ceanothus*)**

13. Leaves are aromatic 14

13. Leaves are not aromatic 15

  14. Leaves are 5 cm (2 in.) to 15 cm (6 in.) long. Flowers occur singly .... **spice bush (*Calycanthus occidentalis*)**

  14. Leaves are 1.2 cm (.5 in.) to 7.5 cm (3 in.) long. Inflorescences have many flowers .... **sage (*Salvia*)**

15. Leaf blade margins are finely serrated to coarsely toothed 16

15. Leaf blade margins are entire, wavy, or irregularly lobed 20

  16. Stems are distinctly 4-angled in cross section .... **western burning bush (*Euonymus occidentalis*)**

  16. Stems are not distinctly 4-angled in cross section ... 17

17. Fruits are single samaras. Plants grow in desert mountains .... **singleleaf ash (*Fraxinus anomala*)**

17. Fruits are drupes or capsules. Plants do not grow in desert mountains 18

  18. Leaf margins are coarsely toothed except at the base. Leaf tips are broad to round .... **western viburnum (*Viburnum ellipticum*)**

  18. Leaf margins are entire to shallowly toothed. Leaf tips are acute 19

19. Plants are spreading or trailing. Flowers are red or yellow .... **keckiella (*Keckiella*)**

19. Plants are erect shrubs. Flowers are white .... **wild mock-orange (*Philadelphus lewisii*)**

  20. Flowers and fruits are set in pairs .... **honeysuckle (*Lonicera*)**

  20. Flowers and fruits are not set in pairs 21

21. Leaves are mostly opposite, with some set in whorls of 3, 4, or 5 22

21. Leaves are all opposite 23

22. Leaves have a broad elliptical to egg shape . . . . **button-willow** (*Cephalanthus occidentalis*)
22. Leaves are linear and crescent shaped . . . . **desert-willow** (*Chilopsis linearis*)
23. Leaves have a lance shape or narrow egg shape. Fruits are capsules . . . . **keckiella** (*Keckiella*)
23. Leaves have a broad elliptical to round shape. Fruits are berries or drupes 24
24. Leaf margins are irregularly lobed or toothed, sometimes entire. Leaves do not have prominent, arched side veins . . . . **snowberry** (*Symphoricarpos*)
24. Leaf margins are entire to wavy. Leaves have prominent, arched side veins . . . . **dogwood** (*Cornus*)

GROUP 5: LEAVES ARE SIMPLE AND ALTERNATE; BRANCHES AND STEMS ARE ARMED

1. Plants are trees 2
1. Plants are shrubs 3
2. Leaves are less than 2 cm (.75 in.) long and are dotted with glands. Fruits are legumes. Plants grow in the deserts of southern California . . . . **smoke tree** (*Psorothamnus spinosus*)
2. Leaves are more than 2 cm (.75 in.) long and do not have glands. Fruits are pomes. Plants grow in northern California . . . . **hawthorn** (*Crataegus*)
3. Leaves are lobed . . . . **gooseberry** (*Ribes*)
3. Leaves are unlobed 4
4. Leaves are evergreen 5
4. Leaves are deciduous 9
5. Leaves have 3 prominent veins originating from their bases . . . . **California-lilac** (*Ceanothus*)
5. Leaves do not have 3 prominent veins 6
6. Leaves are somewhat fleshy . . . . **hop-sage** (*Grayia spinosa*)
6. Leaves are not fleshy 7
7. Leaves are gray and covered with scurfy, gray, inflated hairs. Fruits are utricles . . . . . . **shadscale** (*Atriplex confertifolia*)
7. Leaves are green and not scurfy. Fruits are not utricles . . . 8
8. Leaves are linear to awl shaped. Fruits are legumes . . . . **gorse** (*Ulex europaea*)
8. Leaves are oval. Fruits are drupes . . . . **spiny redberry** (*Rhamnus crocea*)

9. Plants most often appear leafless, as leaves are short-lived. Fruits are legumes . . . . **smoke tree** (*Psorothamnus spinosus*)
9. Plants have long-lasting leaves. Fruits are utricles, berries, or drupes 10
    10. Leaf blade surfaces are scurfy with gray, inflated hairs. Fruits are utricles . . . . **shadscale** (*Atriplex confertifolia*)
    10. Leaf blade surfaces are not scurfy. Fruits are berries or drupes 11
11. Leaves have silvery hairs . . . . **silver buffaloberry** (*Shepherdia argentea*)
11. Leaves are hairless or they have hairs that are not silvery 12
    12. Leaves are somewhat fleshy and are entire. Fruits are berries . . . . **boxthorn** (*Lycium*)
    12. Leaves are not fleshy and are finely serrated. Fruits are drupes . . . . **cherry** (*Prunus*)

GROUP 6: LEAVES ARE SIMPLE, ALTERNATE, AND LOBED

1. Leaves are pinnately lobed and veined 2
1. Leaves are palmately lobed and veined 4
    2. Leaf margins are not rolled under . . . . . . **oak** (*Quercus*)
    2. Leaf margins are rolled under 3
3. Leaves are gray and appear threadlike . . . . **California sagebrush** (*Artemisia californica*)
3. Leaves are green and do not appear threadlike . . . . **bitterbrush** (*Purshia*)
    4. Plants are trees more than 9 m (30 ft) tall . . . . **California sycamore** (*Platanus racemosa*)
    4. Plants are shrubs or small trees less than 7.5 m (25 ft) tall 5
5. Older leaves are set on spur shoots. Margins of lobes are entire 6
5. Older leaves are not set on spur shoots. Margins of lobes are serrated 7
    6. Leaves are evergreen. Fruits are capsules . . . . **flannelbush** (*Fremontodendron californicum*)
    6. Leaves are deciduous. Fruits are berries . . . . **currant** (*Ribes*)

7. Leaves are hairless, or they have only a few hairs on upper surfaces and are hairy on lower surfaces 8
7. Leaves have a dense covering of hair on both surfaces . . . 9
   8. Bark is shredding. Fruits are follicles . . . . **ninebark** (*Physocarpus capitatus*)
   8. Bark is smooth. Fruits are berries . . . . . . **currant** (*Ribes*)
9. Leaves are 5- to 7-lobed, maplelike, and more than 5 cm (2 in.) long. Fruits are "raspberry-like," i.e., sweet drupelets attached to an expanded receptacle . . . . **thimbleberry** (*Rubus parviflorus*)
9. Leaves are 3-lobed at their tips. They are not maplelike and are less than 3 cm (1.2 in.) long. Fruits are not "raspberry-like" and are instead achenes . . . . **big sagebrush** (*Artemisia tridentata*)

GROUP 7: TREES HAVE SIMPLE, ALTERNATE, UNLOBED LEAVES

1. Upper leaf surfaces are sandpapery. Lower leaf surfaces have conspicuous netted veins . . . . **netleaf hackberry** (*Celtis reticulata*)
1. Upper leaf surfaces are not sandpapery. Lower leaf surfaces do not have conspicuous netted veins 2
   2. Leaves are evergreen 3
   2. Leaves are deciduous 14
3. Leaf margins are entire 4
3. Leaf margins are serrated, toothed, or spiny 9
   4. Leaf margins are curled under . . . . **curlleaf mountain-mahogany** (*Cercocarpus ledifolius*)
   4. Leaf margins are not curled under 5
5. Mature leaves are sickle shaped . . . . **blue gum** (*Eucalyptus globulus*)
5. Mature leaves are not sickle shaped 6
   6. Fruits are acorns . . . . **oak** (*Quercus*)
   6. Fruits are not acorns 7
7. Leaves have a broad oval to oblong shape. Bark on small stems and twigs is smooth and reddish brown, and it shreds in sheets . . . . **Pacific madrone** (*Arbutus menziesii*)
7. Leaves are elliptical to lance shaped. Bark is not smooth and does not shred 8
   8. Undersides of leaves are golden or rusty. Crushed leaves are not aromatic . . . . **giant chinquapin** (*Chrysolepis chrysophylla*)

8. Undersides of leaves are green. Crushed leaves are strongly aromatic . . . . **California bay** (*Umbellularia californica*)

9. Leaves have spiny margins 10

9. Leaves do not have spiny margins 11

10. Fruits are acorns . . . . **oak** (*Quercus*)

10. Fruits are drupes . . . . **hollyleaf cherry** (*Prunus ilicifolia*)

11. Leaf blades generally have margins with teeth only on the upper half to two-thirds 12

11. Leaf blades have fully serrated margins 13

12. Blades are 5 cm (2 in.) to 10 cm (4 in.) long, lance shaped, and not found on spur shoots. Undersides of leaf blades lack prominent veins . . . . **Pacific wax-myrtle** (*Myrica californica*)

12. Blades are 1.2 cm (.5 in.) to 4 cm (1.5 in.) long; egg shaped, with the widest part above the middle; and often found on spur shoots. Undersides of leaf blades have prominent parallel veins . . . . **birchleaf mountain-mahogany** (*Cercocarpus betuloides*)

13. Leaves are hairy, especially on their undersides. Bark is blocky and gray and does not shred in sheets . . . . **tanoak** (*Lithocarpus densiflorus*)

13. Leaves are not hairy. Bark is smooth and reddish brown, and it shreds in sheets . . . . **Pacific madrone** (*Arbutus menziesii*)

14. Leaf blades are round and have heart-shaped bases . . . 15

14. Leaf blades are not round and do not have heart-shaped bases 16

15. Leaf blades are entire . . . . **western redbud** (*Cercis occidentalis*)

15. Leaf blades are serrated . . . . **California hazel** (*Corylus cornuta*)

16. Leaves arise from spur shoots 17

16. Leaves do not arise from spur shoots 18

17. Leaves have prominent, pinnately arranged veins. Petioles are less than 12 mm (.5 in.) long. Bark is reddish brown, with conspicuous lenticels. Fruits are nutlets in a cylindrical "cone" . . . . **water birch** (*Betula occidentalis*)

17. Leaves do not have prominent, pinnately arranged veins. Petioles are 2.5 cm (1 in.) to 3 cm (1.5 in.) long. Bark is grayish. Fruits are pomes . . . . . **Oregon crab apple** (*Malus fusca*)

18. Leaves are doubly serrated 19

18. Leaves are singly serrated, scalloped, or entire 20

19. Leaves and twigs are hairy. Buds are not stalked .... **California hazel** (*Corylus cornuta*)
19. Leaves and twigs are not hairy. Buds are stalked .... **alder** (*Alnus*)
    20. Leaves have prominent, pinnately arranged veins 21
    20. Leaves do not have prominent, pinnately arranged veins 22
21. Buds have protective scales. Fruits are set in woody, persistent catkins .... **alder** (*Alnus*)
21. Buds lack protective scales (they are naked). Fruits are drupes .... **cascara** (*Rhamnus purshiana*)
    22. Leaves lack petioles. Blades are entire and usually somewhat sickle shaped ... **desert-willow** (*Chilopsis linearis*)
    22. Leaves have petioles. Blades are serrated or scalloped and not sickle shaped 23
23. Blades are finely serrated and have a pair of glands at their bases or on their petioles .... **cherry** (*Prunus*)
23. Blades are serrated or scalloped and lack glands 24
    24. Leaves are generally much longer than wide. Buds have 1 scale. Petioles are round .... **willow** (*Salix*)
    24. Leaves are usually about as long as wide. Buds have more than 3 scales. Petioles may be flat .... **cottonwood** (*Populus*)

GROUP 8: SHRUBS HAVE SIMPLE, ALTERNATE, UNLOBED, WINTER-DECIDUOUS LEAVES

1. Leaves have 3 teeth at the tips of their blades .... **bitterbrush** (*Purshia*)
1. Leaves do not have 3 teeth at the tips of their blades 2
    2. Leaves are round with heart-shaped bases 3
    2. Leaves are other than round with heart-shaped bases 4
3. Twigs are grayish. Leaves are gray and hairy beneath. Fruits are capsules .... **snowdrop bush** (*Styrax officinalis*)
3. Twigs are brownish. Leaves are green and hairless. Persistent legumes are often seen on the stems .... **western redbud** (*Cercis occidentalis*)
    4. Leaves have glands near the bases of their blades .... **cherry** (*Prunus*)
    4. Leaves do not have glands near the bases of their blades 5

5. Leaves are hairless 6
5. Leaves are hairy on at least 1 surface 14
   6. Leaves have strong, raised vein patterns on their lower surfaces 7
   6. Leaf blades lack raised vein patterns on their lower surfaces 12
7. Leaf blades have 3 veins that originate from their bases; the 2 outside veins are arched . . . . **California-lilac** (*Ceanothus*)
7. Leaf blades have straight veins that originate from their midribs 8
   8. Stipules are present. Buds have single, caplike scales . . . . **willow** (*Salix*)
   8. Stipules are absent or they drop off early. Buds have several scales 9
9. Leaf margins are entire . . . . **oso berry** (*Oemleria cerasiformis*)
9. Leaf margins are not entire 10
   10. Leaf margins are doubly serrated . . . . **alder** (*Alnus*)
   10. Leaf margins are singly serrated or entire 11
11. Branchlets are brown, glandular, and warty . . . . **resin birch** (*Betula glandulosa*)
11. Branchlets are reddish and smooth . . . . **buckthorn** (*Rhamnus*)
   12. Leaf blades are entire or more or less uniformly serrated . . . . **huckleberry** (*Vaccinium*)
   12. Leaf blades have serrated margins on their upper halves 13
13. Leaf blades have sharp, unequal teeth above round bases . . . . **mountain spiraea** (*Spiraea densiflora*)
13. Leaf blades have sharp, equal teeth above somewhat heart-shaped bases . . . . **western serviceberry** (*Amelanchier alnifolia*)
   14. Leaves are densely hairy on both surfaces 15
   14. Leaves are hairy, mainly on their lower surfaces 17
15. Twigs have a strong zigzag pattern . . . . **California hazel** (*Corylus cornuta*)
15. Twigs are straight 16
   16. Leaves have stipules. Leaf margins are hairless. Buds are small . . . . **willow** (*Salix*)
   16. Leaves lack stipules. Leaf margins have small, fine hairs. Buds are large . . . . **western azalea** (*Rhododendron occidentale*)

17. Leaves have a dense covering of white hair beneath . . . . **Douglas spiraea (*Spiraea douglasii*)**
17. Leaves are hairy but lack a dense covering of white hair beneath 18
    18. Leaves have scattered brown hairs on their upper surfaces . . . . **mock-azalea (*Menziesia ferruginea*)**
    18. Leaves are hairy but do not have brown hairs on their upper surfaces 19
19. Leaf hairs are sparse and are found mainly on veins of lower surfaces 20
19. Leaf hairs are abundant on lower surfaces 23
    20. Buds are large and red in winter. Leaf margins have small, fine hairs . . . . **western azalea (*Rhododendron occidentale*)**
    20. Buds are small in winter. Leaf margins are hairless 21
21. Leaf blades are doubly serrated . . . . **snow wreath (*Neviusia cliftonii*)**
21. Leaf blades are entire or singly serrated 22
    22. Leaf blades are entire or have fine serrations all along their margins . . . . **huckleberry (*Vaccinium*)**
    22. Leaf blades have serrations above their upper halves . . . . **western serviceberry (*Amelanchier alnifolia*)**
23. Leaves are deeply lobed . . . . **oak (*Quercus*)**
23. Leaves are coarsely toothed or serrated but not deeply lobed 24
    24. Leaf blades are serrated only above their midpoints . . . . **Sierra sweet-bay (*Myrica hartwegii*)**
    24. Leaf blades have fully serrated margins 25
25. Twigs are hairy and ribbed below leaf scars . . . . **oceanspray (*Holodiscus*)**
25. Twigs are not hairy or ribbed 26
    26. Lateral buds are stalked . . . . **alder (*Alnus*)**
    26. Lateral buds are not stalked . . . . **willow (*Salix*)**

GROUP 9: SHRUBS HAVE SIMPLE, ALTERNATE, UNLOBED, AND EVERGREEN OR DROUGHT-DECIDUOUS LEAVES

1. Leaves are linear or needlelike 2
1. Leaves are other than linear or needlelike 14
    2. Leaves are linear or are narrow and widest above the middle 3

2. Leaves are needles 12

3. Leaves are sickle shaped . . . . **desert-willow** (*Chilopsis linearis*)

3. Leaves are not sickle shaped 4

4. Mature plants have conspicuously peeling reddish bark . . . . **red shank** (*Adenostoma sparsifolium*)

4. Mature plants do not have peeling reddish bark 5

5. Leaf margins are curled under. Flowers are set in clusters. Fruits are utricles 6

5. Leaf margins are not curled under. Flowers form heads. Fruits are achenes 7

6. Leaves are covered with whitish, scurfy hairs . . . . **fourwing saltbush** (*Atriplex canescens*)

6. Leaves are densely covered with star-shaped hairs that are initially whitish and later rusty . . . . **winter fat** (*Krascheninnikovia lanata*)

7. Plants are willowlike and grow in river bottoms and other wet places in southern California (especially in deserts) . . . . **arrow-weed** (*Pluchea sericea*)

7. Plants are not willowlike and do not grow in wet areas in southern California 8

8. Flower heads are daisylike, having both ray and disk flowers . . . . **interior goldenbush** (*Ericameria linearifolia*)

8. Flower heads are pincushion-like, having only disk flowers 9

9. Leaves are threadlike. Seed-bearing heads have 1 flower . . . . **cheese bush** (*Hymenoclea salsola*)

9. Leaves are not threadlike. Seed-bearing heads have many flowers 10

10. Leaves have a dense covering of white hair . . . . **gray horsebrush** (*Tetradymia canescens*)

10. Leaves do not have a dense covering of white hair . . . 11

11. Flower heads have 2 to 20 flowers . . . . **rabbitbrush** (*Chrysothamnus*)

11. Flower heads have 30 to 40 flowers . . . . **yellow-aster** (*Eastwoodia elegans*)

12. Leaves are clustered. Plants are found in low-elevation chaparrals and woodlands . . . . **chamise** (*Adenostoma fasciculatum*)

12. Leaves are not clustered. Plants are not found in low-elevation chaparrals and woodlands 13

13. Flowers are red. Plants are found in high-elevation meadows and woodlands ...... **mountain heather** (*Phyllodoce*)
13. Flowers are white or pink. Plants are found in coastal locations in Humboldt County .... **heath** (*Erica lusitanica*)
    14. Twigs are angled or ridged 15
    14. Twigs are round and not angled or ridged 16
15. Leaves have clearly differentiated blades and petioles .... **California-lilac** (*Ceanothus*)
15. Leaves have wedge-shaped bases that are not clearly differentiated from the petiole .... **coyote brush** (*Baccharis*)
    16. Leaf margins are entire 17
    16. Leaf margins are not entire 34
17. Leaf margins are curled under 18
17. Leaf margins are flat and not curled under 22
    18. Leaves are clustered together on stems and are tightly curled under 19
    18. Leaves are not clustered together on stems. Leaves are variably curled under......................... 20
19. Plants are trees that can grow to be 10 m (33 ft) tall .... **curlleaf mountain-mahogany** (*Cercocarpus ledifolius*)
19. Plants are shrubs from 30 cm (1 ft) to 90 cm (3 ft) tall .... **California buckwheat** (*Eriogonum fasciculatum*)
    20. Leaves are more than 5 cm (2 in.) long. Plants grow in dry habitats .... **mountain balm** (*Eriodictyon californicum*)
    20. Leaves are less than 5 cm (2 in.) long. Plants grow in boggy areas and moist meadows 21
21. Plants are less than 50 cm (1.5 ft) tall. Flowers are solitary and pinkish .... **laurel** (*Kalmia polifolia*)
21. Plants are 60 cm (2 ft) to 150 cm (5 ft) tall. Flowers grow in branched inflorescences and are cream colored .... **western Labrador-tea** (*Ledum glandulosum*)
    22. Leaf surfaces are identical. Leaves are typically vertical .... **manzanita** (*Arctostaphylos*)
    22. Leaf surfaces are different. Leaves are typically horizontal or droopy 23
23. Flowers are set in sunflower-like heads, having both ray and disk flowers .... **brittlebush** (*Encelia*)
23. Flowers are not set in sunflower-like heads 24
    24. Leaves are folded along their midribs 25
    24. Leaves are not folded along their midribs 27

25. Undersides of leaves have golden hairs. Fruits are nuts and located in spiny burs . . . . **golden chinquapin** (*Chrysolepis chrysophylla*)

25. Undersides of leaves do not have golden hairs. Fruits are drupes and are not located in spiny burs 26

 26. Leaves are aromatic. Leaf margins and veins are reddish. Fruits are hairless . . . . **laurel sumac** (*Malosma laurina*)

 26. Leaves are not aromatic. Leaf margins and veins are not reddish. Fruits are hairy . . . . . . **sugarbush** (*Rhus ovata*)

27. Plants are willowlike and grow in river bottoms and other wet places in southern California (especially in deserts) . . . . **arrow-weed** (*Pluchea sericea*)

27. Plants are not willowlike and do not grow in wet areas in southern California 28

 28. Lower leaf surfaces are heavily covered with hairs or scales 29

 28. Lower leaf surfaces are hairless and scaleless 30

29. Lower leaf surfaces have golden to rusty scales . . . . **bush chinquapin** (*Chrysolepis sempervirens*)

29. Lower leaf surfaces have a dense covering of white hair . . . . **hoary coffeeberry** (*Rhamnus tomentella*)

 30. Leaves are more than 6 cm (2.5 in.) long. Petioles are about 2.5 cm (1 in.) long. Buds at the ends of twigs are large . . . . **Pacific rhododendron** (*Rhododendron macrophyllum*)

 30. Leaves are less than 6 cm (2.5 in.) long. Petioles are less than 6 mm (.25 in.) long. Buds at the ends of twigs are small 31

31. Plants are prostrate. Fruits are fleshy or mealy berries . . . . **manzanita** (*Arctostaphylos*)

31. Plants are upright. Fruits are capsules or drupes 32

 32. Plants are found in boggy places in upper montane forests. Leaves are not aromatic when crushed . . . . **black-laurel** (*Leucothoe davisiae*)

 32. Plants are not found in boggy places. Leaves are aromatic when crushed 33

33. Leaves are 5 cm (2 in.) to 15 cm (6 in.) long. Veins are not prominent. Twigs are green. Drupes are hairless . . . . **California bay** (*Umbellularia californica*)

33. Leaves are 2.5 cm (1 in.) to 6 cm (2.5 in.) long. Veins are prominent. Twigs are reddish. Drupes are hairy . . . . **lemonadeberry** (*Rhus integrifolia*)

34. Leaves have 3 teeth at the tips of their blades 35
34. Leaves do not have 3 teeth at the tips of their blades 36
35. Leaves are gray with silky hairs and have flat margins . . . . **big sagebrush** (*Artemisia tridentata*)
35. Leaves are green above, with white hairs below. Margins are curled under . . . . **bitterbrush** (*Purshia*)
36. Leaves have 3 prominent veins that originate from their bases; the 2 outside veins are arched . . . . **California-lilac** (*Ceanothus*)
36. Leaves do not have 3 prominent veins that originate from their bases 37
37. Buds are naked, showing immature leaves . . . . **buckthorn** (*Rhamnus*)
37. Buds have scales that enclose immature leaves 38
38. Leaves are clustered on short spur branches . . . . **birchleaf mountain-mahogany** (*Cercocarpus betuloides*)
38. Leaves are individually placed along the stems 39
39. Buds are clustered at the tips of stems. Leaves and margins are variable . . . . **oak** (*Quercus*)
39. Buds are not clustered. Leaves and margins are uniform 40
40. Leaves have spiny margins 41
40. Leaves have serrated, not spiny, margins 43
41. Leaves are 1 cm (.4 in.) to 5 cm (2 in.) long 42
41. Leaves are more than 5 cm (2 in.) long . . . . **toyon** (*Heteromeles arbutifolia*)
42. Flowers do not have petals. Plants are shrubs less than 4 m (13 ft) tall . . . . **buckthorn** (*Rhamnus*)
42. Flowers have petals. Plants are large shrubs that can grow to be 10 m (33 ft) tall . . . . **hollyleaf cherry** (*Prunus ilicifolia*)
43. Undersides of leaves are hairy 44
43. Undersides of leaves are hairless 45
44. Upper leaf surfaces are not sticky . . . . **tanoak** (*Lithocarpus densiflorus*)
44. Upper leaf surfaces are sticky . . . . **mountain balm** (*Eriodictyon californicum*)
45. Stems zigzag between nodes. Leaves have a broad egg shape . . . . **salal** (*Gaultheria shallon*)
45. Stems are straight and do not zigzag between nodes. Leaves have a narrow egg shape or oblong to lance shape 46

46. Leaves are light green and erect. Flowers are showy, yellow, and 3 cm (1.25 in.) to 6 cm (2.5 in.) in diameter. Plants grow on dry slopes and washes . . . . **bush poppy** (***Dendromecon rigida***)

46. Leaves are dark green and not erect. Flowers are pink, bell shaped, and about 6 mm (.25 in.) long or they form small catkins. Plants grow on moist slopes 47

47. Leaves are less than 3 cm (1.25 in.) long. Flowers are pink, bell shaped, and about 6 mm (.25 in.) long . . . . **evergreen huckleberry** (***Vaccinium ovatum***)

47. Leaves are more than 5 cm (2 in.) long. Flowers form small catkins . . . . **Pacific wax-myrtle** (***Myrica californica***)

DESCRIPTIONS OF GENERA AND SPECIES

ACACIA (ACACIA)

Acacia is a large genus with 1,200 tree and shrub species; 1 species is native to California. The genus is especially diverse in Australia.

CATCLAW ACACIA (Fig. 50) ***Acacia greggii***

DESCRIPTION: A shrub armed with prickles that often grows to be 3.5 m (12 ft) tall. The largest lives in Red Rock, New Mexico, and is 14.9 m (49 ft) tall and 62 cm (24 in.) in diameter. **LEAVES** are alternate and deciduous. Blades are compound, with several blade axes measuring from 2.5 cm (1 in.) to 5 cm (2 in.) long. Each axis has 8 to 12 leaflets that are about 6 mm (.25 in.) long. Blade axes lack terminal leaflets. **INFLORESCENCES** are elongated clusters of many small flowers. Each cluster is from 2 cm (.75 in.) to 5 cm (2 in.) long. **FLOWERS** are yellow and individually inconspicuous. The many yellow stamens provide the conspicuous color of the showy inflorescences. **FRUITS** are legumes with blunt tips and measure 5 cm (2 in.) to 15 cm (6 in.) long. **TWIGS** are gray with curved prickles about 6 mm (.25 in.) long.

HABITAT AND RANGE: Grows along washes below 600 m (2,000 ft) in the Mojave, Sonoran, and Colorado Deserts in California. It ranges east into Texas and south into Mexico.

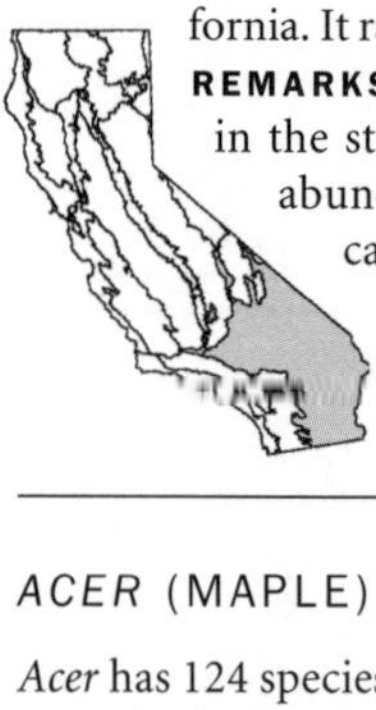

REMARKS: Catclaw acacia is the only native acacia in the state; it is generally uncommon but locally abundant. Thirteen species of acacia have escaped from cultivation in California, some of which are noxious weeds. Another descriptive common name for this species is wait-a-minute.

ACER (MAPLE)

Acer has 124 species, nearly all of which grow in temperate climates in the Northern Hemisphere. Fourteen maple species are found in the United States, and 4 of these grow in California.

Figure 50
Catclaw acacia,
Acacia greggii

Most maples have deciduous, opposite leaves that are either simple or compound. A few evergreen maples grow in Southeast Asia. Simple maple leaves are palmately lobed and veined. The flowers are not showy and can be complete or pistil bearing only or stamen bearing only, on the same or on different plants. Fruits are double samaras that break into segments at maturity.

The California maples usually grow on lower slopes, where soil moisture is more abundant. Maples often provide the primary fall color in California's forests and woodlands. In the fall, bigleaf maple and box-elder are noted for yellow leaves, mountain maple for yellow to orange-yellow leaves, and vine maple for orange-red to red leaves. The genus is renowned as a source of syrups and sweets as well as timber. In California, bigleaf maple can be tapped for syrup if it grows in an environment that is sufficiently cold. Bigleaf maple furniture, cabinets, and flooring are manufactured in northern California. Recently, cancer chemotherapeutic drugs have been made from box-elder. Maples are widely planted as ornamentals in gardens and on streets. The genus has great potential in urban forestry applications.

1. Leaves are simple and palmately lobed 2
1. Leaves are compound, with 3 to 7 leaflets 4
    2. Leaves are generally broader than 12.5 cm (5 in.) in diameter. Lobes are coarsely toothed but not serrated. The central lobe is narrow at its base and wide at its apex. Flowers and samaras form elongated clusters . . . . **bigleaf maple** (*A. macrophyllum*)
    2. Leaves are less than 12.5 cm (5 in.) in diameter. Lobes are serrated. Flowers and samaras form rounded clusters 3
3. Leaves have 5 to 9 major lobes. Lobe sinuses are less than 33% of the length of the lobe. Leaves are almost fanlike in outline. Lobe margins are singly serrated . . . . **vine maple** (*A. circinatum*)
3. Leaves have 3 major lobes. Lobe sinuses are 33% or more of the length of the lobe. Leaves are not almost fanlike in outline. Lobe margins are often doubly serrated . . . . **mountain maple** (*A. glabrum*)
    4. Leaves have 3 to 7 pinnately compound leaflets. Leaflets are highly variable in shape. Leaflet margins are coarsely lobed or toothed. Trees are found at low elevations . . . . **box-elder** (*A. negundo*)
    4. Leaves have 3 leaflets. Leaflets are somewhat uniform in shape. Leaflet margins are often doubly serrated. Shrubs are found at montane elevations . . . . **mountain maple** (*A. glabrum*)

VINE MAPLE (Fig. 51) *Acer circinatum*

DESCRIPTION: A multistemmed, straggly shrub or small tree. Mature plants typically grow to be 10 m (35 ft) tall. The largest lives in Tillamook County, Oregon, and is 14 m (46 ft) tall and 53 cm (21 in.) in diameter. Stems are typically crooked and spreading, and they can root at the nodes. Crowns are uneven and broad to rounded. **LEAVES** are opposite, simple, and deciduous. Blades are pale green, thin, 5 cm (2 in.) to 12.5 cm (5 in.) wide, and palmate, with 5 to 9 lobes. Lobes are acute, with sharply serrated margins. Lobe sinuses are less than 25% of the length of the lobe. The leaf is almost round in outline. **INFLO-**

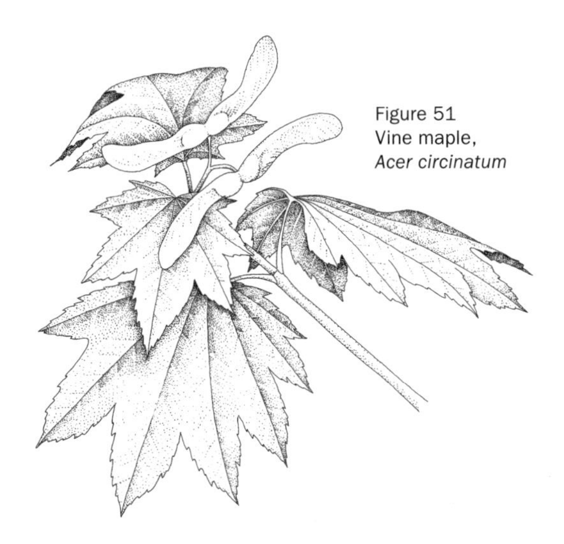

Figure 51
Vine maple,
Acer circinatum

RESCENCES form rounded clusters of 2 to 10 flowers. **FLOWERS** may be complete or may bear stamens only. Petals are greenish white. Plants bloom in April or May as new leaves appear. **FRUITS** are reddish double samaras. Fruit bodies lack hairs. Wings are widely divergent at a nearly 180° angle. Samaras are 1 cm (.4 in.) to 4 cm (1.5 in.) long. **TWIGS** are hairless; are brown, purple, or pale green; and often have white speckles.
HABITAT AND RANGE: Grows in coastal and montane forests and woodlands, where it inhabits stream banks in deep shade. The species occurs from southern Alaska to the mountains of northern California, at elevations from sea level to 1,500 m (5,000 ft).

REMARKS: In the autumn, vine maple leaves turn yellow, orange, or red, bringing patches of color to dark forests. The species is ornamental and easily propagated from seed or cuttings. Leaves are palatable to cattle, deer, elk, and sheep.

MOUNTAIN MAPLE (Fig. 52) *Acer glabrum*

DESCRIPTION: An erect, single-stemmed shrub or small tree. It commonly grows to be 4.5 m (15 ft) tall. The largest lives

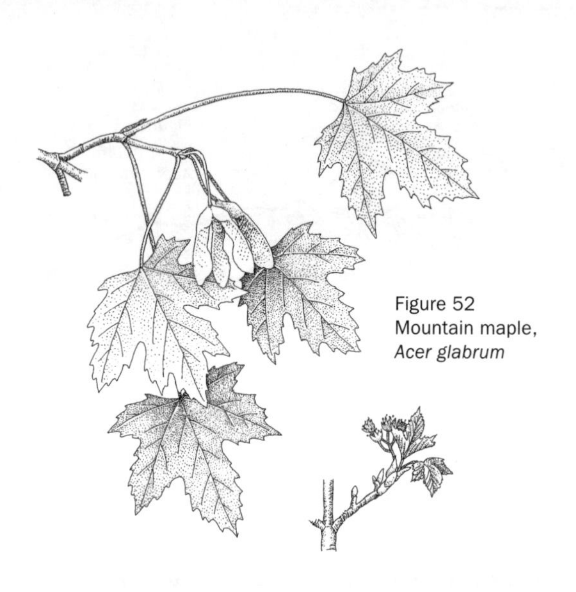

Figure 52
Mountain maple,
Acer glabrum

in Island City, Washington, and is 20.4 m (67 ft) tall and 86 cm (34 in.) in diameter. Crowns are rounded, with branches close to the ground. **LEAVES** are opposite, simple (rarely compound), 2 cm (.75 in.) to 4 cm (1.5 in.) wide, and deciduous. Blades are green on the upper surfaces and gray on the lower surfaces. They are thin and palmate, with 3 major and 2 supplementary lobes. Lobes are acute, with serrated margins. Lobe sinuses are more than 33% of the length of the lobe. **INFLORESCENCES** form rounded clusters of 2 to 10 flowers. **FLOWERS** may be complete or may bear only stamens or pistils. Petals are greenish yellow. Plants bloom in May or June after the new leaves appear. **FRUITS** are yellow or red double samaras. Fruit bodies lack hairs. Wings diverge at oblique angles. Samaras are 1 cm (.4 in.) to 2.5 cm (1 in.) long. **TWIGS** are red or white.

HABITAT AND RANGE: Grows on montane rocky slopes and canyons in western North America from Mexico to Alaska. In California it grows near riverbanks and occurs as undergrowth in montane forests, at elevations from 900 m (3,000 ft) to 2,700 m (9,000 ft).

REMARKS: *A. glabrum* is a complex of forms that have been treated as subspecies and varieties by different botanists. The typical form occurs in the Rocky Mountains. California types differ in leaf size and shape, color of twigs, and fruit characteristics. One variety, *A. g.* var. *diffusum,* has small leaves and white, rather than red, twigs. It grows in the San Jacinto Mountains, other desert mountains, and the eastern Sierra Nevada. The more extensive variety, *A. g.* var. *torreyi,* has 5-lobed leaves and samaras whose wings spread at an approximately 45° angle, whereas *A. g.* var. *greenei* has very small leaves and samaras with overlapping wings. It grows only in the southern Sierra Nevada.

BIGLEAF MAPLE (Fig. 53) ***Acer macrophyllum***

DESCRIPTION: An erect, tall, single-stemmed tree that can grow to be around 30 m (100 ft) tall. The largest lives in Clatsop County, Oregon, and is 30 m (101 ft) tall and 3.4 m (133 in.) in diameter. When the species is open-grown, its crown is extensive and domed. **LEAVES** are opposite, simple, and deciduous. Blades are palmate, with 5 lobes, sometimes 3. Lobes have large coarse teeth. The central lobe is "narrow waisted," i.e., narrow at the base and flared at the apex. Sinuses extend to at least 50% of the lobe length. Leaf size varies greatly, but leaves can be as large as 35 cm (14 in.) wide. Leaves are more pubescent on the lower surfaces than upper ones. Petioles are red and exude a milky sap. **INFLORESCENCES** form elongated clusters of flowers. **FLOWERS** may be complete or may bear stamens only. Petals are yellowish green. Plants bloom in April and May as the new leaves appear. **FRUITS** are tawny when mature. Samaras are covered with stiff hairs, and wings diverge at oblique angles. **TWIGS** are green when young. **BARK** of old trees is thick and deeply fissured.

HABITAT AND RANGE: Commonly grows along stream banks and in canyons, but it can occur in a variety of habitats, even on dry, rocky slopes. It is found from Alaska to California. It inhabits all regions of California except the Central Valley and the deserts. It grows at elevations from sea level to 1,500 m (5,000 ft).

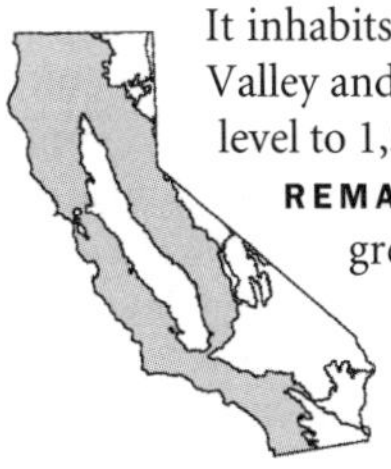

REMARKS: Leaf size and hairiness vary greatly throughout the range of the species, enough so that at least 9 species have been proposed in the past, although only 1 is recognized today. Bigleaf maples of

Figure 53
Bigleaf maple,
Acer macrophyllum

all ages support luxuriant epiphytic growth, especially when living in riparian zones. This species is the most abundant of the native maples growing in California.

BOX-ELDER (Fig. 54) *Acer negundo*

DESCRIPTION: An erect, small- to medium-sized, single-stemmed tree that can grow to be around 24 m (80 ft) tall. The largest lives in Washtenaw County, Michigan, and is 33.5 m (110 ft) tall and 1.7 m (68 in.) in diameter. Crowns are broad and irregular, with branches close to the ground. **LEAVES** are pinnately compound, opposite, and deciduous. The 3 to 7 leaflets are each 5 cm (2 in.) to 11 cm (4.5 in.) long, with coarsely toothed margins. The lower surfaces are paler, with more hairs than on the upper ones. Juvenile leaves are quite hairy and adult leaves are covered with whitish hairs. Terminal leaflets are the largest. **INFLORESCENCES** form elongated, droopy clusters of flowers. **FLOWERS** are incomplete, bearing only stamens on some plants or pistils on others. Petals are absent. Plants bloom in April or May before the leaves appear. **FRUITS** are double samaras and are red when young and yel-

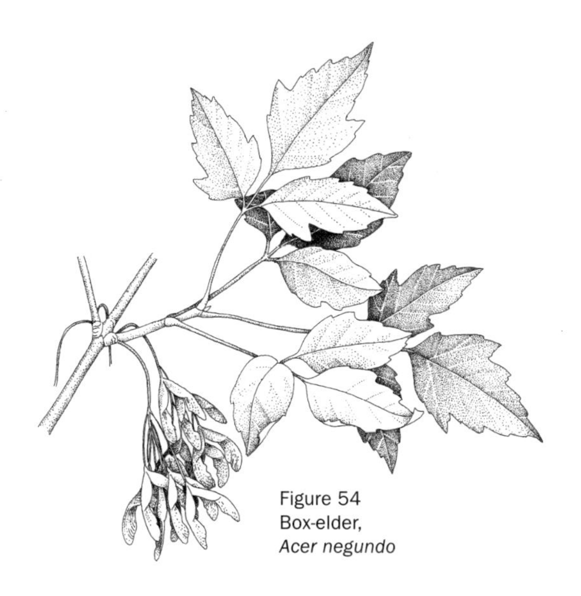

Figure 54
Box-elder,
Acer negundo

low at maturity. Fruit bodies are covered with fine hairs. Wings diverge at oblique angles. **TWIGS** are hairy and green.

HABITAT AND RANGE: Grows in forests and woodlands inhabiting streamsides and bottomlands. It occurs throughout California, except in the deserts, from sea level to 1,800 m (6,000 ft), although it is usually found at relatively low elevations. While it may be locally abundant along a particular stream, it is not found on every stream in a region.

REMARKS: Heavy pubescence on the leaves distinguishes the California subspecies, *A. n.* ssp. *californicum,* from the 3 other recognized subspecies. It is widely planted as an ornamental in California, especially as a street tree in the Central Valley. Box-elder has many cultivars.

ADENOSTOMA (CHAMISE)

The genus *Adenostoma* is composed of 2 species native to California and Baja California.

Adenostoma is a collection of evergreen shrubs with clustered, needlelike or linear leaves that are alternate in arrangement. Flowers are small, white, saucerlike blossoms with many stamens, massed in showy, terminal inflorescences. Fruits are dry, 1-seeded achenes embedded in a hardened calyx.

1. Leaves are clustered, and bark on old stems is dark gray .... **chamise (*A. fasciculatum*)**
1. Leaves are individual, not clustered, and bark on old stems is red .... **red shank (*A. sparsifolium*)**

CHAMISE (Fig. 55) ***Adenostoma fasciculatum***

DESCRIPTION: An erect, rigidly branched shrub that may grow to be 60 cm (2 ft) to 3.5 m (12 ft) tall. Fire-resistant burls can be found at the base of each shrub. **LEAVES** are evergreen, simple, linear or needlelike, sharp pointed, and about 6 mm (.25 in.) long. They are sticky and clustered along the stems. Seedlings and sprouts may have pinnately divided leaves. **INFLORESCENCES** consist of many flowers in terminal, pyramidal clusters that measure 4 cm (1.5 in.) to 10 cm (4 in.) long. **FLOWERS** are small and white. **FRUITS** are 1-seeded achenes.

HABITAT AND RANGE: This species is one of the more common shrubs in the state's chaparrals. It also grows in coastal scrubs, open forests, and woodlands. Chamise occurs locally and extensively at low elevations in the state, where it graces foothills surrounding the Central Valley, in the central Coast Ranges, and throughout southern California, below 1,500 m (5,000 ft).

REMARKS: Chamise is most important as a stabilizer of watersheds, especially in southern California. In addition, it acts as forage and habitat for wildlife. Native Americans had several medicinal uses for the leaves. The common name greasewood is sometimes used for this species, though this causes it to be confused with the common desert shrub *Sarcobatus vermiculatus*, found in alkali habitats.

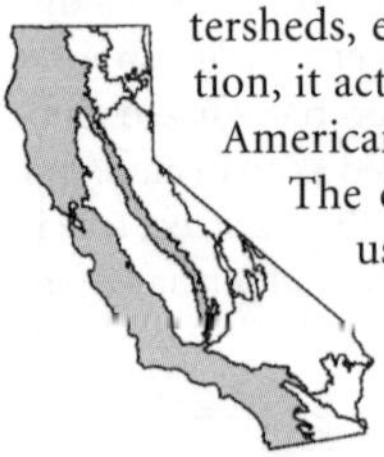

RED SHANK ***Adenostoma sparsifolium***

DESCRIPTION: A large, multistemmed, open-branched shrub or small tree that can grow to be 2 m (6.5 ft) to 6 m (20 ft)

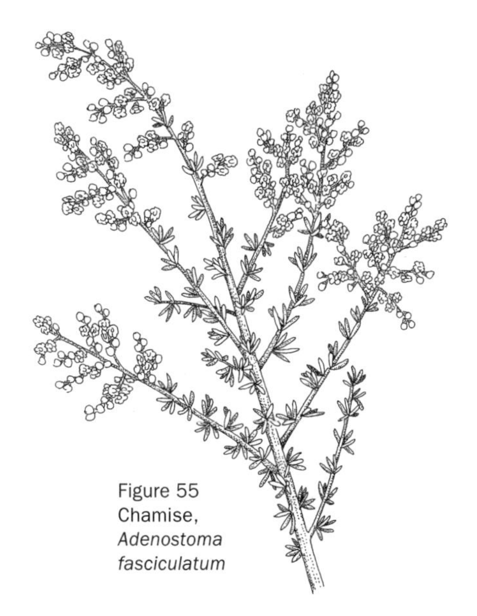
Figure 55
Chamise,
Adenostoma fasciculatum

tall. The largest lives in North Warner Springs and is 7 m (23 ft) tall and 34 cm (13 in.) in diameter. Crowns are rounded, and foliage grows at the ends of long, bare stems. Some plants are more than 100 years old. **LEAVES** are alternate, simple, evergreen, linear, sticky, and 6 mm (.25 in.) to 20 mm (.75 in.) long. Blades are linear and sharp pointed. **INFLORESCENCES** are showy, branched, terminal flower clusters that measure 2 cm (.75 in.) to 6 cm (2.5 in.) long. **FLOWERS** are white to pink and about 2 mm (.1 in.) long. **FRUITS** are small 1-seeded achenes. **TWIGS** are light green and smooth. **BARK** on large stems is reddish brown and thin, and it peels in sheets.

HABITAT AND RANGE: Grows discontinuously in foothill and lower montane chaparrals in the interior southern Coast Ranges and the mountains of southern California from San Luis Obispo County to northern Baja California, at elevations from 500 m (1,600 ft) to 1,800 m (6,000 ft). The species is common in the San Bernardino Mountains and the Peninsular Ranges.

REMARKS: Red shank regenerates primarily from root and burl sprouts. Seedlings are rare. The species provides browse for small mammals and helps protect watersheds. Native Americans used it to treat skin ailments, and early European settlers used red shank for colds and snakebites. This species is also called red shanks and ribbonwood. The term *red shank* also refers to drill bits.

AESCULUS (BUCKEYE)

The genus *Aesculus* has about 13 species of trees and large shrubs. Five are native to North America, and 1 of these is native to California.

CALIFORNIA BUCKEYE *Aesculus californica*

(Fig. 56; Pl. 12)

DESCRIPTION: An erect, single- or multistemmed large shrub or small tree. Mature plants are typically 3.5 m (12 ft) to 9 m (30 ft) tall and 10 cm (4 in.) to 15 cm (6 in.) in diameter. The largest grows in Walnut Creek and is 14.6 m (48 ft) tall and 1.4 m (55 in.) in diameter. Crowns are flat topped to rounded and very broad. Trunks are short, with numerous ascending branches. The species can live to be at least 200 years old. **LEAVES** are deciduous, opposite, and palmately compound. There are between 5 and 7 serrated, oblong to lance-shaped leaflets. Each leaflet is 7.5 cm (3 in.) to 15 cm (6 in.) long. Petioles are 1 cm (.4 in.) to 11 cm (4.5 in.) long. **INFLORESCENCES** form erect, showy clusters that measure 15 cm (6 in.) to 25 cm (10 in.) long. **FLOWERS** bloom from May to July and are pinkish white. **FRUITS** are pear shaped, pendant, and 4 cm (1.5 in.) to 5 cm (2 in.) long. Each leathery capsule contains 1 or 2 large, glossy brown seeds. **BARK** is smooth and grayish white.

HABITAT AND RANGE: A California endemic found in woodlands of the foothills and valleys of the Coast Ranges, Sierra Nevada, and Tehachapi Mountains. The species can grow in riparian zones or on dry, hot slopes below 1,700 m (5,500 ft).

REMARKS: California buckeye is summer-deciduous, losing its leaves in mid- to late summer. In September, California buckeyes can have myriad large, dangling, pear-shaped fruits. All parts of this tree are toxic

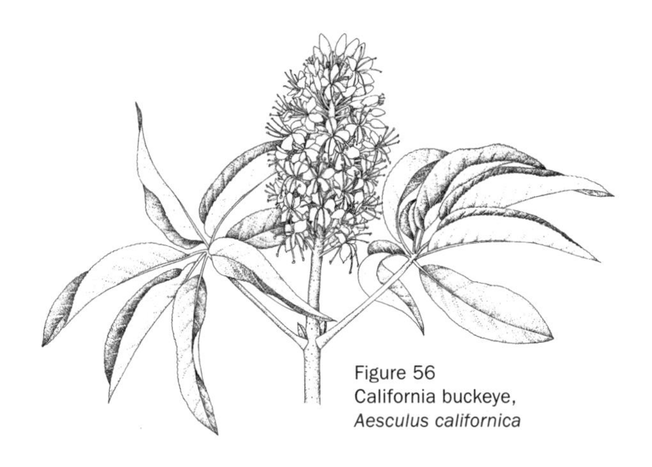

Figure 56
California buckeye,
Aesculus californica

to humans, wildlife, and livestock. They contain glycosidal compounds that affect red blood cells and the central nervous system. Native Americans used the ground seed to stun fish so that they would float to the surface, where they could be harvested. They also ate the mashed, roasted seeds after leaching away their toxins. California buckeye is planted as an ornamental because of its showy flowers and picturesque growth form. It sprouts from its base following injury.

AILANTHUS (AILANTHUS)

The genus *Ailanthus* has about 10 species of trees native to Australia and Asia. One species has become naturalized in California.

TREE-OF-HEAVEN (Fig. 57) ***Ailanthus altissima***

DESCRIPTION: A small- to medium-sized, single-stemmed tree. Mature trees are typically 12 m (40 ft) to 21 m (70 ft) tall and about 30 cm (1 ft) in diameter. The largest grows on Long Island, New York, and is 19.5 m (64 ft) tall and 1.9 m (76 in.) in diameter. Crowns are rounded and spreading, and the trees have slender trunks. **LEAVES** are deciduous, alternate, pinnately compound (usually 13 to 25 leaflets), and 30 cm (1 ft) to 90 cm (3 ft) long. Leaflets are egg shaped to lance shaped and 5 cm (2 in.) to

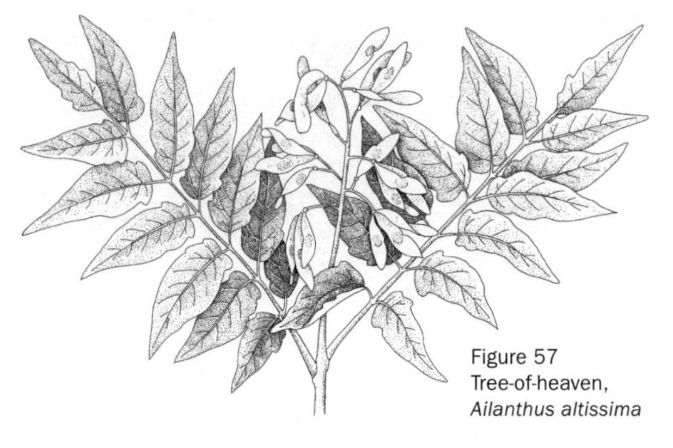
Figure 57
Tree-of-heaven, *Ailanthus altissima*

15 cm (6 in.) long. Leaflet margins are entire, with 1 to 4 glandular teeth at their bases. Leaflet tips are sharp pointed and bases are horizontal. **INFLORESCENCES** are multibranched flower clusters that measure 15 cm (6 in.) to 30 cm (12 in.) long. **FLOWERS** are whitish or yellowish green. Pollen-bearing and seed-bearing flowers are on separate trees, or flowers are complete with pollen-bearing and seed-bearing parts. **FRUITS** are samaras about 4 cm (1.5 in.) long, with the seed in the middle of the papery, twisted wing. Samaras form in clusters.

HABITAT AND RANGE: A native to China that has become naturalized along streams in river canyons in California, except in the deserts. It is especially abundant in the Sierra Nevada and Klamath Mountains foothills. Remnant populations can be found near old Chinese and other early-settler living areas. In California the species is found below 1,200 m (4,000 ft).

REMARKS: Tree-of-heaven grows rapidly on harsh sites. It can reproduce as seedlings, root suckers, and stump sprouts. It is an aggressive weed in disturbed areas. The foliage, flowers, and twigs of tree-of-heaven emit an unpleasant odor when bruised.

ALNUS (ALDER)

The genus *Alnus* has about 30 species; 10 are native to North America, and 4 of these are native to California. All but 1, found in South America, are native to the Northern Hemisphere. The genus is made up of trees and shrubs.

Leaves are deciduous, alternate, simple, elliptical to egg shaped, and 2.5 cm (1 in.) to 15 cm (6 in.) long. Margins are serrated. Leaf tips are sharp pointed. Small pollen-bearing flowers are aggregated in droopy, slender catkins that measure 5 cm (2 in.) to 20 cm (8 in.) long. Catkins form clusters of 3 to 5. The small seed-bearing flowers form catkins 5 mm (.2 in.) to 20 mm (.8 in.) long. They occur in clusters of 2 or 3. Fruits are winged nutlets enclosed in a woody catkin that resembles a conifer cone. Male and female catkins are found on the same tree. Twigs are hairless to hairy and reddish gray; they usually have stalked lateral buds. Lenticels are evident. Bark is grayish and often mottled with grayish blue and tan patches.

Alders typically grow in cool, moist woodlands and forests, generally near streams. Nitrogen-fixing bacteria form root nodules in alders, adding considerable nitrogen to ecosystems via decaying plant parts. Alder wood is used in furniture and novelty items and to smoke fish and meat. Leaves and twigs provide important browse for ungulates, and birds eat alder buds and seeds.

1. Plants are trees more than 12 m (40 ft) tall at maturity and do not grow in thickets 2
1. Plants are shrubs or small trees less than 9 m (30 ft) tall at maturity; they grow in thickets 3
   2. Leaves are doubly serrated and have tightly rolled-under leaf margins . . . . **red alder** (*A. rubra*)
   2. Leaves are generally singly serrated and have flat leaf margins . . . . **white alder** (*A. rhombifolia*)
3. Lateral buds do not grow on stalks. "Cone" stalks are as long as or longer than the "cone" and usually have at least 1 basal leaf. Leaves are thin and more or less translucent . . . . **Sitka alder** (*A. viridis*)
3. Lateral buds grow on stalks. "Cone" stalks are shorter than the "cone" and are leafless. Leaves are thick and opaque . . . . **mountain alder** (*A. incana*)

MOUNTAIN ALDER *Alnus incana* ssp. *tenuifolia*

DESCRIPTION: A large, multistemmed shrub or small tree that often grows in thickets. Mature plants are typically 2 m (6.6

ft) to 9 m (30 ft) tall and 10 cm (4 in.) to 20 cm (8 in.) in diameter. The largest grows in Umatilla National Forest in Oregon and is 21.6 m (71 ft) tall and 76 cm (30 in.) in diameter. Crowns are broad and spreading. The species is short-lived. **LEAVES** are deciduous, alternate, simple, oval to egg shaped, 5 cm (2 in.) to 10 cm (4 in.) long, and 2.5 cm (1 in.) to 6 cm (2.5 in.) wide. Margins are doubly serrated. Leaf tips are round to sharp pointed, and bases are round to heart shaped. Upper surfaces are dark green and hairless. Lower surfaces are yellowish green and may be hairless. **FRUITS** occur in conelike catkins that measure 1 cm (.4 in.) to 1.5 cm (.6 in.) long and have many woody scales. Each scale has a small, round, narrow-winged nutlet. The "cone" stalks are shorter than the "cones" and are leafless. Flowers open before leaves appear. **TWIGS** are slender, light green, and hairy. Lateral buds are stalked. Lenticels are evident. **BARK** is thin, gray to reddish gray, and somewhat scaly on older plants.

HABITAT AND RANGE: Outside of California, mountain alder grows in moist forests and woodlands in the Rocky Mountains and the Cascades. In California it is found in the Klamath Mountains, the Cascades, and the Sierra Nevada, from 1,200 m (4,000 ft) to 2,400 m (8,000 ft). It is commonly found along streams and in and around montane to upper montane meadows.

REMARKS: Mountain alder is shade intolerant and is eventually overtopped by conifers on most sites that lack recurring disturbances. Seasonal flooding is the usual disturbance that permits mountain alder to reseed itself. The species' primary value lies in its ability to protect stream banks and watersheds, as well as provide wildlife habitat. Deer and elk eat its foliage, and small mammals eat the twigs and buds. Other subspecies of *A. incana* grow in Canada, Alaska, and Europe.

WHITE ALDER (Fig. 58) *Alnus rhombifolia*

DESCRIPTION: A small- to medium-sized, single-stemmed tree. Mature trees are typically 15 m (50 ft) to 24 m (80 ft) tall and 30 cm (1 ft) to 60 cm (2 ft) in diameter and often have several stems arising from a clump. The largest grows in Camp Nelson, California, and is 24 m (79 ft) tall and 1.2 m (46 in.) in diameter. Crowns are broadly rounded on long, clear trunks. This species lives to be only about 100 year old. **LEAVES** are decid-

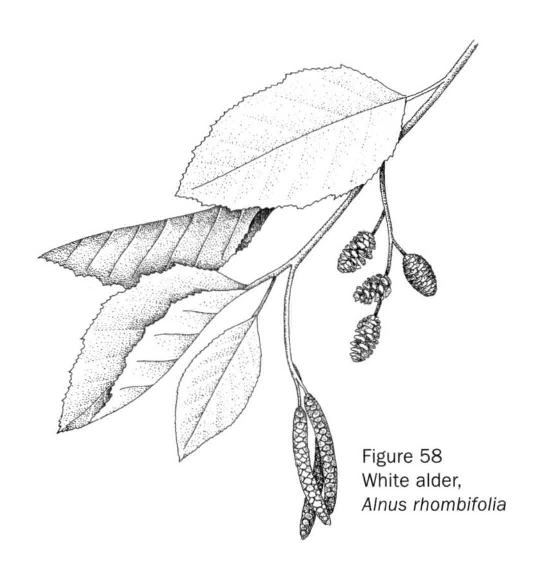

Figure 58
White alder,
Alnus rhombifolia

uous, alternate, simple, thick, egg shaped, oval to rhombic, and 4 cm (1.5 in.) to 8 cm (3.2 in.) long. Margins are usually singly serrated (sometimes doubly) and are not rolled under. Serrations are generally smaller than those of red alder. Leaf tips are round to sharp pointed, and bases are tapered to round. Upper surfaces are green and lower surfaces are yellowish green, with prominent parallel veins. **FRUITS** occur in conelike catkins that measure 6 mm (.25 in.) to 12 mm (.5 in.) long and have many scales. Each scale has a small, round, unwinged nutlet. The "cone" stalks are shorter than the "cones" and are leafless. Flowers open before leaves appear. **TWIGS** are slender and orange-red. Lenticels are evident. Lateral buds are stalked. **BARK** is thin and gray, with mottled, whitish gray patches. Older trees' bark is furrowed and separated by grayish brown plates.

HABITAT AND RANGE: Grows in riparian woodlands and forests from southern British Columbia to California. In California it occurs from 90 m (300 ft) to 2,400 m (7,900 ft) along permanent streams and lakes. It is found on the Modoc Plateau but is considered uncommon.

REMARKS: White alder is shade intolerant. It needs a continual supply of moisture and is a pioneering species in riparian areas. It is susceptible to fire but often escapes damage because of its moist habitat. White alder is used for firewood, somewhat as lumber, and as an ornamental. Native Americans extracted a red dye from the bark and made a sweat-inducing tea from it. White alder's greatest value lies in its ability to protect watersheds and provide wildlife habitat. These roles are especially critical in areas subject to development.

RED ALDER (Fig. 59) *Alnus rubra*

DESCRIPTION: An erect, medium-sized, single-stemmed tree. Mature trees are typically 24 m (80 ft) to 30 m (100 ft) tall and 30 cm (1 ft) to 60 cm (2 ft) in diameter. The largest grows in Clatsop County, Oregon, and is 31.7 m (104 ft) tall and 2 m (78 in.) in diameter. Crowns are narrow to broad on long, clear trunks. The species lives to be only about 100 years old. **LEAVES** are deciduous, alternate, simple, thick, egg shaped to elliptical, and 7.5 cm (3 in.) to 15 cm (6 in.) long. Margins are doubly serrated and tightly rolled under. Serrations are generally larger than those of white alder. Leaf tips are sharp pointed and bases are tapered to round. Upper surfaces are dark green and hairless. Lower surfaces are pale green and have prominent, parallel, rust-colored veins. **FRUITS** occur in conelike catkins that measure 1 cm (.4 in.) to 3 cm (1.2 in.) long and have many scales. Each scale has a small, round, broad-winged nutlet. The "cone" stalks are shorter than the "cones" and are leafless. Flowers open before leaves appear. **TWIGS** are bright red to reddish brown. Lenticels are evident. Lateral buds are stalked. **BARK** is thin and gray, with bluish gray, tan, and grayish white mottling.

HABITAT AND RANGE: Grows in coastal and montane woodlands and forests from southeastern Alaska to San Luis Obispo County. It occurs below 900 m (3,000 ft) on riparian and moist upland slopes. The species usually grows on lower slopes on deep, fertile soils. In the southern part of its range, it is mostly confined to coastal habitats. In northwestern California, it grows on moist, cool inland sites as well.

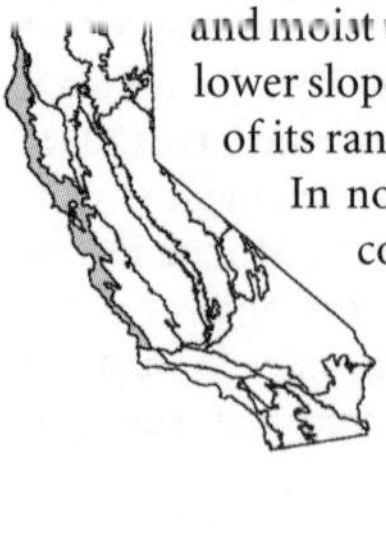

REMARKS: Red alder is shade intolerant. It often becomes established on disturbed sites, only to be overtopped by

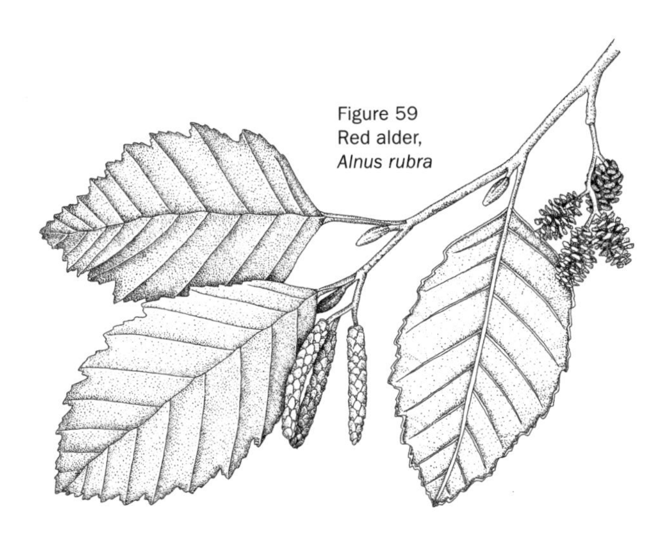
Figure 59
Red alder,
Alnus rubra

more shade-tolerant conifers. Although its bark is thin, mature red alders can survive light surface fires. The species' wood is used for furniture, novelty items, cabinets, firewood, and numerous other items. Deer and elk eat its foliage, and beavers eat its bark. Native Americans used the wood to smoke salmon and for medicinal purposes.

SITKA ALDER *Alnus viridis* ssp. *sinuata*

(Fig. 60; Pl. 13)

DESCRIPTION: A tall, multistemmed shrub or small tree that grows in thickets. Mature plants range in height from 2 m (6.5 ft) to 4.5 m (15 ft). The largest grows on Maury Island, Washington, and is 11.3 m (37 ft) tall and 22 cm (8.5 in.) in diameter. Crowns are broadly rounded and spreading. The species is short-lived. **LEAVES** are deciduous, alternate, simple, thin, oval to egg shaped, 6 cm (2.5 in.) to 15 cm (6 in.) long, and 4 cm (1.5 in.) to 9 cm (3.5 in.) wide. Margins are doubly serrated. Leaf tips are sharp pointed and bases are round to tapered. Upper surfaces are yellowish green and shiny; lower surfaces are green, with hairs restricted to, or occurring more densely in, the major vein axils. **FRUITS** occur in conelike catkins about 13 mm (.5 in.) long and have many scales. Each scale has a small, round,

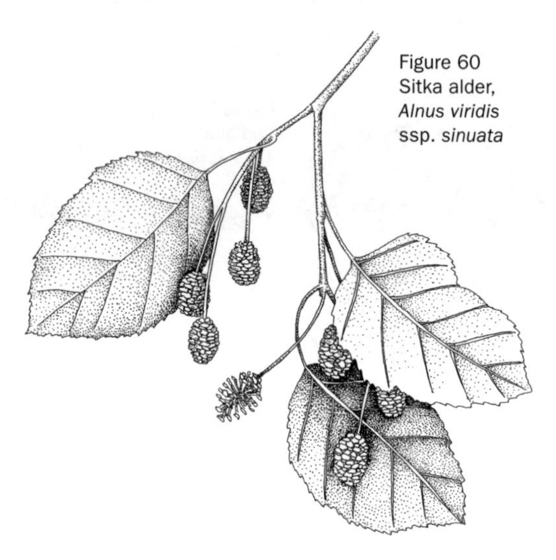

Figure 60
Sitka alder,
Alnus viridis
ssp. *sinuata*

broad-winged nutlet. The "cone" stalks are longer than the "cones" and usually have at least 1 leaf at the base. Flowers appear as leaves reach full size. **TWIGS** are slender, reddish to yellowish brown, and smooth. Lateral buds are not stalked. Lenticels are evident. **BARK** is thin and reddish to grayish brown; it has prominent lenticels.

HABITAT AND RANGE: Outside of California, Sitka alder grows in moist forests and woodlands from Alaska, western Canada, and the northern Rocky Mountains and into Washington and Oregon. In California it grows at elevations from 1,000 m (3,200 ft) to 2,700 m (8,800 ft) in the Klamath Mountains and at higher elevations in the northern Coast Ranges. The species occurs along streams, seeps, and meadow margins.

REMARKS: Sitka alder is shade intolerant and may eventually become overtopped by conifers on most sites that lack recurring disturbances. Usual disturbances that permit this species to reseed itself include flooding, avalanches, and landslides. It quickly invades habitats disturbed by these events. Sitka alder's greatest value lies in its ability to protect watersheds and stabilize slopes. It is an indicator of high water tables. The other subspecies of *A. viridis* grow in Canada, Alaska, and Eurasia.

Figure 61
Western serviceberry,
Amelanchier alnifolia

AMELANCHIER (SERVICEBERRY)

The genus *Amelanchier* has about 10 species growing in temperate North America, Eurasia, and northern Africa. Two species are native to California.

WESTERN SERVICEBERRY *Amelanchier alnifolia* (Fig. 61)

DESCRIPTION: A loosely branched shrub or small tree that may grow to be between 1 m (3.3 ft) and 8 m (25 ft) tall. The largest lives in Beacon Rock State Park, Washington, and is 12.8 m (42 ft) tall and 30 cm (1 ft) in diameter. **LEAVES** are deciduous, simple, alternate, and mostly borne on short lateral twigs. Leaf blades are elliptical or round, 2 cm (.75 in.) to 4 cm (1.5 in.) long, and hairless, with teeth on the upper half of the margins only. **INFLORESCENCES** consist of 1 to several flowers borne on short lateral twigs. Clusters are 2.5 cm (1 in.) to 5 cm (2 in.) long. **FLOWERS** have 5 separate white petals. **FRUITS**

are small, hairless or hairy pomes. **TWIGS** are hairy or hairless and reddish brown.

HABITAT AND RANGE: Grows in forests, woodlands, scrublands, and grasslands in northern California's mountains, from 60 m (200 ft) to 2,600 m (8,500 ft). Outside of California it occurs throughout northwest North America.

REMARKS: Serviceberry plants show a great deal of local and regional variation in leaf and fruit characteristics, to the point that books report different numbers of species and varieties for California. The 2 California serviceberry species can be distinguished by leaf hairiness. Western serviceberry, which has hairless leaves, is more commonly encountered than Utah serviceberry (*A. utahensis*), which has hairy leaves. Both species may grow in the same region and are important forage and habitat plants for wildlife. Outside of California this species is often called Saskatoonberry.

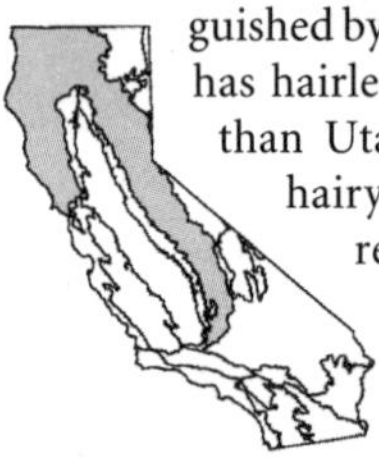

ARBUTUS (MADRONE)

The genus *Arbutus* has about 20 species; 3 are native to the United States, and 1 of these is native to California. The genus is made up of trees and shrubs growing in North America, the Mediterranean basin, Central America, the Canary Islands, and Asia.

PACIFIC MADRONE (Fig. 62) ***Arbutus menziesii***

DESCRIPTION: An erect, medium-sized, single- or multistemmed tree. On harsh sites it can be a shrub. Mature trees are typically 8 m (25 ft) to 40 m (130 ft) tall and 30 cm (1 ft) to 90 cm (3 ft) in diameter. The largest grows in Humboldt County and is 24.4 m (80 ft) tall and 3.4 m (133 in.) in diameter. Stems typically are crooked and slanted and have large, skewed, meandering branches. Crowns are broad, irregular, and sparse. Trees can reach 400 to 500 years in age. **LEAVES** are evergreen, simple, alternate, leathery, oval to oblong, 7.5 cm (3 in.) to 12.5 cm (5 in.) long, and 4 cm (1.5 in.) to 8 cm (3 in.) wide. Margins have fine serrations or are entire. Upper surfaces are dark green and lower surfaces are light to silvery green. Petioles are grooved

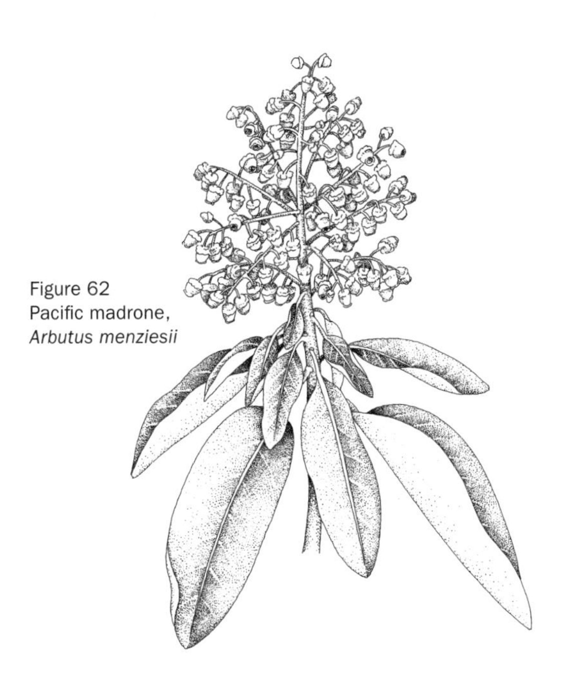

Figure 62
Pacific madrone,
Arbutus menziesii

and about 2.5 cm (1 in.) long. **INFLORESCENCES** are sweet smelling, hairy-branched clusters measuring 7.5 cm (3 in.) to 15 cm (6 in.) long. **FLOWERS** are urn shaped, about 6 mm (.25 in.) long, and white to pink. **FRUITS** are spherical, fleshy, orange-red berries that measure 6 mm (.25 in.) to 12 mm (.5 in.) in diameter. Each berry has about 20 hard, dark brown seeds. **BARK** is thin and flaky, and it exfoliates in strips, revealing smooth, reddish brown inner bark. Older bark is fissured and reddish brown, while younger bark can be green to reddish brown.

HABITAT AND RANGE: Grows in broadleaved and coniferous forests, woodlands, and chaparrals, from Vancouver Island south to Mexico. In California it occurs along the coast and west of the Sierra Nevada and Cascades, at elevations between 100 m (328 ft) and 1,500 m (5,000 ft). It is generally found on relatively dry sites that experience mild, oceanic winters.

REMARKS: Pacific madrone is moderately shade tolerant. It is susceptible to fire, various leaf spot diseases, and the fungal "madrone canker." It sprouts following injury from fire or cutting. Pacific madrone foliage is not very palatable to browsing wildlife. Deer devour its flowers, however, and small mammals and birds, especially the band-tailed pigeon, favor its fruits. Pacific madrone bark has been used to tan leather. Its wood is used for cabinets, novelty items, and firewood.

ARCTOSTAPHYLOS (MANZANITA)

The genus *Arctostaphylos* has about 60 species, and 56 of these are native to California. One species is circumpolar, and the other non-Californian species are native to Mexico and Central America. The genus is composed of shrubs and small trees.

Leaves are simple, evergreen, leathery, and alternate in arrangement. Flowers are showy and pink to white, and they appear in winter or spring. Each is an urn topped by 4 or 5 small lobes. Flowers are arranged in droopy or upturned inflorescences. Fruits are drupes with 2 to 10 stones embedded in a dry, mealy pulp. Branches and stems are typically smooth and red. Some species have fire-resistant burls at the bases of their stems.

Several manzanita species have many subspecies. Plants are easily recognized at the genus level, but many species are locally distributed and hybridization may make identification difficult in areas where species' ranges overlap. Before keying a plant, note its habit and whether it has a burl. Not all species are included in this book, but we have included the more commonly seen species that grow in California's forests and woodlands.

Young fruits look like small apples, hence the Spanish-based common name (*manzana* is the Spanish word for apple). The generic name means "bear berries": manzanita fruits are crops that animals feed on throughout the summer. Young fruits of some species are used to make jellies. *Arctostaphylos* species have hard seed coats that must be scarified before germination can take place. Many stands of *Arctostaphylos* have become established as a result of their seed coats being cracked by the heat of a fire. Abrasions on the seed coats can also allow germination to be initiated.

1. Plants are prostrate or mounding shrubs that root when branches come in contact with the ground 2
1. Plants are upright shrubs or trees 5

2. Young twigs are covered with long, bristly hairs . . . . **glossyleaf manzanita** (*A. nummularia*)
2. Young twigs are not covered with long, bristly hairs 3
3. Leaves are slightly cupped, and upper surfaces are darker than the lower ones . . . . **bearberry** (*A. uva-ursi*)
3. Leaves are flat, and both surfaces are the same color 4
4. Leaf blades are widest above the middle . . . . **pinemat manzanita** (*A. nevadensis*)
4. Leaf blades are widest at or below the middle . . . . **Parry manzanita** (*A. parryana*)
5. Inflorescence bracts are large and leaflike 6
5. Inflorescence bracts are small and scalelike, but the lower ones may be larger 9
6. Shrubs have burls that sprout after fire. Immature inflorescences are nodding 7
6. Shrubs lack burls that sprout after fire. Immature inflorescences are upturned 8
7. Leaf blades have flat or heart-shaped bases, and upper surfaces are darker than lower ones . . . . **woollyleaf manzanita** (*A. tomentosa*)
7. Leaf blades have tapered or round bases, and both surfaces are the same color . . . . **Eastwood manzanita** (*A. glandulosa*)
8. Young twigs have long, bristly hairs that may be gland tipped . . . . **hairy manzanita** (*A. columbiana*)
8. Young twigs lack long, bristly hairs. Instead, they have many short, white, downy hairs . . . . **hoary manzanita** (*A. canescens*)
9. Young twigs are covered with long, bristly hairs . . . . **glossyleaf manzanita** (*A. nummularia*)
9. Young twigs are not covered with long, bristly hairs 10
10. Leaf blades are grayish green or whitish green 11
10. Leaf blades are green 12
11. Leaves are grayish green . . . . **bigberry manzanita** (*A. glauca*)
11. Leaves are whitish green . . . . **whiteleaf manzanita** (*A. viscida*)
12. Young twigs, inflorescence branches, and petioles have golden, glistening, gland-tipped hairs . . . . **greenleaf manzanita** (*A. patula*)
12. Young twigs, inflorescence branches, and petioles lack golden, glistening, gland-tipped hairs 13

13. Inflorescences are round in shape. Leaves are 2 cm (.75 in.) to 2.5 cm (1 in.) long . . . . **Mexican manzanita (*A. pungens*)**
13. Inflorescences are oblong in shape. Leaves are 2.5 cm (1 in.) to 6.5 cm (2.5 in.) long . . . . **common manzanita (*A. manzanita*)**

HOARY MANZANITA *Arctostaphylos canescens*

DESCRIPTION: An erect shrub that lacks burls and can grow to be 2 m (6.5 ft) tall. **LEAVES** are erect and have petioles that measure 3 mm (.12 in.) to 10 mm (.4 in.) long. Blades are oval to elliptical and 2 cm (.75 in.) to 5 cm (2 in.) long. Blades have wedge-shaped or round bases with smooth margins. Upper and lower surfaces of a leaf are similar in color and texture: dull gray covered with fine, small hairs. **INFLORESCENCES** are compact, 1 cm (.4 in.) to 2 cm (.75 in.) long, and droopy. Most bracts are large, leafy, lance shaped, and covered with fine white hairs that may end in glands. **FLOWERS** are white to pink. **FRUITS** are spherical, 5 mm (.2 in.) to 10 mm (.4 in.) in diameter, and covered with fine hairs. **TWIGS** have a dense covering of short gray hair. **BARK** is smooth and dark red.

HABITAT AND RANGE: Grows in coniferous forests and chaparrals at low elevations of the coastal ranges, from southwestern Oregon south to the Santa Cruz Mountains, at elevations from 300 m (1,000 ft) to 1,500 m (5,000 ft).

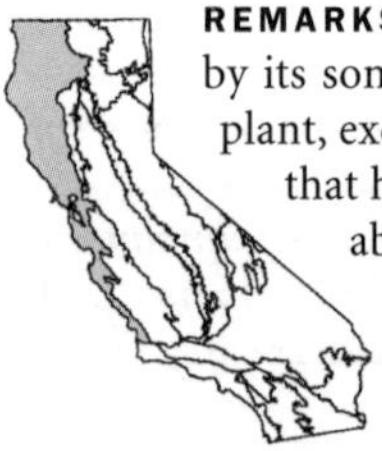

REMARKS: Hoary manzanita is easily recognized by its somber, ashy (hoary) hue on all parts of the plant, except its red bark. As with other manzanitas that have leafy inflorescences, there is considerable variation in hairiness among populations. It does not sprout from a burl after fires. *A. canescens* ssp. *sonomensis* is considered rare.

HAIRY MANZANITA *Arctostaphylos columbiana*

DESCRIPTION: An erect shrub or small tree that lacks burls and can grow to be 3 m (10 ft) tall. **LEAVES** have petioles that measure 4 mm (.15 in.) to 10 mm (.4 in.) long, oval to elliptical blades, wedge-shaped or round bases, and smooth margins. Leaf blades are 4 cm (1.5 in) to 6 cm (2.4 in.) long. Both blade surfaces are similar in color and texture: dull gray-green covered with fine, small hairs. **INFLORESCENCES** are open, with flower stalks that measure 15 mm (.6 in.) to 25 mm (1 in.) long. Most bracts are 10 mm (.4 in.) to 18 mm (.7 in.) long, leafy, lance shaped, and covered with bristles or fine hairs. **FLOWERS** are white to pink. **FRUITS** are spherical, 8 mm (.3 in.) to 10 mm (.4 in.) in diameter, bright red, slightly hairy, and sticky, with a deep dimple. **TWIGS** are densely covered with fine hairs and longer white bristles that may end in glands. **BARK** is smooth and dark reddish brown.

HABITAT AND RANGE: Grows in coniferous forests, chaparrals, and coastal scrubs, from sea level to 750 m (2,500 ft), in the western Klamath Mountains and northern Coast Ranges, and into the Pacific Northwest.

REMARKS: Hairy manzanita resembles Eastwood manzanita and woollyleaf manzanita. However, it lacks the plant-to-plant variation seen in those species, and, unlike those species, it does not sprout from a burl after fires. This species is also called Columbia manzanita.

EASTWOOD MANZANITA (Fig. 63) — *Arctostaphylos glandulosa*

DESCRIPTION: An erect shrub with burls. The species can grow to be 2 m (6.5 ft) tall. **LEAVES** have hairy petioles that measure 5 mm (.2 in.) to 10 mm (.4 in.) long, elliptical to lance-shaped blades, round bases, and smooth or toothed margins. Leaf blades are 2 cm (.75 in.) to 4.5 cm (1.75 in.) long. Both blade surfaces are similar in color and texture. Surfaces are bright green to gray and vary in degree of hairiness, becoming less hairy with age. Leaf characteristics differ greatly among plants. **INFLORESCENCES** are dense; flower stalks measure 1 cm (.4 in.) to 3 cm (1.2 in.) long. Most bracts are 8 mm (.3 in.) to 15 mm (.6 in.) long, leafy, lance shaped, and covered with fine hairs. **FLOWERS** are white. **FRUITS** are spherical, 6 mm (.25 in.) to 10 mm (.40 in.) wide, sticky, reddish brown, and hairy, with a dimple. **TWIGS** are covered with erect, even-sized, bristly hairs that often end in glands. **BARK** is smooth and red.

HABITAT AND RANGE: This species, in all its forms, grows

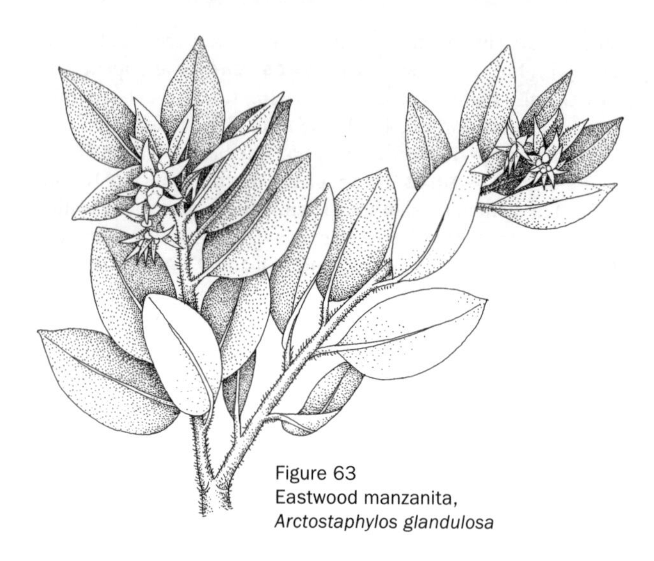

Figure 63
Eastwood manzanita,
Arctostaphylos glandulosa

in lower-elevation forests, woodlands, and chaparrals in coastal ranges from southwestern Oregon to Baja California. It is found from sea level to 2,100 m (7,000 ft).

REMARKS: Many populations have characteristics shared among Eastwood manzanita, woolly manzanita, and other manzanitas, probably as a result of hybridization. Eastwood manzanita is a complex of 6 subspecies; 1 is considered rare (*A. g.* ssp. *crassifolia*). All forms sprout from the burl after fires. Birds and small mammals eat Eastwood manzanita fruits but find its leaves less palatable. Dense stands provide valuable hiding, resting, and nesting sites. Native Americans ground Eastwood manzanita's dried fruits into flour, and jelly can be made from the fruits. The foliage has chemicals that inhibit ponderosa and knobcone pine seedlings.

BIGBERRY MANZANITA *Arctostaphylos glauca* (Fig. 64)

DESCRIPTION: An erect shrub or small tree that lacks burls and typically grows to be 7 m (23 ft) tall. The largest lives in

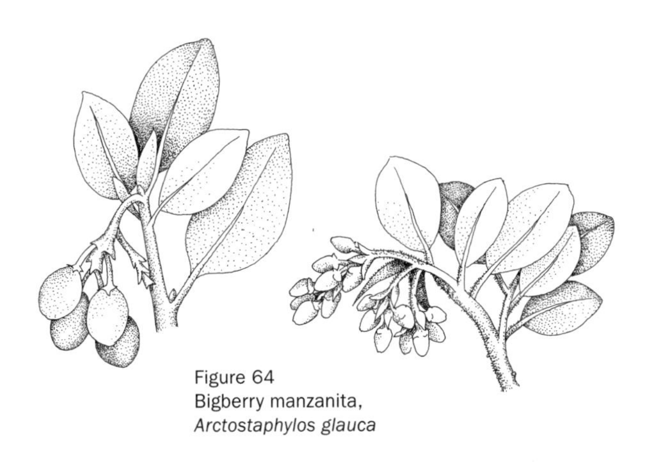

Figure 64
Bigberry manzanita,
Arctostaphylos glauca

Sunol Regional Park, Alameda County, and is 10 m (33 ft) tall and 42 cm (16.5 in.) in diameter. **LEAVES** are erect, with petioles measuring 6 mm (.25 in.) to 15 mm (.6 in.) long. Leaf blades are 2.5 cm (1 in.) to 5 cm (2 in.) long. The oval blades have round bases and smooth or toothed margins. Both blade surfaces are similar in color and texture: dull, hairless, waxy, and grayish green. **INFLORESCENCES** are open and they hang down. Most bracts are small scales that measure 3 mm (.1 in.) to 6 mm (.25 in.) long. The lowest bract is 10 cm (4 in.) to 15 cm (6 in.) and leaflike. **FLOWERS** are white and tinged with pink. **FRUITS** are spherical, about 13 mm (.5 in.) in diameter, and very sticky. **TWIGS** are covered by greenish wax and sometimes with small bristles. **BARK** is smooth and dark red.

HABITAT AND RANGE: Grows in woodlands and chaparrals at low and mid elevations, from the San Francisco Bay area south to the mountains of southern California and Baja California. Populations are found in the desert mountains. The species grows from sea level to 1,400 m (4,500 ft).

REMARKS: Bigberry manzanita is easily recognized by its large fruits and massive stature. It does not sprout from a burl after fires. The species is planted as an ornamental in hot, dry environments. Birds and small mammals eat its fruits. Bigberry manzanita foliage has chemicals that inhibit growth of annual plants near the canopy's drip line.

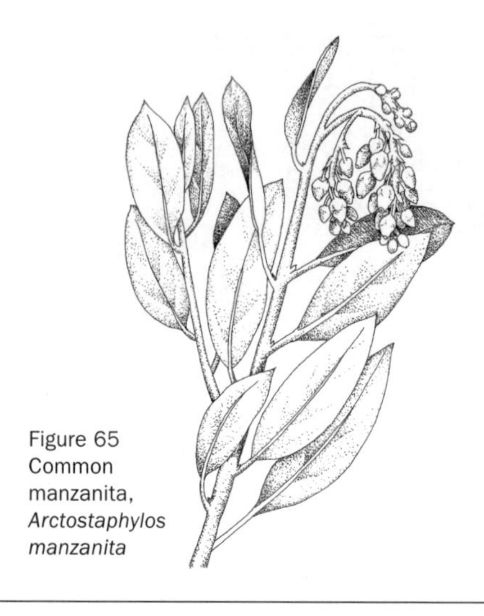

Figure 65
Common manzanita, *Arctostaphylos manzanita*

COMMON MANZANITA *Arctostaphylos manzanita*

(Fig. 65)

DESCRIPTION: An erect shrub or tree that lacks burls and typically grows to be 7 m (23 ft) tall. The largest lives in Guerneville, California, and is 6.7 m (22 ft) tall and 74 cm (29 in.) in diameter. **LEAVES** are erect, smooth margined, and oval to oblong, with petioles measuring 6 mm (.25 in.) to 12 mm (.5 in.) long. Leaf blades are 2 cm (.75 in.) to 5 cm (2 in.) long. Bases are round to wedge shaped. Both blade surfaces are similar in color and texture on the same plant: dull or shiny, hairless, and green or gray-green. **INFLORESCENCES** are open, droopy, and 15 mm (.6 in.) to 45 mm (1.75 in.) long. Most bracts are small scales that are appressed to the stalks. **FLOWERS** are white. **FRUITS** are spherical, 8 mm (.3 in.) to 12 mm (.5 in.) in diameter, deep red, and hairless or covered with fine hairs. **TWIGS** are covered with small, fine white hairs. **BARK** is smooth and reddish brown.

HABITAT AND RANGE: Grows in forests, woodlands, and chaparrals at lower elevations surrounding the Central Valley north of Contra Costa and Mariposa Counties, and into the Klamath Mountains and northern Coast Ranges. It is found at elevations from 100 m (300 ft) to 1,200 m (4,000 ft).

REMARKS: Common manzanita varies in leaf color and hairiness, so it is not surprising that it has 6 subspecies. Like other manzanitas, it hybridizes with other species. One subspecies, *A. m.* ssp. *insulicola,* has a limited range and is considered uncommon. Most forms fail to sprout from a burl after fires. Birds and small mammals eat the species' fruits. Common manzanita leaves have been used to treat urinary tract infections, jelly can be made from its fruits, and novelty items have been carved from the wood of large stems.

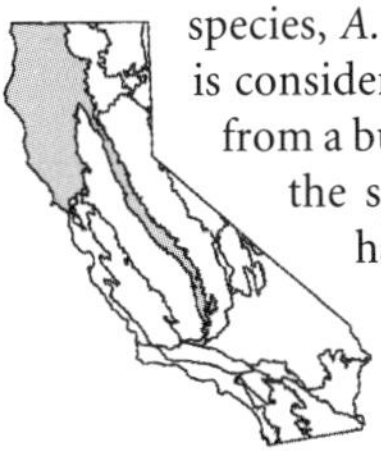

PINEMAT MANZANITA *Arctostaphylos nevadensis*

DESCRIPTION: A prostrate or mounded shrub that lacks burls and can grow to be 60 cm (2 ft) tall. **LEAVES** are erect, with petioles measuring 3 mm (.12 in.) to 6 mm (.25 in.) long. Blades are wider above the middle and lance shaped to egg shaped, with smooth margins. Blades are 1 cm (.4 in.) to 3 cm (1.2 in.) long. Both blade surfaces are similar in color and texture: bright green and either shiny or dull. **INFLORESCENCES** are dense; flower stalks measure 5 mm (.2 in.) to 10 mm (.4 in.) long. Most bracts are small, linear scales, but the lowest bract is larger and leaflike. **FLOWERS** are white to pink. **FRUITS** are spherical, about 6 mm (.25 in.) wide, mealy, brown, and hairless. **TWIGS** have fine hairs. **BRANCHES** are spreading, and often they root when in contact with the ground. **BARK** is red under curls of exfoliating bark.

HABITAT AND RANGE: Grows in montane and subalpine coniferous forests and chaparrals of northern California, from 300 m (1,000 ft) to 3,000 m (10,000 ft). It also grows in the Pacific Northwest.

REMARKS: A commonly occurring prostrate, spreading shrub in the mountains of northern California. Beginners may confuse this species with bearberry, another prostrate manzanita. The leaves and range of the two species are very different. Pinemat manzanita hybridizes with the equally common upright greenleaf manzanita and whiteleaf manzanita, forming plants that are intermediate in habit and leaf characteristics.

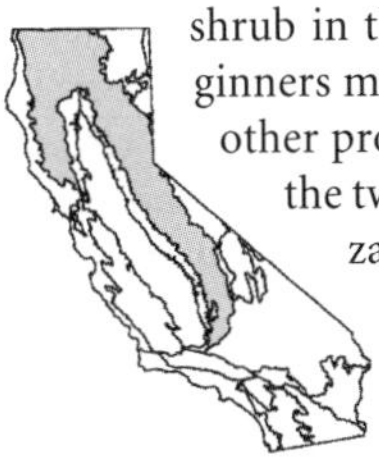

Figure 66
Glossyleaf manzanita, *Arctostaphylos nummularia*

GLOSSYLEAF MANZANITA (Fig. 66) *Arctostaphylos nummularia*

DESCRIPTION: A normally prostrate shrub lacking burls, it can grow to be from 15 cm (.5 ft) to 60 cm (2 ft) high. Erect shrubs can grow to be 2 m (7 ft) tall. **LEAVES** are crowded on the stems, round or heart shaped, cupped, and covered with hairs. Petioles are only 1 mm (.04 in.) to 3 mm (.12 in.) long, and leaf blades are 1 cm (.4 in.) to 2.2 cm (.87 in.) long. Upper leaf surfaces are dark green and shiny, and lower surfaces are light green, with bristles on their midribs. **INFLORESCENCES** are dense; flower stalks measure 5 mm (.2 in.) to 10 mm (.4 in.) long. Bracts are small triangular scales. **FLOWERS** are small and white. **FRUITS** are oblong, about 3 mm (.12 in.) long, about 2 mm (.08 in.) wide, and green. **TWIGS** are densely hairy and covered with white, gland-tipped bristles. **BRANCHES** are ascending or spreading and often root when in contact with the ground. **BARK** is smooth and red.

HABITAT AND RANGE: Grows in coniferous forests and chaparrals of the north coast and in the Santa Cruz Moun-

tains. Plants are often seen in seasonally wet sites on acid soils near the coast. The species grows from sea level to 750 m (2,500 ft).
REMARKS: The size of glossyleaf manzanita is apparently related to soil characteristics: prostrate plants are associated with infertile, acidic soils found on Mendocino County terraces near Fort Bragg, which support pygmy conifer forests; plants growing inland are erect. Glossyleaf manzanita can reproduce vegetatively from branch-to-ground contact, but it does not sprout following fire.

PARRY MANZANITA *Arctostaphylos parryana*

DESCRIPTION: A prostrate or mounded shrub that lacks burls and can grow to be 2 m (7 ft) tall. **LEAVES** are erect, with petioles measuring 5 mm (.2 in.) to 10 mm (.4 in.) long. Blades are round or elliptical with smooth margins; they measure 1.5 cm (.6 in.) to 5 cm (2 in.) long and 1.5 cm (.6 in.) to 2.5 cm (1 in.) wide. Both blade surfaces are similar in color and texture: bright green and shiny. **INFLORESCENCES** are dense; flower stalks measure 5 mm (.2 in.) to 15 mm (.6 in.) long. Most bracts are small scales, but the lowest bract is larger and leaflike. **FLOWERS** are white. **FRUITS** are spherical, 9 mm (.35 in.) wide, dark reddish brown, and hairless. **TWIGS** lack hairs or have many fine white hairs. **BRANCHES** are spreading and often root when they are in contact with the ground. **BARK** is dark red.
HABITAT AND RANGE: Grows in montane coniferous forests and chaparrals of the Transverse Ranges. It is found from 1,200 m (4,000 ft) to 2,300 m (7,500 ft).
REMARKS: Parry manzanita most closely resembles greenleaf manzanita, but habits of the two differ. Greenleaf manzanita is erect, while Parry manzanita is a spreading shrub. In addition, Parry manzanita leaves are lighter in color.

GREENLEAF MANZANITA *Arctostaphylos patula*

(Fig. 67)

DESCRIPTION: An erect shrub that grows to be 2 m (6.5 ft) tall. It may have burls. **LEAVES** are erect and have petioles measuring 6 mm (.25 in.) to 15 mm (.6 in.) long. The oval blades

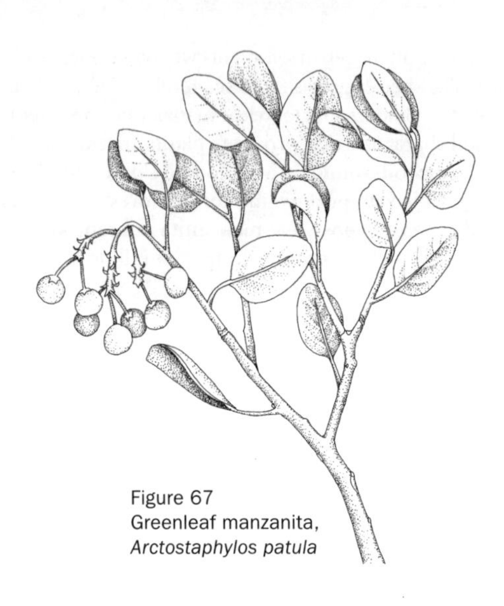

Figure 67
Greenleaf manzanita,
Arctostaphylos patula

are 2.5 cm (1 in.) to 6 cm (2.4 in.) long, with round bases and smooth margins. Both blade surfaces are similar in color and texture: dull, hairless, and shiny green. **INFLORESCENCES** are open; flower stalks measure 15 mm (.6 in.) to 30 mm (1.2 in.) long. Most bracts are small scales, although the lowest one is larger and leaflike. **FLOWERS** are pink. **FRUITS** are spherical, about 9 mm (.35 in.) wide, hairless, and dark chestnut-red. **TWIGS** are slender and covered with golden, glistening hairs topped with glands. **BARK** is smooth and deep chestnut-brown.

HABITAT AND RANGE: Grows in montane and subalpine forests, woodlands, and chaparrals. Outside of California it is found in the Pacific Northwest, the Rocky Mountains, and Baja California. In California it grows in the mountains north of Lake County in the coastal ranges, and southward in inland mountains as far as the San Jacinto Mountains in southern California, at elevations from 600 m (2,000 ft) to 3,300 m (11,000 ft).

REMARKS: Greenleaf manzanita is a common manzanita easily recognized by its habitat and sticky twigs. Plants in some populations have burls and many trunks, while plants in other populations have single trunks and lack burls. Greenleaf manzanita may sprout from its burls after fires. Its seeds can remain dormant in the soil for decades. Birds and small mammals eat greenleaf manzanita fruits but find the leaves less palatable. Native Americans ate the fruits, made cider from them, and used them as a diuretic. The species hybridizes with the equally common, upright, whiteleaf manzanita and the prostrate pinemat manzanita, forming plants intermediate in habit and leaf characteristics.

MEXICAN MANZANITA *Arctostaphylos pungens*

DESCRIPTION: An erect or spreading shrub that lacks burls and can grow to be 3 m (10 ft) tall. **LEAVES** are erect and have petioles measuring 4 mm (.15 in.) to 8 mm (.3 in.) long. Blades are 1.5 cm (.6 in.) to 4 cm (1.5 in.) long and elliptical or lance shaped, with wedge-shaped bases and smooth margins. Both blade surfaces are similar in color and texture on the same plant: dull, hairless, and shiny green. **INFLORESCENCES** are a dense ball of flowers. Most bracts are large, leafy, and lance shaped. **FLOWERS** are white. **FRUITS** are spherical, about 6 mm (.25 in.) wide, dimpled, hairless, and brownish red. **TWIGS** are covered with small, fine white hairs. **BARK** is smooth and reddish brown.
HABITAT AND RANGE: Grows mainly in montane woodlands and forests in southern California's mountains. The species is found from 600 m (2,000 ft) to 2,100 m (7,000 ft). Outside of California it is found in Nevada, Utah, Arizona, New Mexico, Texas, and Mexico. 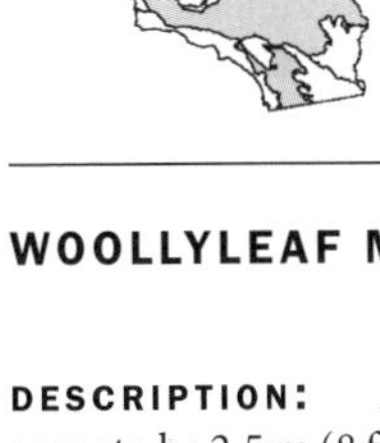
REMARKS: Outside of southern California's mountains, localized populations of Mexican manzanita can be found in the mountains of Monterey and San Benito Counties, and in the New York and Providence Mountains of the Mojave Desert. It does not sprout from a burl after fires.

WOOLLYLEAF MANZANITA *Arctostaphylos tomentosa*

DESCRIPTION: An erect shrub with burls. The species can grow to be 2.5 m (8 ft) tall. **LEAVES** are spreading, and they have petioles that measure 2 mm (.08 in.) to 5 mm (.2 in.) long. Leaf blades are 2 cm (.8 in.) to 5 cm (2 in.) long and oblong, with flat or heart-shaped bases and smooth margins. Upper leaf surfaces are convex, dark or bright green, and shiny; lower surfaces are white due to a dense covering of fine hair (i.e., blade surfaces are not similar). **INFLORESCENCES** are dense; flower stalks measure 10 mm (.4 in.) to 25 mm (1 in.) long. Most bracts are large, leafy, and lance shaped. **FLOWERS** are white. **FRUITS** are spherical, 6 mm (.25 in.) to 10 mm (.4 in.) wide, dimpled, brown, and hairy. **TWIGS** have a dense covering of white, fine or bristly hair. **BARK** is shreddy and evergreen.

HABITAT AND RANGE: This species, in all its forms, grows in chaparrals and coniferous forests and woodlands of the northwestern San Francisco Bay area, the outer central Coast Ranges, the western Transverse Ranges, and the Channel Islands. It grows at elevations from sea level to 1,000 m (3,500 ft).

REMARKS: Many populations have intermediate characteristics shared with woollyleaf manzanita, Eastwood manzanita, and other manzanitas, probably as a result of hybridization. Woollyleaf manzanita is a complex of 9 subspecies. Six of these have limited ranges. Rare subspecies are *A. t.* ssp. *daciticola* and *A. t.* ssp. *eastwoodiana*. Subspecies considered uncommon are *A. t.* ssp. *insulicola* and *A. t.* ssp. *subcordata*. Restricted forms include *A. t.* ssp. *rosei* and *A. t.* ssp. *tomentosa*. All forms sprout from their burls after fires.

BEARBERRY (Fig. 68; Pl. 14) *Arctostaphylos uva-ursi*

DESCRIPTION: A prostrate or mounded shrub that generally lacks burls and can grow to be 60 cm (2 ft) tall. Branches are spreading, and they often root when in contact with the ground. **LEAVES** are inversely lance shaped or oblong, cupped, and hairless. Leaf blades are 1 cm (.4 in.) to 2.5 cm (1 in.) long. Their upper surfaces are dark green and shiny. Lower surfaces are light green. **INFLORESCENCES** are dense; flower stalks measure 3 mm (.12 in.) to 10 mm (.4 in.) long. Bracts are small, triangular scales. **FLOWERS** are white to pink. **FRUITS** are spherical, 6 mm (.25 in.) to 12 mm (.5 in.) wide, bright red, and hairless. **TWIGS** may have small, bristly hairs. **BARK** is dark brown.

HABITAT AND RANGE: Grows in coniferous forests, chap-

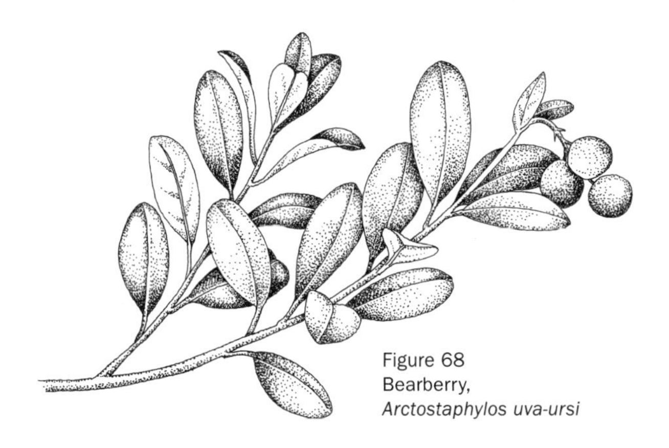

Figure 68
Bearberry,
Arctostaphylos uva-ursi

arrals, and open habitats such as dunes, coastal bluffs, scrubs, and rock outcrops, along the north coast from the Oregon border south to the Point Reyes Peninsula. It is found from sea level to 60 m (200 ft). In addition, isolated populations occur above Convict Lake (2,400 m [7,800 ft] to 3,200 m [10,500 ft]) on the eastern slopes of the Sierra Nevada. Outside of California the species grows from Oregon to Alaska and in Canada, Eurasia, the Rocky Mountains, and the eastern United States.

REMARKS: Botanists believe that hybridization has occurred between bearberry and woollyleaf manzanita in populations that have small burls, and between bearberry and hairy manzanita in populations of erect shrubs with intermediate leaf characteristics. Bearberry leaves have a medicinal use in treating sores and urinary diseases. Birds and small mammals eat the berries in late winter. The species has been used for erosion control along highways and as a ground cover. Native Americans smoked its leaves. Another common name for the species is kinnikinnick.

WHITELEAF MANZANITA *Arctostaphylos viscida*

(Fig. 69; Pl. 15)

DESCRIPTION: An erect shrub that lacks burls and typically grows to be 3.5 m (12 ft) tall. The largest lives in Springville, Cali-

Figure 69
Whiteleaf manzanita,
Arctostaphylos viscida

fornia, and is 9.5 m (31 ft) tall and 35 cm (14 in.) in diameter. **LEAVES** are erect and have petioles measuring 2 mm (.08 in.) to 4 mm (.16 in.) long. Blades are 10 mm (.4 in.) to 25 mm (1 in.) long and oval, with round bases and smooth or fine-toothed margins. Both blade surfaces are similar in color and texture: dull, hairless, white, and waxy. **INFLORESCENCES** are open, and their branches are covered with small, spiky hairs. Most bracts are small scales, although the lowest one is larger and leaflike. **FLOWERS** are white to pink. **FRUITS** are spherical, 6 mm (.25 in.) to 8 mm (.30 in.) wide, dimpled, and hairless or covered with small sticky hairs. **TWIGS** are slender and may have a dense covering of fine, bristly, glandular hair. **BARK** is smooth and deep red-brown.

HABITAT AND RANGE: Grows in montane coniferous forests, woodlands, and chaparrals in the inner northern Coast Ranges, the Klamath Mountains, Cascades, and Sierra Nevada. It ranges from 120 m (400 ft) to 1,800 m (6,000 ft) in elevation. Outside of California it is found in southwestern Oregon.

REMARKS: Whiteleaf manzanita is common and easily recognized by its distinctive white leaves. Three geographic subspecies are recognized based on the stickiness of inflorescences and fruits and the hairiness of twigs: *A. v.* ssp. *mariposa* in the Sierra Nevada, *A. v.* ssp. *pulchella* in the Klamath and northern Coast Ranges, and *A. v.* ssp. *viscida* in the Klamath Mountains, Cascades, and Sierra Nevada. It does not sprout from a burl after fires. Birds and small mammals eat whiteleaf manzanita fruits but find its leaves less palatable. Jelly can be made from the fruits. Whiteleaf manzanita has chemicals that inhibit conifer seedlings, although its root fungi aid the survival of Douglas-fir seedlings.

ARTEMISIA (SAGEBRUSH)

The genus *Artemisia* grows throughout the temperate parts of the Northern Hemisphere and South America. It has more than 300 species, and 21 of these are native to California.

Artemisia is a collection of aromatic annual or perennial herbs or shrubs with simple, often cleft or dissected, leaves that are alternate in arrangement. Flowers are small, inconspicuous, and collected into heads that form erect inflorescences. Fruits are dry, 1-seeded achenes.

The shrubby species are called sagebrushes and the herbaceous ones are called mugworts, sandworts, or wormwoods. The generic name honors Artemis, the Greek goddess of the hunt and a noted herbalist. Many species have medicinal uses.

1. Leaves are pinnately divided into threadlike lobes . . . . **California sagebrush** (*A.* ***californica***)
1. Leaves are wedge shaped with 3 broad lobes . . . . **big sagebrush** (*A.* ***tridentata***)

CALIFORNIA SAGEBRUSH *Artemisia californica*

(Fig. 70)

DESCRIPTION: A sprawling shrub that may grow to be between 60 cm (2 ft) and 2 m (6.5 ft) tall. **LEAVES** are drought-deciduous, thin, and clustered at axillary nodes. Blades are aromatic, 1 cm (.4 in.) to 10 cm (4 in.) long, and divided into 2 to 4 narrow lobes. The leaves are gray from many appressed hairs, and the leaf margins are rolled under, making them appear threadlike. **INFLORESCENCES** are terminal, and flower

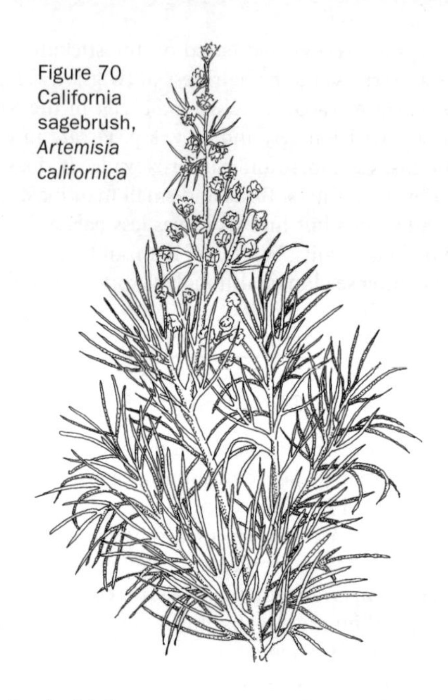
Figure 70 California sagebrush, *Artemisia californica*

heads are mixed with leaves. **TWIGS** are slender, flexible, and wandlike.

HABITAT AND RANGE: Grows in the coastal hills from Marin County south into southern California and Baja California. It is characteristic of coastal scrubs and chaparrals, but it can also be found in forests and woodlands. It occurs below 750 m (2,500 ft).

REMARKS: California sagebrush grows on the mainland and the Channel Islands. A closely related species, *A. nesiotica*, which has wider leaf blades, is restricted to the southern Channel Islands.

BIG SAGEBRUSH (Fig. 71) ***Artemisia tridentata***

DESCRIPTION: A large gray shrub that typically grows to be between 45 cm (1.5 ft) and 3 m (10 ft) tall. The largest lives in Crooked River National Grassland in Oregon and is 4 m (13 ft) tall and 15 cm (6 in.) in diameter. **LEAVES** are evergreen and

Figure 71
Big sagebrush,
Artemisia tridentata

clustered at axillary nodes. Blades are 1 cm (.4 in.) to 3 cm (1.2 in.) long and wedge shaped, with tips divided into 3 lobes. The leaves are gray from a dense covering of hair. **INFLORESCENCES** are terminal, with flower heads extending above the leaves.

HABITAT AND RANGE: Grows throughout the intermountain West in California's inland mountains, the San Joaquin Valley, and the Mojave Desert. It ranges from 450 m (1,500 ft) to 3,200 m (10,500 ft).

REMARKS: Big sagebrush is further divided into 5 subspecies, each of which grows in a distinctive habitat. There are several other common gray sagebrush species with 3-lobed leaves, which makes identification difficult. The value of the species to wildlife as browse varies among the subspecies.

ATRIPLEX (SALTBUSH)

The genus *Atriplex* has over 250 species that commonly grow in arid, temperate, or tropical environments with alkaline or saline soils. Thirty-one of these species are native to California and typically do not grow in forests and woodlands.

The herbs and shrubs have leaves with hairs that expand as salt is collected and then deflate with age, becoming scaly or powdery—many books describe the leaves as scurfy. The wind-pollinated flowers are small and not showy. Fruits are utricles, each surrounded by 2 distinct or fused bracts.

Many saltbush species are common in California's deserts, dunes, and grasslands, but some occur in woodlands. They are among the most variable and rapidly evolving groups in North America. Numerous forms have been recognized, each adapted to a specific habitat. Common names include allscale, desert-holly, lenscale, quail bush, scalebush, shadscale, and spinescale.

1. Leaf blades are round or have a wide lance shape. . . . . **shadscale** (*A. confertifolia*)
1. Leaf blades are linear or somewhat broader above the center . . . . **fourwing saltbush** (*A. canescens*)

FOURWING SALTBUSH *Atriplex canescens*

DESCRIPTION: An erect shrub that may grow to be 1.5 m (5 ft) tall. **LEAVES** are alternate and evergreen. Blades are 8 mm (.3 in.) to 50 mm (2 in.) long, gray, scurfy, and linear or somewhat broader above the center; they have 1 vein from the base. Margins are entire. Leaves lack petioles. **INFLORESCENCES** are 2 kinds of terminal spikes, one containing pollen-bearing flowers, and the other seed-bearing flowers, which are most commonly found on separate plants. **FLOWERS** are small and drab yellow or brown. **FRUITS** are utricles with bracts forming 4 wings. **TWIGS** may have fine gray hairs.

HABITAT AND RANGE: In California this species grows in the southern Great Basin and Mojave Deserts and their adjoining mountain slopes, between 600 m (2,000 ft) and 2,400 m (7,800 ft). It mixes with other saltbush species in the interior woodlands; it is also found in desert scrubs.

REMARKS: Fourwing saltbush is an important browse plant for livestock and wildlife, and it provides cover for upland gamebirds. The species has been widely used for rehabilitating mine spoils and roadsides. Fourwing saltbush grown in selenium-enriched soil

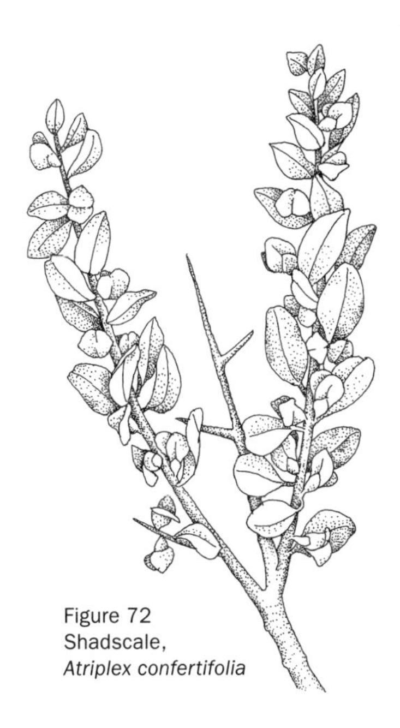

Figure 72
Shadscale,
Atriplex confertifolia

can accumulate large amounts of this element, making the shrubs toxic.

SHADSCALE (Fig. 72) *Atriplex confertifolia*

DESCRIPTION: A rigidly branched shrub that may grow to be 1 m (3.3 ft) tall. **LEAVES** are alternate, crowded, and short-lived. Blades are 8 mm (.3 in.) to 25 mm (1 in.) long, are round or elliptical, and have 1 or 3 veins originating from the base. Upper and lower surfaces are scurfy from gray, inflated hairs that cover them. **INFLORESCENCES** are 2 kinds of terminal spikes, one containing pollen-bearing flowers, and the other seed-bearing flowers, which are found on separate plants. **FLOWERS** are small, drab, and yellow or brown. **FRUITS** are utricles. **TWIGS** are straw colored and stiff, becoming spinelike with age.

HABITAT AND RANGE: In California this species grows in the southern Great Basin and Mojave Deserts and their adjoining mountain slopes, between 600 m (2,000 ft) and 2,400 m (7,800 ft). Outside of California it is found throughout the Great Basin and the Southwest. It mixes with other species in the interior coniferous woodlands and desert scrubs.

REMARKS: Shadscale is an important browse plant for livestock and wildlife. Its wind-borne pollen can cause hay fever in pollen-sensitive people.

BACCHARIS (COYOTE BRUSH)

The genus *Baccharis* grows throughout the temperate parts of the Western Hemisphere and is especially diverse in South America. The genus has more than 400 species, with 9 of these growing in California.

The aromatic, often sticky, perennial herbs or shrubs have simple, evergreen leaves that are alternate in arrangement. Flowers are small, inconspicuous, and clustered into flower heads. Seed-bearing and pollen-bearing flowers are found on separate plants. Fruits are dry achenes.

The generic name honors Bacchus, the Greek god of wine.

1. Leaves are linear or have a narrow lance shape .... **mule fat** (*B. salicifolia*)
1. Leaves have a wedge shape or a wide lance shape .... **coyote brush** (*B. pilularis*)

COYOTE BRUSH (Fig. 73) ***Baccharis pilularis***

DESCRIPTION: A large, often sticky shrub that may grow to be between 60 cm (2 ft) and 3.5 m (11.5 ft) tall. Plants are "leggy," with leaves clustered at the ends of long, bare stems and branches. One variety is a prostrate shrub ranging in height from 15 cm (6 in.) to 30 cm (12 in.). **LEAVES** are evergreen and green, with 3 principal veins originating from the base. Leaves are wider above the middle, egg shaped, and 12 mm (.5 in.) to 38 mm (1.5 in.) long. Blades have a wide wedge shape and short petioles. Margins are usually irregularly toothed but are sometimes entire. **INFLORESCENCES** are tight clusters of flower heads.

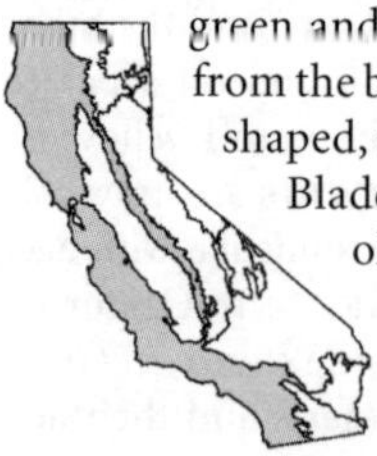

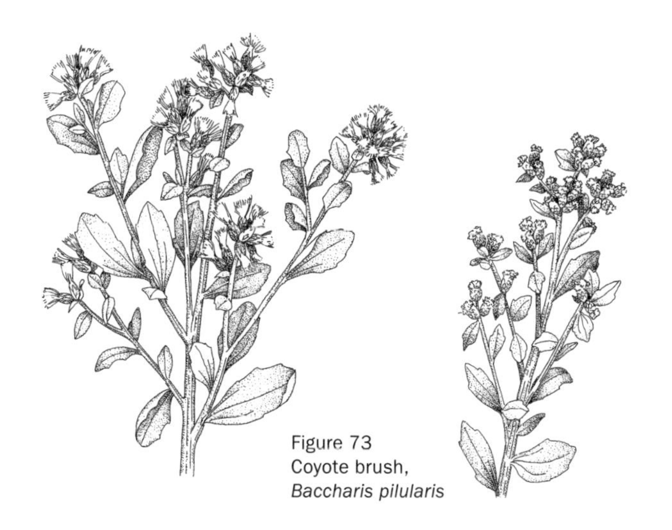

Figure 73
Coyote brush,
Baccharis pilularis

HABITAT AND RANGE: Grows throughout California's Coast Ranges and the Sierra Nevada foothills and south to northern Mexico. It occurs below 1,500 m (5,000 ft) and is associated with coastal scrubs, chaparrals, and recently disturbed areas within forests.

REMARKS: On windswept headlands along the central coast, coyote brush forms dense mats. Away from the coast, the plants are erect.

MULE FAT *Baccharis salicifolia*

DESCRIPTION: A large, dark, often sticky shrub that may grow to be 2 m (6.5 ft) to 3.5 m (11.5 ft) tall, with straight, ascending stems. **LEAVES** are evergreen, dark green, and sticky. Blades have a linear or narrow lance shape and are 2.5 cm (1 in.) to 9 cm (3.5 in.) long. Petioles are winged. Margins may be toothed or smooth. **INFLORESCENCES** are open, terminal clusters of heads or tight, lateral clusters of heads.

HABITAT AND RANGE: Grows throughout much of California and ranges south into South America. It occurs below 750 m (2,500 ft) and is associated with streamsides, where it forms large thickets. The species can be found in riparian forests and woodlands.

REMARKS: Mule fat has the interesting habit of having 2 phenological forms. Summer forms have terminal inflorescences and toothed leaves. Winter forms have lateral inflorescences and toothless leaves.

BERBERIS (OREGON-GRAPE)

The genus *Berberis* has about 600 species, 6 of which grow in California. Three of these are frequently found in forests and woodlands: the ranges of 2 species, *B. haematocarpa* and *B. fremontii,* are more extensive in the southwestern United States than in California; and 1 species, *B. nevinii,* is an endangered California endemic.

In California, *Berberis* is a collection of rhizomatous, viny, or upright shrubs with pinnately compound, evergreen leaves. Leaves are alternate in arrangement. Leaflets have spiny margins. Flowers are yellow and have 3 whorls that can be interpreted as bracts, sepals, and petals. The flowers are arranged in showy, open or closed inflorescences. Fruits are berries.

In some books, barberries, which have simple leaves and spiny stems, and Oregon-grapes, which have compound leaves and spiny blades, have been treated as separate genera: *Berberis* and *Mahonia,* respectively. Horticulturists often recognize *Mahonia,* but botanists notice that many species show different combinations of characteristics associated with spine location, and so treat the group of species as a single genus.

Many species are cultivated as ornamentals. Some, but not all, species harbor black-stem wheat rust (*Puccinia graminis*). Wildlife species eat the berries throughout the summer. Jams and jellies can be made from the fruits. Native Americans made medicines from the often-toxic leaves and roots.

1. Terminal buds are covered with large evergreen scales that mingle with upper leaf bases. Leaves generally have 11 to 23 leaflets . . . . **little Oregon-grape (*B. nervosa*)**
1. Terminal buds are not covered with large scales, and upper leaf bases are separate. Leaves generally have 5 to 11 leaflets 2

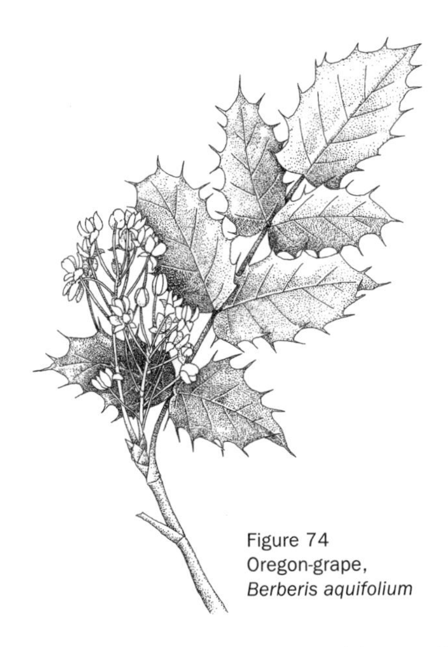

Figure 74
Oregon-grape,
Berberis aquifolium

2. Leaf petioles are over 2.5 cm (1 in.) long and leaflets are not crowded . . . . **Oregon-grape (*B. aquifolium*)**
2. Leaf petioles are less than 2.5 cm (1 in.) long and leaflets are overlapping or crowded . . . . **California barberry (*B. pinnata*)**

OREGON-GRAPE (Fig. 74; Pl. 16) ***Berberis aquifolium***

DESCRIPTION: A creeping or erect shrub that may grow to be 3 m (10 ft) tall. **LEAVES** are evergreen, alternate, pinnately compound (usually 5 to 9 leaflets), and 10 cm (4 in.) to 25 cm (10 in.) long. Leaflets are elliptical to round and 2 cm (.8 in.) to 7.5 cm (3 in.) long. Each glossy green leaflet has 10 to 20 spine-tipped teeth on its margins. Scales are less than 2 cm (.75 in.) long and are deciduous. **INFLORESCENCES** are rounded clusters of 30 to 60 yellow flowers. **FRUITS** are blue to purple with a waxy covering. **BRANCHES** are grayish brown to purple.
HABITAT AND RANGE: Grows throughout much of the

western United States. In California it is absent from the deserts. It occurs below 2,100 m (7,000 ft) in forests, woodlands, and chaparrals.

REMARKS: Oregon-grape takes many forms, which grade into each other. The forms have been treated differently by botanists. Some recognize 6 species, while we recognize 1. Some of the variation appears to be environmentally induced. We recognize 3 varieties of *B. aquifolium:* the wavy, dull, thick-leaved *B. a.* var. *dictyota* occurs in California's low-elevation woodlands and chaparrals; the typical, glossy-leaved *B. a.* var. *aquifolium* grows in coniferous forests; and the short, few-leaved *B. a.* var. *repens* is most commonly seen in open woodlands. This species is resistant to rust infection.

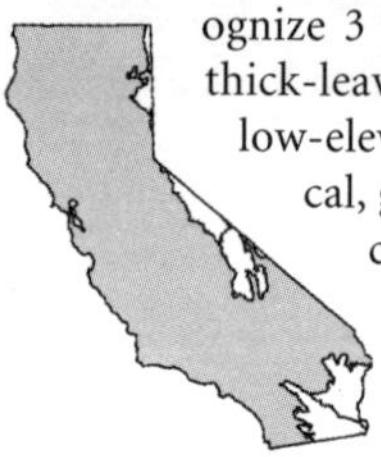

LITTLE OREGON-GRAPE (Fig. 75) *Berberis nervosa*

DESCRIPTION: A creeping or erect shrub that may grow to be 2 m (6.5 ft) tall. **LEAVES** are evergreen, alternate, pinnately compound (usually 11 to 23 leaflets), and 20 cm (8 in.) to 45 cm (18 in.) long. Leaflets are lance shaped to egg shaped and 2.5 cm (1 in.) to 7.5 cm (3 in.) long. They are dull green and have 6 to 13 spine-tipped teeth on each margin. Veins on lower leaflet surfaces are obscure. Leaves clustered at the tops of branches, just below the terminal buds, arise among evergreen scales, which measure 2 cm (.75 in.) to 4 cm (1.5 in.) long. **INFLORESCENCES** are elongated clusters of 30 to 70 yellow flowers. **FRUITS** are blue with a waxy covering. **BRANCHES** are brown or yellowish brown.

HABITAT AND RANGE: This species is common in coastal forests from British Columbia south to Monterey County. It also is found in Sierra County. It occurs below 2,000 m (6,500 ft).

REMARKS: The generally short, dull-leaved little Oregon-grape (which measures less than 30 cm [1 ft]) is easily differentiated from the generally taller, glossy-leaved Oregon-grape (*B. aquifolium*) (which measures up to 2 m [6.5 ft]), which grows in its range. Little Oregon-grape is resistant to rust infection.

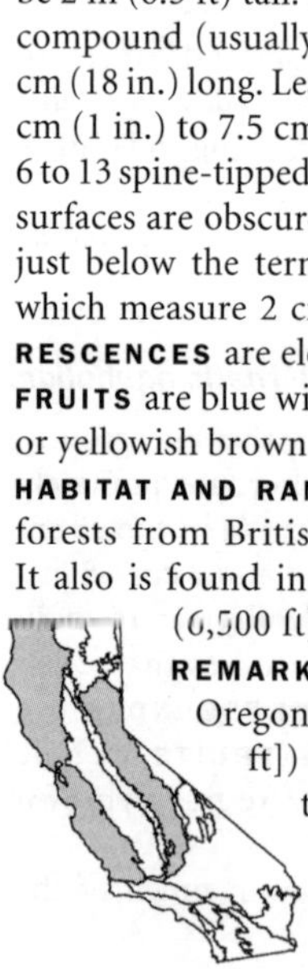

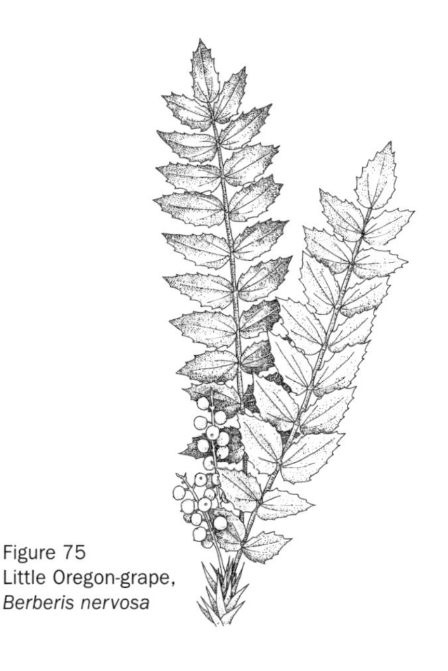

Figure 75
Little Oregon-grape,
Berberis nervosa

CALIFORNIA BARBERRY *Berberis pinnata*

DESCRIPTION: A vinelike or erect shrub that may grow to be 2 m (6.5 ft) tall. **LEAVES** are evergreen, alternate, pinnately compound (usually 7 to 11 leaflets), and 9 cm (3.5 in.) to 20 cm (8 in.) long. Leaflets are egg shaped to elliptical and 3 cm (1.2 in.) to 7 cm (2.8 in.) long. They are crowded and glossy green and have 10 to 20 spine-tipped teeth on each undulating margin. Scales are small (less than 2 cm [.75 in.]) and deciduous. **INFLORESCENCES** are dense clusters of 20 to 50 yellow flowers. **FRUITS** are blue. **BRANCHES** are grayish brown.

HABITAT AND RANGE: Found in California's Coast Ranges and the southern California mountains. It occurs below 1,200 m (4,000 ft) in forests, woodlands, and chaparrals.

REMARKS: California barberry takes 2 forms: the mainland subspecies, *B. p.* ssp. *pinnata,* is an erect shrub, whereas the Channel Islands' endangered *B. p.* ssp. *in-*

sularis is a vine. Many mainland populations share characteristics with Oregon-grape and are intermediate in appearance. California barberry is resistant to rust infection.

BETULA (BIRCH)

The genus *Betula* is restricted to the Northern Hemisphere and has about 50 species. Twelve are native to North America, and 2 of these are native to California. The genus is made up of shrubs and small- to medium-sized trees growing in temperate and boreal forests.

Leaves are deciduous; alternate; simple; egg shaped, oval, or triangular; and 2 cm (.8 in.) to 5 cm (2 in.) long. Margins are serrated to lobed. Leaves are set on spur shoots. Small pollen-bearing and seed-bearing flowers are arranged separately in erect or droopy catkins. Both kinds of catkins form on the same tree. Pollen-bearing catkins hang in clusters of 2 to 3, and seed-bearing catkins are usually solitary. Mature seed catkins resemble conifer cones and are composed of many small, winged nutlets. The "cones" fall apart when mature. Twigs are slender, greenish to reddish brown, and hairy or hairless; they have conspicuous lenticels. Bark is smooth, papery, and colorful when young. Lenticels are horizontal and prominent. Older bark has roughened patches.

Birches are used as lumber, to make pulp and furniture, and in landscaping. Sweet syrups and fermented drinks can be made from birch sap. Native Americans fashioned canoes, houses, and clothing from birches.

1. Leaves are thick and their margins have round teeth; leaf tips are round. Plants are small shrubs less than 3 m (10 ft) tall .... **resin birch (*B. glandulosa*)**
1. Leaves are thin and their margins have sharp teeth; leaf tips are sharp pointed. Plants are tall shrubs to small trees more than 3 m (10 ft) tall .... **water birch (*B. occidentalis*)**

RESIN BIRCH ***Betula glandulosa***

DESCRIPTION: A small, erect or nearly prostrate shrub that grows to be between 30 cm (1 ft) and 2 m (6.5 ft) tall. **LEAVES** are deciduous, alternate, simple, and less than 2.5 cm (1 in.) long. They have a broad egg shape to round shape. Margins are scalloped. Leaf tips are round and bases are round to wedge shaped. Upper surfaces are dark green and have glands, while lower surfaces are pale green and have glands. Leaves occur on spur shoots. **FRUITS** are winged nutlets in conelike catkins about 2 cm (.75 in.) long. **TWIGS** are waxy and gray and have resin glands. **BARK** is brown to gray and does not peel.

HABITAT AND RANGE: Grows in riparian woodlands. Outside of California it is widely distributed from Alaska to New England and south into the Rocky Mountains and east of the Cascades and Sierra Nevada crest. In California it is found at around 2,100 m (7,000 ft) in the Warner Mountains and Cascades.

REMARKS: Resin birch is found in California but is considered uncommon.

WATER BIRCH (Fig. 76) ***Betula occidentalis***

DESCRIPTION: A single- or multistemmed large shrub or small tree. Mature trees are typically 3 m (10 ft) to 12 m (40 ft) tall and 10 cm (4 in.) to 30 cm (12 in.) in diameter. The largest grows in Wallowa County, Oregon, and is 16 m (53 ft) tall and 89 cm (35 in.) in diameter. The species typically has several stems arising from its base and looks superficially like a willow. Crowns are broad and spreading and trunks are curved to crooked. **LEAVES** are deciduous, alternate, simple, and 2 cm (.8 in.) to 5 cm (2 in.) long. They have a broad egg shape. Margins are sharply doubly serrated. Leaf tips are sharp pointed and bases are horizontal to tapered. Upper surfaces are shiny and yellowish green, while lower surfaces are paler and glandular. Petioles are flat on upper surfaces and hairy. Leaves occur on spur shoots. **FRUITS** are winged nutlets in conelike catkins about 2.5 cm (1 in.) long. **TWIGS** are reddish brown and glandular. **BARK** is reddish brown and does not readily peel. Lenticels are conspicuous.

HABITAT AND RANGE: Outside of California the species grows in riparian woodlands in the Rocky Mountains and Cascades. In California it is found in the

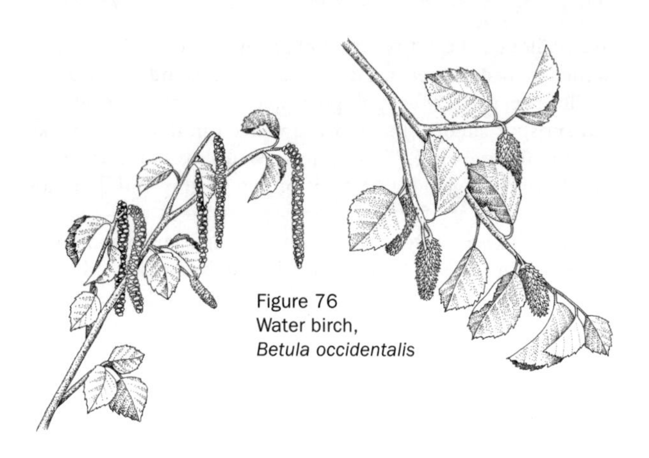

Figure 76
Water birch,
Betula occidentalis

southern Sierra Nevada (mostly east of the crest), the Warner Mountains, and the eastern Klamath Mountains. It is found along streams and springs from 600 m (2,000 ft) to 2,700 m (9,000 ft).

REMARKS: Water birch tolerates flooding, and beavers use its stems and branches to make dams. Birds eat its catkins, buds, and seeds. Water birch's dense habit provides good thermal and hiding cover. It sprouts from its base and can develop stem clumps numbering over 100.

CALYCANTHUS (SPICE BUSH)

The genus *Calycanthus* has 3 species that occur in the United States; 1 is native to California.

SPICE BUSH — *Calycanthus occidentalis*

(Fig. 77; Pl. 17)

DESCRIPTION: A shrub with large fragrant leaves and brown bark. Shrubs may grow to be 3.5 m (12 ft) tall. **LEAVES** are opposite and deciduous. Blades have a broad lance shape and are 5 cm (2 in.) to 15 cm (6 in.) long, rough to the touch, and hairless. **INFLORESCENCES** consist of single flowers at the ends of branches. **FLOWERS** are showy, reddish brown, and about 5 cm (2 in.) in diameter. Sepals and petals are numerous

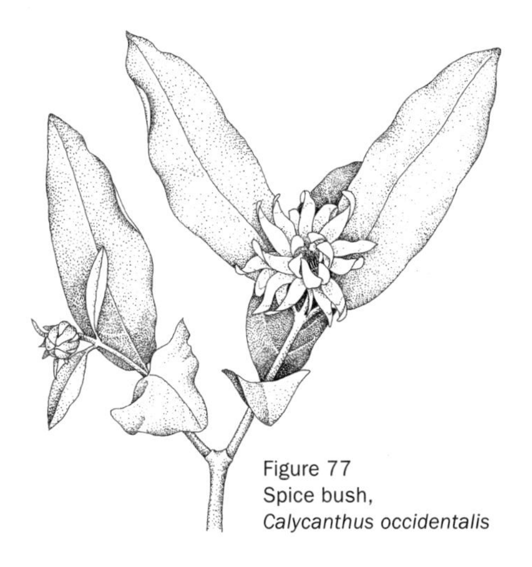

Figure 77
Spice bush,
Calycanthus occidentalis

and the same color. **FRUITS** are about 2.5 cm (1 in.) long, consisting of an urn-shaped receptacle containing many velvety achenes.

HABITAT AND RANGE: Grows along streams in many of northern California's foothills, between 200 m (600 ft) and 1,000 m (3,500 ft).

REMARKS: One other species of *Calycanthus* grow in the eastern United States, and closely related genera occur in China. A disjunct pattern such as this suggests that spice bush is a relict plant that once had a larger range.

CASSIOPE (WHITE HEATHER)

The genus *Cassiope* has about 14 shrub species that grow at high latitudes (are circumboreal) and on high-elevation mountains. One species is native to California.

WHITE HEATHER (Fig. 78) ***Cassiope mertensiana***

DESCRIPTION: An erect or prostrate shrub with dark green, scaly leaves. Shrubs may grow to be 30 cm (1 ft) tall. **LEAVES**

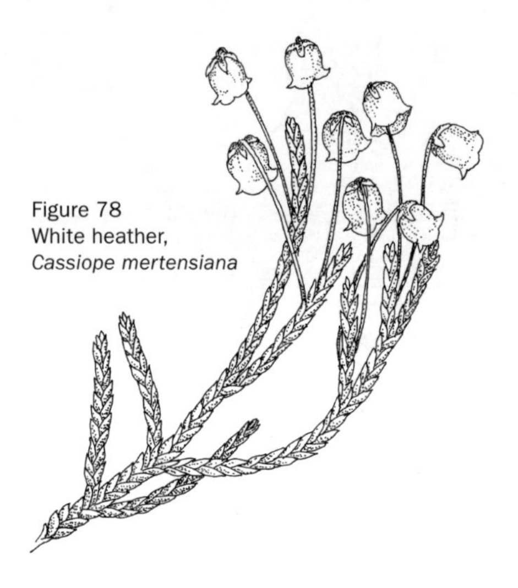

Figure 78
White heather,
Cassiope mertensiana

are opposite, simple, evergreen, and leathery. Blades, which measure less than 6 mm (.25 in.) long, are keeled scales that overlap like shingles on a roof. **INFLORESCENCES** consist of single flowers arising from leaf axils at the tops of the stems. **FLOWERS** are 4- or 5-parted, have a wide bell shape, and are white to pink. **FRUITS** are spherical capsules.

HABITAT AND RANGE: Grows in the Klamath Mountains, Cascades, and Sierra Nevada in open subalpine forests and alpine meadows, at elevations from 1,800 m (5,900 ft) to 3,500 m (11,500 ft). It occurs on moist slopes, among rocks, and near areas of late snowmelt.

REMARKS: White heather was reported to be John Muir's favorite plant. It is one of the many ericaceous (of the Ericaceae) shrubs found at high elevations in California's mountains.

CEANOTHUS (CALIFORNIA-LILAC)

The genus *Ceanothus* has about 45 species of shrubs and small trees, which are found in the Northern Hemisphere from Canada to Guatemala. Of these, 43 are native to California. Many California species have several subspecies.

Leaves are simple, deciduous or evergreen, and alternate or opposite in arrangement. Leaves may have prominent stipules, or blades with 1 or 3 obvious veins. Flowers are arranged in open, showy inflorescences. Flowers are individually small, but flower stems are the same color as the flowers, making the inflorescences showy and the shrubs striking. Flowers are commonly blue or white, with 5 sepals and petals. Fruits are round, 3-parted capsules. Branches are flexible or rigid and may be thorny.

Plants are easily recognized at the genus level; but many species are locally distributed, and hybridization may make identification difficult in areas where species ranges overlap. This books treats only about a third of the species that grow in California.

Ceanothus is often encountered as individual plants or as large populations in the state's forests, woodlands, and chaparrals. Many *Ceanothus* stands have become established as a result of their seed coats being cracked by the heat of a fire. Abrasions on the seed coats can also allow germination to be initiated.

Ceanothus species are highly valued as browse for wildlife and livestock and as ornamentals. Many species form nitrogen-fixing root nodules.

1. Leaf blades are thin and evergreen or deciduous; stipules are thin and fall soon after blades expand 2
1. Leaf blades are thick and evergreen, and stipules are thick, corky, and evergreen 12
   2. Lower sides of leaves have 1 prominent vein 3
   2. Lower sides of leaves have 3 prominent veins 6
3. Plants form mats. Leaves are blue-green .... **pine mat** (*C. diversifolius*)
3. Plants form upright shrubs. Leaves are not blue-green . . . 4
   4. Twigs are yellow to pale green .... **deerbrush** (*C. integerrimus*)
   4. Twigs are gray or gray-brown 5
5. Twigs are covered with rusty hairs .... **woollyleaf ceanothus** (*C. tomentosus*)
5. Twigs have gray hairs .... **Lemmon ceanothus** (*C. lemmonii*)
   6. Twigs are rigid and thorny 7
   6. Twigs are flexible and not thorny 9

7. Shrubs are round and less than 1.2 m (4 ft) tall; they grow at montane elevations .... **mountain whitethorn** (*C. cordulatus*)

7. Shrubs are erect and over 1.2 m (4 ft) tall; they grow at lower elevations 8

 8. Leaf blades are less than 3 cm (1.25 in.) long .... **chaparral whitethorn** (*C. leucodermis*)

 8. Leaf blades are greater than 3 cm (1.25 in.) long .... **coast whitethorn** (*C. incanus*)

9. Leaves are deciduous and blades are entire .... **deerbrush** (*C. integerrimus*)

9. Leaves are evergreen and blades are not entire 10

 10. Twigs are angled and flowers are blue .... **blueblossom** (*C. thyrsiflorus*)

 10. Twigs are round and flowers are blue, purple, or white 11

11. Flowers are white .... **tobacco brush** (*C. velutinus*)

11. Flowers are blue or purple .... **hairy ceanothus** (*C. oliganthus*)

 12. Leaves are alternately, or both alternately and oppositely, arranged on the same plant .... **bigpod ceanothus** (*C. megacarpus*)

 12. Leaves are oppositely arranged 13

13. Lower sides of leaf blades are densely covered with white hairs, and leaf margins are turned under .... **hoaryleaf ceanothus** (*C. crassifolius*)

13. Lower sides of leaf blades are not densely covered with white hairs, and leaf margins are flat 14

 14. Flowers are white. Most plants are upright shrubs 15

 14. Flowers are blue. Most plants are shrubby mat 16

15. Leaves are cupped upward .... **cupleaf ceanothus** (*C. greggii*)

15. Leaves are flat .... **wedgeleaf ceanothus** (*C. cuneatus*)

 16. Upper sides of leaf blades are dull .... **glory bush** (*C. gloriosus*)

 16. Upper sides of leaf blades are shiny 17

17. Leaf blades have teeth at tips only .... **Siskiyou mat** (*C. pumilus*)

17. Leaf blades have teeth on sides and on tips .... **Mahala mat** (*C. prostratus*)

Figure 79 Mountain whitethorn, *Ceanothus cordulatus*

MOUNTAIN WHITETHORN *Ceanothus cordulatus* (Fig. 79)

DESCRIPTION: A flat-topped shrub that may grow to be 1.5 m (5 ft) tall. **LEAVES** are thin, alternate, evergreen, and less than 3 cm (1.25 in.) long. Stipules are thin. Blades are elliptical and grayish green with scattered hairs. Three prominent mid veins are evident on the undersides of blades. Petioles are less than 6 mm (.25 in.) long. **INFLORESCENCES** are small flower clusters less than 4 cm (1.5 in.) long. **FLOWERS** are white. **FRUITS** are sticky and have minute crests. **TWIGS** are yellowish green, with a covering of fine hair. **BRANCHES** are round and rigid and have thorns.

HABITAT AND RANGE: Grows in montane and subalpine forests, woodlands, and chaparrals in most mountain ranges in the state. It occurs between 900 m (3,000 ft) and 3,000 m (9,500 ft). Outside of California it grows in Oregon, Nevada, and Baja California.

REMARKS: Mountain whitethorn is easily recognized as the gray, thorny ceanothus of the mountains. It is abundant and extensive in areas following fire or logging. The resulting stands are transitional to various kinds of forests and woodlands.

HOARYLEAF CEANOTHUS *Ceanothus crassifolius*

DESCRIPTION: An open-branched, erect shrub that may grow to be 3.5 m (12 ft) tall. **LEAVES** are opposite, evergreen, and less than 3 cm (1.25 in.) long. Stipules are thick and evergreen. Blades are elliptical. Petioles are 2 mm (.1 in.) to 6 mm (.25 in.) long. Upper blade surfaces are olive green and hairless. Lower blade surfaces are covered with fine white hairs and have 1 prominent mid vein. **INFLORESCENCES** are set on short lateral branches and are less than 3 cm (1.25 in.) long. **FLOWERS** are white. **FRUITS** are sticky and have short horns on the shoulders of the capsules. **TWIGS** are gray, white, or brown, with rusty hairs. **BRANCHES** are opposite and rigid.

HABITAT AND RANGE: Grows on mountain slopes of southern California and Baja California. It is characteristic of chaparrals but can be found in forests and woodlands as well, at elevations below 1,000 m (3,500 ft).

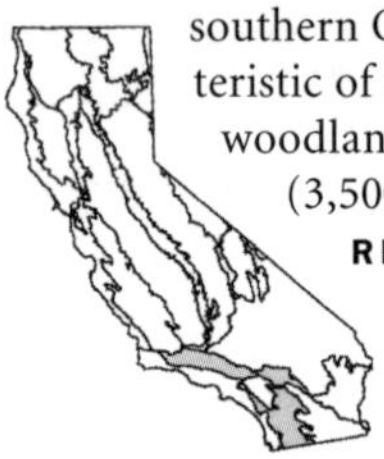

REMARKS: Hoaryleaf ceanothus hybridizes with cupleaf ceanothus, producing plants lacking hoaryleaf ceanothus's horned fruits.

WEDGELEAF CEANOTHUS *Ceanothus cuneatus*

(Fig. 80)

DESCRIPTION: An intricately branched, prostrate or erect shrub that may grow to be 2.5 m (8 ft) tall. **LEAVES** are evergreen, opposite, flat, and less than 3 cm (1.25 in.) long. Stipules are thick and evergreen. Blades are wedge shaped (broad at the tips and tapering to narrow bases). Petioles are absent to 3 mm (.12 in.) long. Upper blade surfaces are hairless and dull to shiny green. Lower blade surfaces have 1 prominent mid vein and a covering of fine hair. **INFLORESCENCES** are set on short lateral branches and are less than 2.5 cm (1 in.) long. **FLOWERS** are white, blue, or lavender. **FRUITS** have horns near the tops

Figure 80
Wedgeleaf ceanothus,
Ceanothus cuneatus

of the capsules. **TWIGS** are gray to brown and are hairless. **BRANCHES** are opposite and rigid.

HABITAT AND RANGE: Grows throughout California, except in the deserts; southern Oregon; and Baja California, at elevations below 1,800 m (6,000 ft). It is characteristic of chaparrals, woodlands, and forests.

REMARKS: This species is broken up into 3 varieties: *C. c.* var. *cuneatus*, *C. c.* var. *fascicularis*, and *C. c.* var. *rigidus* (which is uncommon). Wedgeleaf ceanothus is thought to hybridize with several ceanothus species, giving rise to great diversity in habits and flowers. It is a preferred wildlife browse plant, and birds and small mammals eat its seeds. Leaves and flowers may be used for teas and tonics. Wedgeleaf ceanothus is also known as buck brush.

PINE MAT — *Ceanothus diversifolius*

DESCRIPTION: A low, trailing shrub that may grow to be 30 cm (1 ft) tall. **LEAVES** are thin, alternate, evergreen, and less than 5 cm (2 in.) long. Stipules are deciduous. Blades are round to egg shaped. Petioles are less than 12 mm (.5 in.) long. Upper blade surfaces are bluish green, with a scattering of hairs. Lower blade surfaces are pale green and hairy, with 1 prominent mid vein. **INFLORESCENCES** are small clusters of a few flowers; they measure less than 3 cm (1.25 in.) long. **FLOWERS** range in color from deep blue to white. **FRUITS** have 3 crests on the upper sides of the round capsules. **TWIGS** are hairy and greenish to reddish brown. **BRANCHES** are flexible.

HABITAT AND RANGE: Grows in the montane forests and woodlands of California's northwestern mountains south to the southern Sierra Nevada. It occurs between 1,400 m (4,500 ft) and 2,100 m (7,000 ft).

REMARKS: Pine mat is not abundant over its large range but is easily identified as the prostrate mat with thin, hairy leaves.

GLORY BUSH *Ceanothus gloriosus*

DESCRIPTION: A prostrate or erect shrub that may grow to be 2 m (6.5 ft) tall. **LEAVES** are evergreen, opposite, and less than 5 cm (2 in.) long. Stipules are thick and evergreen. Blades are round to oblong, with spiny margins. Petioles are less than 4 mm (.15 in.) long. Upper blade surfaces are dark green and hairless. Lower blade surfaces have 1 prominent mid vein and a covering of fine hair. **INFLORESCENCES** are set on short lateral branches and are less than 2.5 cm (1 in.) long. **FLOWERS** are deep blue or purple. **FRUITS** are sticky and have horns near the tops of the capsules. **TWIGS** are angled, are green or brown, and have a covering of hair. **BRANCHES** are opposite and reddish brown.

HABITAT AND RANGE: Grows along the coast in Mendocino, Sonoma, and Marin Counties on dunes, on headlands, and in forests, below 500 m (1,600 ft).

REMARKS: Of the 3 varieties, the erect *C. g.* var. *exaltatus* (which means "the glorious and exalted one") is relatively common and is associated with open forests. The other, prostrate, varieties, *C. g.* var. *gloriosus* (which is uncommon) and *C. g.* var. *porrectus* (which is rare), are localized and associated with coastal habitats.

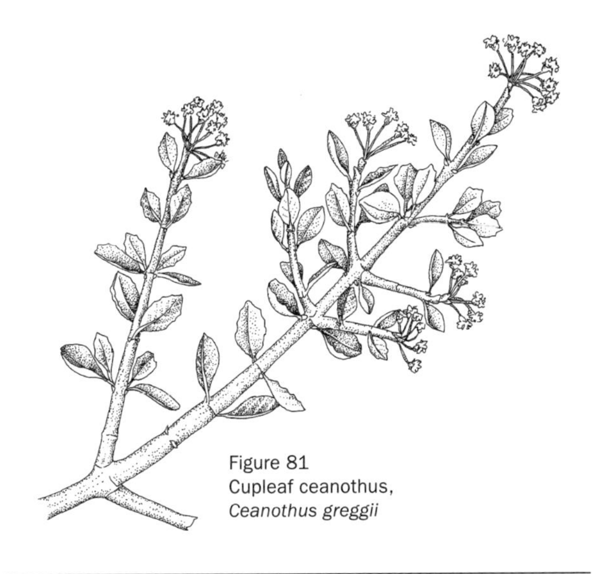

Figure 81
Cupleaf ceanothus,
Ceanothus greggii

CUPLEAF CEANOTHUS (Fig. 81) *Ceanothus greggii*

DESCRIPTION: An intricately branched, erect shrub that may grow to be 2 m (6.5 ft) tall. **LEAVES** are evergreen, opposite, and 9 mm (.37 in.) to 20 mm (.75 in.) long. Stipules are thick and evergreen. Petioles are less than 3 mm (.12 in.) long. Blades are variably shaped on the same plant and may be cupped. Blade surfaces are yellowish green or grayish green, with fine hairs and 1 prominent mid vein. **INFLORESCENCES** are set on short lateral branches and are less than 2 cm (.75 in.) long. **FLOWERS** are white. **FRUITS** have short horns on the shoulders of the capsules. **TWIGS** are gray with fine white hairs. **BRANCHES** are opposite and rigid.

HABITAT AND RANGE: Grows on inland mountain slopes of southern California, southwestern United States, and Mexico, at elevations from 1,000 m (3,500 ft) to 2,300 m (7,500 ft). It is characteristic of chaparrals and pinyon-juniper woodlands but can be found in forests as well.

REMARKS: Cupleaf ceanothus is broken into 4 geographical varieties, 2 of which grow in California. Plants of the eastern Transverse and Peninsular Ranges have yellowish green leaves (*C. g.* var. *perplexans*). Plants from the western Transverse Ranges and east through the desert mountains have grayish green leaves (*C. g.* var. *vestitus*). Cupleaf ceanothus is a preferred browse plant for deer and bighorn sheep.

COAST WHITETHORN *Ceanothus incanus*

DESCRIPTION: An erect shrub that may grow to be 4 m (13 ft) tall. **LEAVES** are thin, alternate, evergreen, and less than 6 cm (2.4 in.) long. Stipules are deciduous. Blades are elliptical to egg shaped and are grayish green. They have 3 prominent mid veins and a covering of fine hair (but are sometimes hairless). Petioles are less than 12 mm (.5 in.) long. **INFLORESCENCES** are large flower clusters less than 7 cm (2.75 in.) long. **FLOWERS** are white. **FRUITS** are triangular and have roughened warts. **TWIGS** are gray and waxy, with a covering of fine hair. **BRANCHES** are round and rigid and have thorns.

HABITAT AND RANGE: Grows in the outer Coast Ranges from Humboldt County south to Santa Cruz County, below 900 m (3,000 ft). Plants are most commonly seen in logged forests and chaparrals.

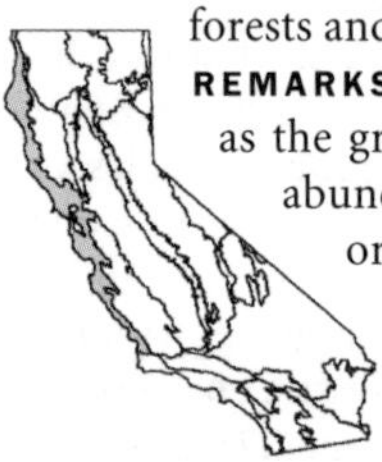

REMARKS: Coast whitethorn is easily recognized as the gray, thorny ceanothus near the coast. It is abundant and extensive in areas following fire or logging. The resulting stands are transitional to various coniferous forest types. The town of Whitethorn in southern Humboldt County is named in its honor.

DEERBRUSH (Fig. 82) *Ceanothus integerrimus*

DESCRIPTION: A highly branched, erect shrub that may grow to be 4 m (13 ft) tall. **LEAVES** are thin, alternate, deciduous, and less than 7.5 cm (3 in.) long. Stipules are thin and deciduous. Blades are elliptical to oblong. Upper blade surfaces are pale green, with scattered hairs. Lower blade surfaces are light green, have 1 or 3 prominent mid veins, and may have hairs. **INFLORESCENCES** are clusters of flowers that form at the ends of branches; they measure less than 15 cm (6 in.) long. **FLOW-**

Figure 82
Deerbrush,
Ceanothus integerrimus

ERS are white, deep to pale blue, or even pink. **FRUITS** are sticky and have minute crests. **TWIGS** are pale green and may have hairs. **BRANCHES** are round and flexible.

HABITAT AND RANGE: Grows in the lower-elevation and montane forests, woodlands, and chaparrals of most of the state's mountains, at elevations between 150 m (500 ft) and 2,100 m (7,000 ft). Outside of California it occurs in Washington, Oregon, Arizona, and New Mexico.

REMARKS: Deerbrush is abundant and extensive in areas following fire or logging. The resulting stands are transitional to various kinds of forest. Four poorly defined varieties of *C. integerrimus* are recognized but are difficult to identify. Deerbrush is widely regarded as one of the most valuable and abundant browse plants for California's wildlife and livestock. Birds and small mammals eat its seeds. Native Americans made soap from the flowers. Deerbrush is planted as an ornamental.

LEMMON CEANOTHUS *Ceanothus lemmonii*

DESCRIPTION: A low, much-branched, rounded shrub that may grow to be 1 m (3.3 ft) tall. **LEAVES** are thin, alternate, evergreen, and less than 3 cm (1.25 in.) long. Stipules are deciduous. Blades are elliptical to oblong. Upper blade surfaces are dull green and waxy. Lower blade surfaces have 1 or 3 prominent mid veins and are white from a dense covering of hair. Petioles are less than 8 mm (.3 in.) long. **INFLORESCENCES** are small clusters of a few flowers; they measure less than 5 cm (2 in.) long. **FLOWERS** are blue. **FRUITS** have 3 crests on the upper sides of the round capsules. **TWIGS** are hairy and greenish gray. **BRANCHES** are flexible.

HABITAT AND RANGE: Grows in the low-elevation woodlands and chaparrals of California's northwestern mountains south to the central Sierra Nevada, at elevations from 350 m (1,200 ft) to 1,000 m (3,500 ft).

REMARKS: Lemmon ceanothus is often seen in disturbed areas along roadsides. It is especially noticeable in the spring, when the shrubs are covered with bright blue flowers.

CHAPARRAL WHITETHORN *Ceanothus leucodermis*

DESCRIPTION: A large, erect shrub that may grow to be 4 m (13 ft) tall. **LEAVES** are thin, alternate, evergreen, and less than 4 cm (1.5 in.) long. Stipules are deciduous. Blades are egg shaped to elliptical. Upper blade surfaces are dull green, with a scattering of long hairs. Lower blade surfaces are pale green and hairy and have 3 prominent mid veins. Petioles are about 3 mm (.12 in.) long. **INFLORESCENCES** are tight clusters of flowers, they measure less than 11 cm (4.3 in.) long. **FLOWERS** are pale blue. **FRUITS** are sticky, have 3 lobes, and lack crests. **TWIGS** are gray and waxy, with a covering of fine hair. **BRANCHES** are round and rigid and have thorns.

HABITAT AND RANGE: Grows at lower elevations in the Sierra Nevada, the Coast Ranges south of San Francisco, southern

California's mountains, and Baja California. The species is most commonly seen in chaparrals, but it can be a component of open forests and woodlands. It occurs below 1,800 m (6,000 ft).

REMARKS: Chaparral whitethorn is easily recognized as the green-leaved, thorny ceanothus of lower-elevation chaparrals. It provides cover for wildlife and is a preferred food for deer.

BIGPOD CEANOTHUS *Ceanothus megacarpus*

DESCRIPTION: An erect shrub that may grow to be 4 m (13 ft) tall. **LEAVES** are evergreen, alternate, and usually less than 2.5 cm (1 in.) long. Stipules are thick and evergreen. Blades are elliptical to egg shaped. Upper blade surfaces are dull green and hairless. Lower blade surfaces are covered with fine gray hairs and have 1 prominent mid vein. Petioles are less than 7 mm (.28 in.) long. **INFLORESCENCES** form on short lateral branches and are less than 2 cm (.75 in.) long. **FLOWERS** are white to pale lavender. **FRUITS** are spherical and 8 mm (.3 in.) to 12 mm (.5 in.) in diameter. They have prominent horns on the tops of their capsules. **TWIGS** are gray and have fine hairs. **BRANCHES** are alternate (opposite on *C. m.* var. *insularis*), roughened, and rigid.

HABITAT AND RANGE: Grows in the foothills and on mountain slopes from Santa Barbara County south into southern California. It is characteristic of chaparrals but can be found in forests and woodlands as well. It occurs below 600 m (2,000 ft).

REMARKS: Bigpod ceanothus grows on the mainland and the Channel Islands. Some island populations lack horns on the capsules (*C. m.* var. *insularis*), unlike mainland populations (*C. m.* var. *megacarpus*). *Ceanothus megacarpus* var. *insularis* is uncommon.

HAIRY CEANOTHUS *Ceanothus oliganthus*

DESCRIPTION: A large shrub that can be treelike when mature; grows to be 5 m (15 ft) tall. **LEAVES** are thin, alternate, evergreen, and less than 4 cm (1.5 in.) long. Stipules are deciduous and blades elliptical. Upper blade surfaces are dark green. Lower blade surfaces are pale green, with scattered hairs

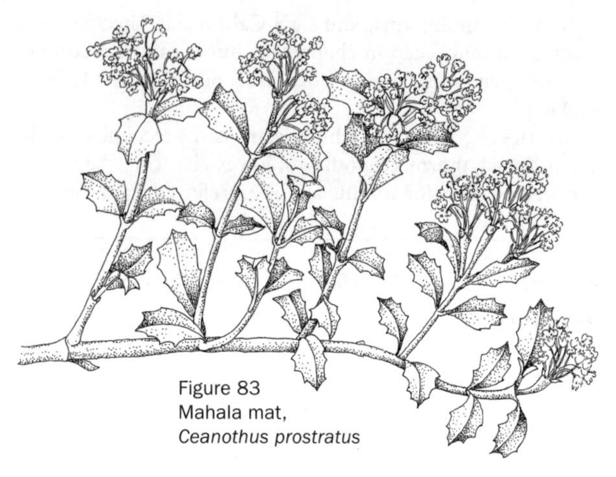
Figure 83
Mahala mat,
Ceanothus prostratus

and 3 prominent mid veins. **INFLORESCENCES** are small clusters of a few flowers. **FLOWERS** are blue or purple to nearly white. **FRUITS** have 3 crests on the upper side of a round capsule. **TWIGS** are round, ridged, and warty or smooth. **BRANCHES** are flexible or rigid, sometimes spiny and sticky. **HABITAT AND RANGE:** Grows in the low-elevation woodlands and chaparrals of the coastal mountains of both northern and southern California, at elevations below 1,300 m (4,500 ft). **REMARKS:** Hairy ceanothus has 2, rather geographically separated, varieties, both of which occur in the Transverse Ranges and southern Coast Ranges. Hairy ceanothus (*C. o.* var. *oliganthus*) has hairy leaf blades and twigs densely covered with stiff or soft hair. It grows in southern California. Jim bush (*C. o.* var. *sorediatus*) has mostly hairless twigs, and leaf blades that are hairy only on the lower surfaces. It grows in northern California.

MAHALA MAT (Fig. 83) ***Ceanothus prostratus***

DESCRIPTION: A prostrate or erect shrub that may grow to be 2.5 m (8 ft) tall. **LEAVES** are evergreen, opposite, and less than 3 cm (1.25 in.) long. Stipules are thick and evergreen. Blades are generally wider above the middle and oblong to egg shaped, with 3 to 9 marginal spines. Upper blade surfaces are pale green and hairless. Lower blade surfaces have 1 prominent mid vein and a covering of fine white hair. Petioles are less than 3 mm (.12 in.) long. **INFLORESCENCES** form on short lateral branches that are less than 2 cm (.75 in.) long. **FLOWERS** are deep or pale blue. **FRUITS** have erect or spreading horns that project from the sides of the capsules. **TWIGS** are angled, brown, and hairy. **BRANCHES** are opposite and reddish brown.

HABITAT AND RANGE: Grows in the montane forests and woodlands of most of northern California's mountains, at elevations between 900 m (3,000 ft) and 2,000 m (6,500 ft). Outside of California it grows in western Nevada, Oregon, and Washington.

REMARKS: Mahala mat's more common form (*C. p.* var. *prostratus*) is a prostrate mat, while *C. p.* var. *laxus* is erect; the latter is found in the northern part of the species' range.

SISKIYOU MAT *Ceanothus pumilus*

DESCRIPTION: A prostrate shrub that may grow to be 30 cm (1 ft) tall. **LEAVES** are evergreen, opposite, and less than 15 mm (.6 in.) long. Stipules are thick and evergreen. Blades are wider above the middle and oblong to egg shaped, with 3 terminal spines. Upper blade surfaces are pale green and hairless. Lower blade surfaces have 1 prominent mid vein and a covering of fine white hair. Petioles are less than 2 mm (.08 in.) long. **INFLORESCENCES** form on short lateral branches that are less than 2 cm (.75 in.) long. **FLOWERS** are blue to white. **FRUITS** have small horns on the sides of their capsules. **TWIGS** are angled and green or brown, with a covering of hair. **BRANCHES** are opposite and reddish brown.

HABITAT AND RANGE: Grows in the Siskiyou Mountains and northern Coast Ranges, often on serpentine soils, at elevations from 600 m (2,000 ft) to 1,700 m (5,700 ft). Outside of California it grows in southwestern Oregon.

REMARKS: Siskiyou mat is commonly encountered among grasses in Jeffrey pine woodlands and in chaparrals growing on serpentine soils. The smaller leaf

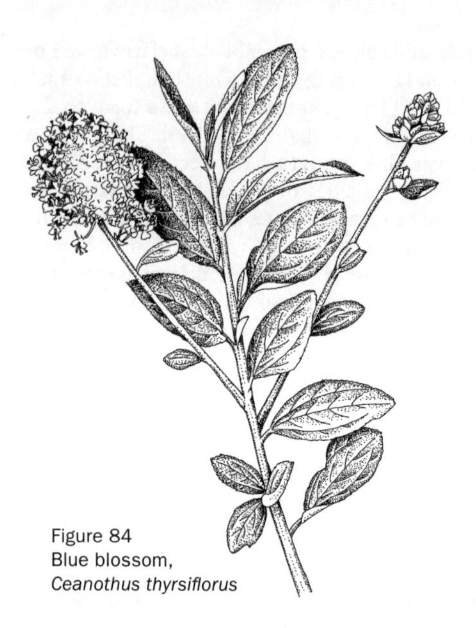

Figure 84
Blue blossom,
Ceanothus thyrsiflorus

with 3 terminal teeth differentiates it from the more wide-ranging mahala mat (with 3 to 9 marginal spines) that it resembles.

BLUE BLOSSOM *Ceanothus thyrsiflorus*

(Fig. 84; Pl. 18)

DESCRIPTION: A prostrate or erect shrub that typically grows to be 9 m (30 ft) tall. The largest lives in Curry County, Oregon, and is 12.5 m (41 ft) tall and 25 cm (10 in.) in diameter. **LEAVES** are alternate, evergreen, and less than 5 cm (2 in.) long. Stipules are deciduous. Blades are oblong to elliptical, with serrated margins. Upper blade surfaces are dark green and hairless. Lower blade surfaces have 3 prominent mild veins. Surfaces are light green and may have hairs. Petioles are less than 12 mm (.5 in.) long. **INFLORESCENCES** are narrow flower clusters less than 8 cm (3 in.) long. **FLOWERS** are deep to pale blue or even white. **FRUITS** are sticky and round. **TWIGS** are green and ribbed. **BRANCHES** are flexible.

HABITAT AND RANGE: Grows in forests, woodlands, and coastal scrubs of northern California's Coast Ranges, at elevations below 600 m (2,000 ft). Outside of California it grows in southwestern Oregon.
REMARKS: Blue blossom is abundant and extensive in areas following fire or logging. The resulting stands are transitional to various kinds of forest. It is planted as an ornamental.

WOOLLYLEAF CEANOTHUS *Ceanothus tomentosus*

DESCRIPTION: A medium-sized or large shrub that may grow to be 2.5 m (8 ft) tall. **LEAVES** are thin, alternate, evergreen, and less than 3 cm (1.25 in.) long. Stipules are deciduous and blades elliptical. Upper blade surfaces are dark green with fine hairs, and margins have fine, gland-tipped teeth. Lower blade surfaces have 3 prominent mid veins and are either hairless or have a dense covering of white or brownish hair. **INFLORESCENCES** are open clusters of flowers. **FLOWERS** are white. **FRUITS** are round with 3 small crests. **TWIGS** are gray to red and may be covered with fine hairs. **BRANCHES** are round, flexible, and sometimes warty.
HABITAT AND RANGE: Grows in chaparrals and woodlands in the foothills and montane elevations of the Sierra Nevada, from Placer to Mariposa Counties and again in southern California, below 1,100 m (3,600 ft). 
REMARKS: Sierra Nevada plants represent the typical form, *C. t.* var. *tomentosus,* with hairy leaves and large teeth. In southern California, plants have hairless leaf blades with small teeth (*C. t.* var. *olivaceus*). Woollyleaf ceanothus is most important as a stabilizer of watersheds, especially in southern California. In addition, it acts as forage and habitat for wildlife.

TOBACCO BRUSH *Ceanothus velutinus*

DESCRIPTION: A large shrub or small tree that may grow to be 6 m (20 ft) tall. **LEAVES** are alternate, evergreen, aromatic, and less than 8 cm (3 in.) long. Stipules are deciduous. Blades have a broad elliptical to egg shape. Leaves have a strong walnut, cinnamon, or balsam odor. Upper blade surfaces are dark green, shiny, and hairless. Lower blade surfaces are pale green, with 3 prominent mid veins. **INFLORESCENCES** are open flower clusters less than 12 cm (4.75 in.) long. **FLOWERS** are white. **FRUITS** are sticky, pyramidal, and rough. **TWIGS** are light brown, with a covering of fine hair. **BRANCHES** are round and flexible.

HABITAT AND RANGE: This species in its typical form, *C. v.* var. *velutinus* is a common and abundant ceanothus in the coniferous forests of northern California's inland mountains, at elevations below 3,000 m (10,000 ft). Here, lower leaf blades of the plants are covered with velvety hairs. A north coast form, *C. v.* var. *hookeri,* has hairless lower leaf blades. Outside of California the species ranges north to Washington and east to Colorado, with an isolated occurrence in South Dakota.

REMARKS: Tobacco brush is easily recognized by its dark green, shiny, aromatic leaves. The odor is overwhelming to a person walking through fields on warm days. Tobacco brush is abundant and extensive in areas following fire or logging. The resulting stands are transitional to various kinds of forest.

CELTIS (NETLEAF HACKBERRY)

The genus *Celtis* has about 70 species of trees and shrubs; 7 are native to North America, and 1 of these is native to California. *Celtis* is found in dry, temperate, and tropical forests.

NETLEAF HACKBERRY (Fig. 85) ***Celtis reticulata***

DESCRIPTION: A single-stemmed small tree or large shrub. Mature plants typically range in height from 1 m (3.3 ft) to 8 m (26 ft). The largest lives in Catron County, New Mexico, and is 21 m (69 ft) tall and 1.4 m (57 in.) in diameter. Crowns are expansive and irregular on spreading, twisting branches. Trunks are short and crooked. **LEAVES** are deciduous, alternate, simple, thick, leathery, lance shaped to egg shaped, and 2 cm (.75 in.) to 7.5 cm (3 in.) long. Margins are entire or have a few teeth. Leaf tips are pointed and bases are tapered to heart shaped. Upper surfaces are dark green and rough, while lower surfaces are yellowish green, hairy, and obviously net veined. **FRUITS** are brown, orange, or purple spherical drupes that measure about

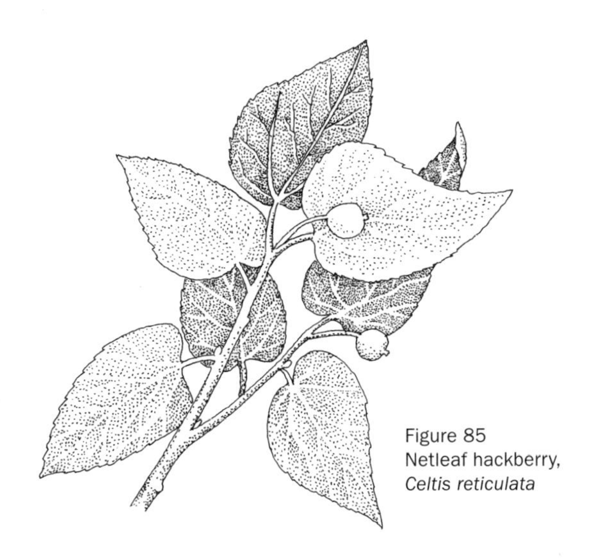

Figure 85
Netleaf hackberry,
Celtis reticulata

6 mm (.25 in.) in diameter. **TWIGS** are reddish brown and hairy. **BARK** is reddish brown to gray, thick, and warty.

HABITAT AND RANGE: Grows in riparian woodlands near streams, springs, and ephemeral creeks. Outside of California it is found throughout the southwestern United States and north to eastern Washington. In California it is found in dry, hot habitats in southern California and in the southeastern deserts but is considered uncommon. It ranges from 450 m (1,500 ft) to 1,700 m (5,500 ft).

REMARKS: Netleaf hackberry has a deep, spreading root system, which helps to make it drought tolerant. It is planted as an ornamental and used for fence posts and firewood. Wildlife species utilize stands of netleaf hackberry for food and cover. Its fruits are an important winter food for birds. Native Americans ate the sweet fruits.

CEPHALANTHUS (BUTTON-WILLOW)

The genus *Cephalanthus* is composed of about 17 species growing in North America, Asia, and southern Africa. One species occurs in California.

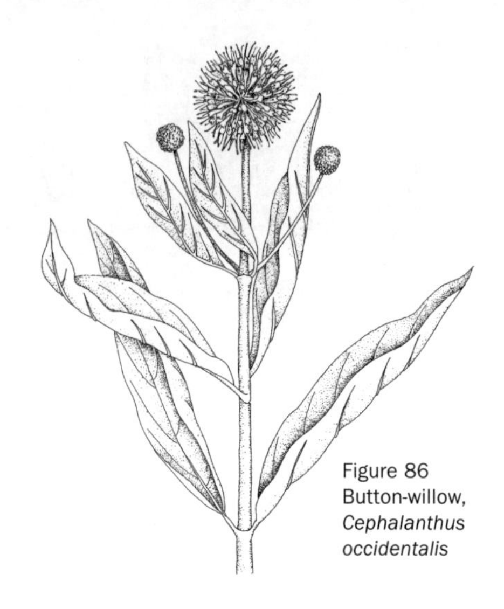

Figure 86
Button-willow,
Cephalanthus occidentalis

BUTTON-WILLOW *Cephalanthus occidentalis*

(Fig. 86)

DESCRIPTION: A shrub with large leaves and gray or brown bark. It typically grows to be between 2 m (6.5 ft) and 10 m (33 ft) tall. The largest lives in Buttonwillow, California, and is 6 m (20 ft) tall and 51 cm (20 in.) in diameter. **LEAVES** are simple, whorled or opposite, and deciduous; they usually have 3 leaves at each node (sometimes 2, 4, or 5 leaves per node). Blades have a broad elliptical to egg shape and are 7.5 cm (3 in.) to 20 cm (8 in.) long and hairless. **INFLORESCENCES** are located at the ends of branches and consist of round heads of flowers that are about 3 cm (1.25 in.) in diameter. **FLOWERS** are not individually showy. Each flower is small, 4-parted, and tubular, with an inferior pistil. **FRUITS** are dry, hard capsules that break into 2 to 4 achenelike fruits.

HABITAT AND RANGE: Grows along streams and around lakes below 1,000 m (3,300 ft). In California it occurs in the Central Valley and surrounding foothills. Outside of California it ranges east to New Brunswick and Florida.

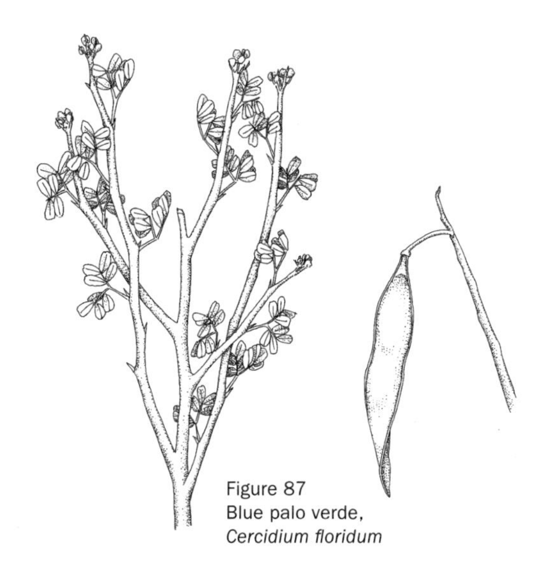
Figure 87
Blue palo verde,
Cercidium floridum

REMARKS: This species grows in the eastern United States; another grows in Texas and Mexico, and 4 others grow in the tropics of both the Old and New Worlds. Such a pattern suggests that button-willow is a relict that once had a larger range. The family of this species is mainly one of small trees and shrubs, including coffee and quinine, which grow in tropical forests.

CERCIDIUM (PALO VERDE)

The genus *Cercidium* has 4 species that grow in the deserts of the southwestern United States and northwestern Mexico. Two species grow in California.

BLUE PALO VERDE (Fig. 87) ***Cercidium floridum***

DESCRIPTION: An erect tree with blue-green bark that typically grows to be 8 m (25 ft) tall. The largest lives in Riverside County and is 16.1 m (53 ft) tall and 68 cm (27 in.) in diameter. Trees are leafless most of the year. **LEAVES** are alternate, compound, deciduous, and short-lived. Primary leaflets are about 1 cm (.4 in.) long and usually found in pairs (sometimes

4 or 6). Each primary leaflet stalk can have 2 to 4 pairs of secondary leaflets that are each about 6 mm (.25 in.) long. The primary leaflet stalks lack terminal leaflets. **INFLORESCENCES** are loose, rounded clusters of flowers. **FLOWERS** are yellow and about 12 mm (.5 in.) broad. **FRUITS** are legumes that measure 3 cm (1.2 in.) to 11 cm (4.3 in.) long. **TWIGS** have thorns in many leaf axils, are blue-green, and have pointed tips.

HABITAT AND RANGE: Grows along washes and in floodplains below 350 m (1,200 ft) in the Colorado and Sonoran Deserts in southeastern California, and in the Sonoran Desert of Arizona and northwestern Mexico. In California it is considered uncommon.

REMARKS: Blue palo verde is one of the taller and more easily identified trees in the Colorado and Sonoran Deserts. It is easily recognized by its showy yellow flowers and green (*verde*) stems (*palo*). A second species of palo verde, foothill palo verde (*C. microphyllum*), which is common in Arizona, grows in California only in the Whipple Mountains. Some botanists place all *Cercidium* species in the genus *Parkinsonia*. Mexican palo verde (*P. aculeata*) is planted along highways in the Colorado Desert for its yellow flowers. Unlike blue palo verde, Mexican palo verde plants have obviously flat and showy blade axes.

CERCIS (REDBUD)

The genus *Cercis* has about 7 species of shrubs and trees in the Northern Hemisphere, and 1 of these is native to California.

WESTERN REDBUD *Cercis occidentalis*

(Fig. 88; Pl. 19)

DESCRIPTION: A tall shrub or small tree that typically grows to be about 8 m (25 ft) tall. The largest lives in Santa Rosa and is 8.8 m (29 ft) tall and 58 cm (23 in.) in diameter. Crowns are rounded on clustered, erect branches. **LEAVES** are deciduous, alternate, simple, round, somewhat leathery, and 5 cm (2 in.) to 9 cm (3.5 in.) in diameter. Margins are entire. Leaf bases are heart shaped. Upper surfaces are dark green, shiny, and hair-

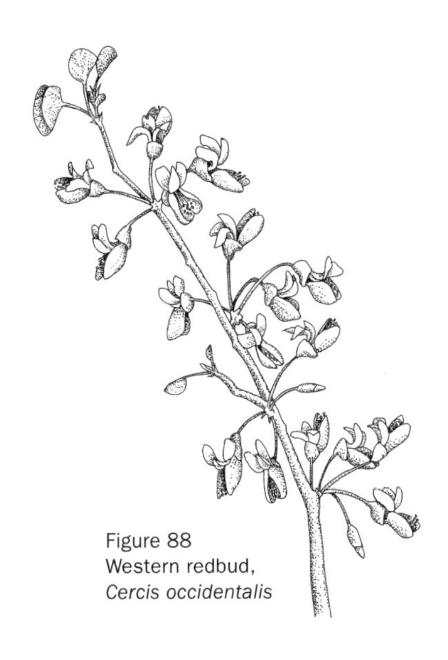

Figure 88
Western redbud,
Cercis occidentalis

less, while lower surfaces are green and hairless and have 7 to 9 fanlike veins radiating from their bases. Petioles are 12 mm (.5 in.) to 25 mm (1 in.) long. **INFLORESCENCES** are unbranched and showy. **FLOWERS** resemble pea flowers, are a striking magenta to red, and measure about 12 mm (.5 in.) long. They appear before leaf emergence. **FRUITS** are legumes, 4 cm (1.5 in.) to 7.5 cm (3 in.) long, about 12 mm (.5 in.) wide, flat, and reddish purple. The clustered pods persist through the winter. **TWIGS** are slender, erect, and hairless.

HABITAT AND RANGE: Grows in foothill woodlands and chaparrals. It is often found on dry, lower canyon slopes near stream banks below 1,000 m (3,300 ft). It ranges east to Texas.

REMARKS: Western redbud provides color to what otherwise can be a drab landscape. From February to April, western redbud's resplendent magenta floral displays are harbingers of spring. In the summer, its green leaves provide a welcome contrast to the golden, dry grasslands and woodlands. And in the fall and winter, its reddish brown pods are striking on bare branches. Nitrogen-fixing bacteria form root nodules in western redbud, adding considerable nitrogen to ecosystems via decaying plant parts.

CERCOCARPUS (MOUNTAIN-MAHOGANY)

The genus *Cercocarpus* has 13 species growing in the western United States and Mexico. Five of these live in California, 2 of which are common.

Cercocarpus is a collection of shrubs and small trees with hard wood and simple leaves that are alternate in arrangement and clustered on short twigs. Flowers are greenish and lack petals. The many stamens of each are attached to the top of a funnel-like cup surrounding a single pistil. The flowers are not showy but instead are sweet with nectar. The show comes in the summer, when the plants are in fruit. Each fruit is an achene ending in a long style that is covered with shiny hairs at maturity. Branches and trunks are gray or reddish brown.

The shrubs glisten in the sun from the mass of silvery fruits, each one a "tailed fruit" as indicated by the generic name.

1. Leaf blades have smooth margins . . . . **curlleaf mountain-mahogany** (*C. ledifolius*)
1. Leaf blades have toothed or scalloped margins . . . . **birchleaf mountain-mahogany** (*C. betuloides*)

BIRCHLEAF MOUNTAIN-MAHOGANY

(Fig. 89) ***Cercocarpus betuloides***

DESCRIPTION: An erect shrub or small tree that typically grows to be between 2 m (6.5 ft) and 8 m (25 ft) tall. The largest lives in Central Point, Oregon, and is 10.4 m (34 ft) tall and 35 cm (14 in.) in diameter. **LEAVES** are evergreen, dark green, and leathery. Leaf blades are elliptical and 12 mm (.5 in.) to 25 mm (1 in.) long. Upper surfaces are hairless and sticky and lower surfaces are covered with white hairs. Margins are toothed or scalloped and are rolled toward the lower surface. **INFLORESCENCES** are clusters of 1 to 8 flowers. **FLOWERS** have funnel-like cups covered with white hairs. **FRUITS** are achenes with twisted, feathery styles that measure from 5 cm (2 in.) to 7.5 cm (3 in.) long. **TWIGS** are gray and smooth. **BARK** is gray and breaks into squarish segments.

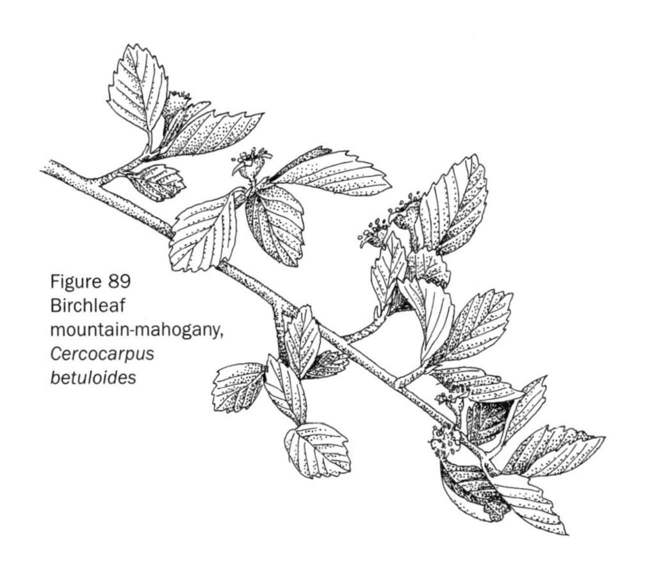

Figure 89 Birchleaf mountain-mahogany, *Cercocarpus betuloides*

HABITAT AND RANGE: Grows in chaparrals, dry forests, and woodlands below 2,400 m (8,000 ft) in the foothills and mountain slopes in most of California's mountains.

REMARKS: Birchleaf mountain-mahogany has 3 varieties. Typical birchleaf mountain-mahogany (*C. betuloides* var. *betuloides*) has leaves smaller than those of the other varieties and fewer lateral veins. On the Channel Islands, *C. betuloides* var. *blancheae* has leaves with toothed margins. In more northerly mountains, *C. betuloides* var. *macrourus* has elongated leaves. To add to the confusion, botanists outside of California consider birchleaf mountain-mahogany's varieties to be part of a westernwide species (*C. montanus*) that includes 7 varieties. Birchleaf mountain-mahogany is preferred forage for wildlife.

CURLLEAF MOUNTAIN-MAHOGANY *Cercocarpus ledifolius*

(Fig. 90; Pl. 20)

DESCRIPTION: A shrub or small tree that usually grows to be between 1 m (3.3 ft) and 9 m (30 ft) tall. The largest lives in Great Basin National Park, Nevada, and is 7.9 m (26 ft) tall and

Figure 90
Curlleaf
mountain-mahogany,
Cercocarpus ledifolius

1.3 m (4 ft) in diameter. **LEAVES** are evergreen, dark green, and leathery. Leaf blades are elliptical and 12 mm (.5 in.) to 25 mm (1 in.) long. Upper surfaces are sticky and hairless and lower surfaces are covered with white hairs. Margins are entire and rolled toward the lower surface. **INFLORESCENCES** are clusters of 1 to 8 flowers. **FLOWERS** have funnel-like cups covered with white hairs. **FRUITS** are achenes with twisted, feathery styles that measure from 5 cm (2 in.) to 7.5 cm (3 in.) long. **TWIGS** are gray and smooth. **BARK** is gray and breaks into squarish segments.

HABITAT AND RANGE: Grows in the intermountain West. In California it is found between 1,000 m (3,300 ft) and 2,700 m (9,000 ft) in shrublands, forests, and woodlands of all the major mountain ranges except those of the central coast.

REMARKS: Two forms of curlleaf mountain-mahogany grow in California. In *C. l.* var. *ledifolius* the midrib on the lower leaf bade is obscured by hair, and in *C. l.* var. *intermontanus* the midrib is not. The first form is more common in the Great Basin. A similar species, *C. intricatus*, which has leaves more tightly

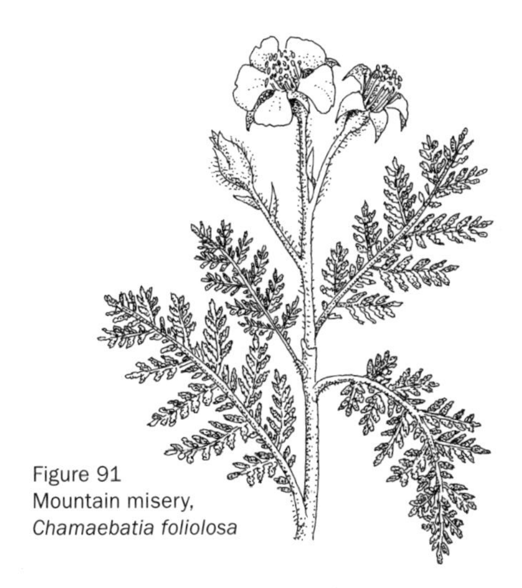

Figure 91
Mountain misery,
Chamaebatia foliolosa

rolled, grows in the eastern Sierra Nevada and nearby mountains of the Great Basin and Mojave Deserts. Curlleaf mountain-mahogany is valuable forage for wildlife. Stems have been used as arrow shafts, digging tools, and spears. The inner bark has medicinal properties.

CHAMAEBATIA (MOUNTAIN MISERY)

The genus *Chamaebatia* has 2 species growing in California and Baja California.

MOUNTAIN MISERY *Chamaebatia foliolosa*

(Fig. 91)

DESCRIPTION: A densely branched, strong-smelling shrub that may grow to be 1 m (3.3 ft) tall. **LEAVES** are alternate, evergreen, fernlike, and clustered at the ends of the stems. Blades are 2 or 3 times pinnately compound and are 12 mm (.5 in.) to 7.5 cm (3 in.) long. Each leaflet is broken into notched segments. **INFLORESCENCES** consist of a few flowers in terminal, elongated clusters that are 3 cm (1.2 in.) to 10 cm (4 in.) long. **FLOWERS** have 5 petals and are showy, white, and about 12 mm (.5 in.) broad. **FRUITS** consist of a single achene per flower. **HABITAT AND RANGE:** Grows in coniferous forests on the western slopes of the Cascades and Sierra Nevada, between 600 m (2,000 ft) and 2,100 m (7,000 ft). Its clones can cover extensive areas under forest canopies.

REMARKS: Most people notice that mountain misery and fern bush look similar since both have fernlike leaves. They need not be confused, however, as the plants have different ranges. Fern bush grows east of the Cascade and Sierra Nevada crest, while mountain misery grows on the west side. *C. australis,* a second, uncommon mountain misery with narrower leaves, occurs in southern San Diego County.

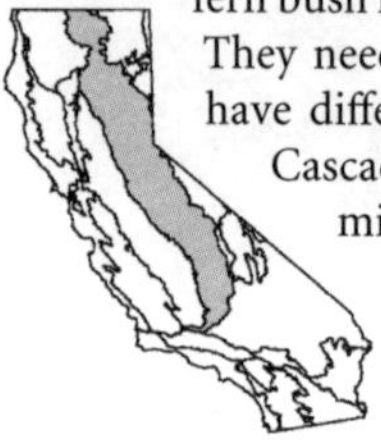

CHAMAEBATIARIA (FERN BUSH)

The genus *Chamaebatiaria* has 1 species that grows in the intermountain West.

FERN BUSH (Fig. 92) ***Chamaebatiaria millefolium***

DESCRIPTION: A densely branched, fragrant shrub that may grow to be 2 m (6.5 ft) tall. **LEAVES** are alternate, deciduous, fernlike, and clustered at the ends of the stems. Blades are 1 or 2 times pinnately compound and are 5 cm (2 in.) to 10 cm (4 in.) long. Each oblong leaflet has many deep notches on the margins. **INFLORESCENCES** consist of many flowers in terminal, elongated clusters that measure 2.5 cm (1 in.) to 10 cm (4 in.) long. **FLOWERS** have 5 showy white petals. **FRUITS** are 4 or 5 leathery follicles per flower.

HABITAT AND RANGE: In California this species primarily grows in the Great Basin and its adjoining mountain slopes, between 900 m (3,000 ft) and 2,100 m (7,000 ft). It occurs as individuals or small populations in eastside woodlands and scrubs.

REMARKS: The generic name of fern bush, *Chamaebatiaria,* is meant to suggest that fern bush resembles *Chamaebatia,* mountain misery, another genus that grows in California. Fern bush fruits, however, are follicles and those of mountain misery are achenes. Their ranges also distinguish them: mountain misery grows on the western slopes of the Cas-

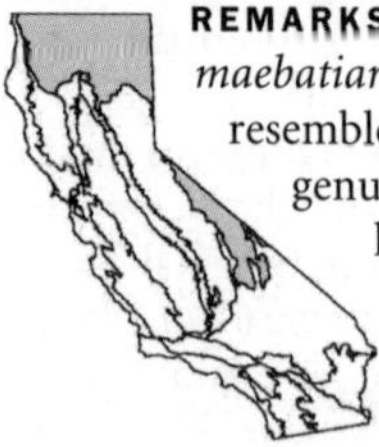

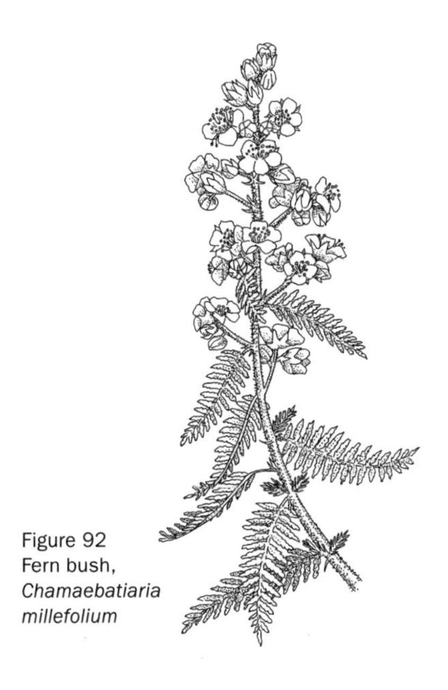

Figure 92
Fern bush,
Chamaebatiaria millefolium

cades and Sierra Nevada, and fern bush generally grows to the east.

CHILOPSIS (DESERT-WILLOW)

The genus *Chilopsis* has 1 species that grows in the warm deserts of the southwestern United States and northwestern Mexico.

DESERT-WILLOW (Fig. 93) *Chilopsis linearis*

DESCRIPTION: A willowlike shrub or small tree that usually grows to be 6 m (20 ft) tall. The largest lives in Gila City, Arizona, and is 20.7 m (68 ft) tall and 1.3 m (51 in.) in diameter. **LEAVES** are most often alternate but can be opposite or even whorled on the same plant, and they are deciduous. Blades are simple, 10 cm (4 in.) to 25 cm (10 in.) long, linear, and crescent shaped. **INFLORESCENCES** are terminal clusters of showy flowers. **FLOWERS** are sweetly fragrant, 2-lipped, and lavender, pink, or white, with purplish lines or markings. **FRUITS** are

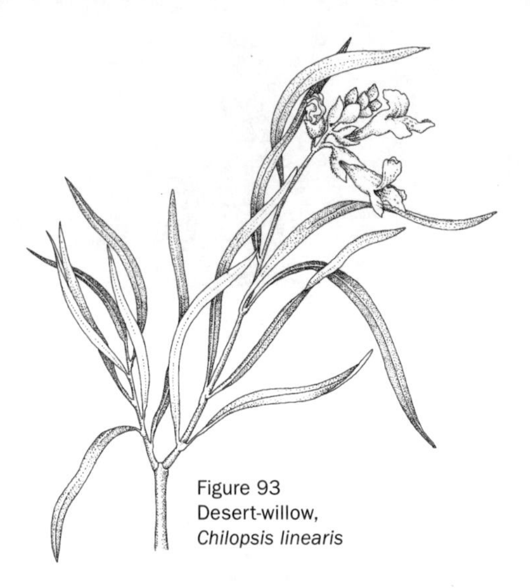

Figure 93
Desert-willow,
Chilopsis linearis

long, thin capsules that measure up to 38 cm (15 in.) long. **TWIGS** are gray.

HABITAT AND RANGE: Grows along washes and watercourses below 1,500 m (5,000 ft) in the eastern Mojave, Colorado, and Sonoran Deserts, as well as on adjoining slopes of the Transverse and Peninsular Ranges. Outside of California its range extends east to Texas and south into northern Mexico.

REMARKS: Desert-willow is one of the easily recognized small trees of California's warm deserts. It is identified by its sweet, showy flowers and distinctively curved leaves.

CHRYSOLEPIS (CHINQUAPIN)

The genus *Chrysolepis* has 2 species that occur in the United States, both of these in California. One is a shrub and the other is either a tree or a shrub.

Leaves are evergreen, alternate, simple, leathery, somewhat elliptical, and 2.5 cm (1 in.) to 12.5 cm (5 in.) long. Margins are entire and wavy. Upper surfaces are dark green and hairless and lower surfaces are golden or rusty. Pollen-bearing flowers are clustered in erect catkins 2.5 cm (1 in.) to 7.5 cm (3 in.) long. Seed-bearing flowers are clustered either below the pollen-bearing catkins or on a separate stalk. Fruits are nuts enclosed in a spiny bur. Bark is thin and smooth or thick and furrowed.

California's chinquapins were once considered as part of the Old World genus *Castanopsis*. Chinquapins and chestnuts (*Castanea*) are closely related and their fruits are very similar.

1. Leaf tips are round and somewhat broad. Bark is thin and smooth. Plants are shrubs .... **bush chinquapin** (*C. sempervirens*)
1. Leaf tips are pointed and tapered. Bark is thick and furrowed. Plants are trees or shrubs .... **golden chinquapin** (*C. chrysophylla*)

GOLDEN CHINQUAPIN *Chrysolepis chrysophylla*

DESCRIPTION: An erect, medium-sized, single-stemmed tree or a shrub. Mature trees are typically 15 m (50 ft) to 36 m (120 ft) tall and 30 cm (1 ft) to 1 m (3.3 ft) in diameter. Shrubs can grow to be 4.5 m (15 ft) tall. The largest golden chinquapin lives in Mendocino County and is 37 m (122 ft) tall and 1.2 m (49 in.) in diameter. Crowns are conical, with large, spreading branches. Trees live to be 400 or 500 years old. **LEAVES** are evergreen, simple, alternate, leathery, lance shaped to elliptical, and 5 cm (2 in.) to 15 cm (6 in.) long. Margins are entire and slightly wavy. Shrub forms have leaves that are folded, with margins pointed upward. Tree forms have flat leaves. Leaf tips are pointed and tapered and bases are tapered. Upper surfaces are dark green and hairless, while lower surfaces are golden and slightly hairy. **FRUITS** are nuts contained in a spiny, spherical bur that measures 2.5 cm (1 in.) to 4 cm (1.5 in.) in diameter. **BARK** on small trees is smooth and gray. On larger trees, it is thick, furrowed, and grayish brown.

HABITAT AND RANGE: Grows in forests and woodlands in California and the Pacific Northwest. In California it is mostly found in the Klamath Mountains and Coast Ranges, growing below 1,800 m

Figure 94
Bush chinquapin,
Chrysolepis sempervirens

(6,000 ft). Some populations occur in the Sierra Nevada near Georgetown.

REMARKS: Golden chinquapin is moderately shade tolerant. It sprouts from burls following injury from fire or cutting. Its wood is used to make cabinets, furniture, and novelty items. Two varieties are recognized: a tree form, *C. c.* var. *chrysophylla*, and a shrub form, *C. c.* var. *minor*. The tree form is found north of San Francisco Bay and in the Sierra Nevada. The shrub form usually grows on infertile soils or at higher elevations throughout the range of the tree form, as well as in the central Coast Ranges south of the Bay Area. *C. c.* var. *minor* hybridizes with *C. sempervirens* in Siskiyou County.

BUSH CHINQUAPIN *Chrysolepis sempervirens*

(Fig. 94)

DESCRIPTION: A prostrate to spreading, multistemmed shrub that can grow to be 2.5 m (8 ft) tall. It typically reaches from 30 cm (1 ft) to 1.5 m (5 ft) tall and has a rounded crown. **LEAVES** are evergreen, simple, alternate, flat, leathery, elliptical, and 2.5 cm (1 in.) to 7.5 cm (3 in.) long. Margins are entire. Leaf tips are round and somewhat broad; bases are tapered. Upper surfaces are dull green and hairless, while lower surfaces are golden to rusty and slightly hairy. **FRUITS** are nuts in a spiny, spherical bur that measures 2 cm (.75 in.) to 4 cm (1.5 in.) in diameter. **BARK** is thin, smooth, and brownish gray.

HABITAT AND RANGE: Grows in coniferous forests and chaparrals, usually on exposed, rocky slopes. It is found from southern Oregon to the mountains of California, from 600 m (2,000 ft) to 3,600 m (12,000 ft).

REMARKS: Bush chinquapin is similar to the shrub form of golden chinquapin. Bush chinquapin has flat leaves with broad, round leaf tips. The shrub form of golden chinquapin, in contrast, has upturned, folded leaves with pointed tips. Many bush chinquapin leaves are yellowish to rusty on their lower surfaces, although not as golden as golden chinquapin leaves, even in its shrub form. Bush chinquapin sprouts from burls following a fire. Rodents and birds eat its seeds. Native Americans ate raw and roasted chinquapin seeds.

CHRYSOTHAMNUS (RABBITBRUSH)

The genus *Chrysothamnus* has 16 species found from southwestern Canada to northern Mexico. Eight of these grow in California.

Chrysothamnus is a collection of perennial herbs or shrubs with simple, drought-deciduous leaves that are alternate in arrangement. Flowers are small, yellow, and set in showy masses of flower heads. Fruits are dry achenes.

1. Stems are hairless and leaves are twisted .... **yellow rabbitbrush (*C. viscidiflorus*)**
1. Stems have a dense covering of fine hair, and leaves are straight 2
   2. Bracts surrounding the flower heads are weakly keeled, and bract tips are long and tapering .... **Parry rabbitbrush (*C. parryi*)**
   2. Bracts surrounding the flower heads are strongly keeled, and bract tips are acute .... **rubber rabbitbrush (*C. nauseosus*)**

Figure 95
Rubber rabbitbrush, *Chrysothamnus nauseosus*

RUBBER RABBITBRUSH (Fig. 95) — *Chrysothamnus nauseosus*

DESCRIPTION: A many-branched, gray, ill-scented shrub that may grow to be 3 m (10 ft) tall. **LEAVES** are drought-deciduous and grayish green. Blades are threadlike to linear and 1 cm (.4 in.) to 7 cm (2.75 in.) long. **INFLORESCENCES** have yellow flowers in open, flat-topped clusters of heads. Bracts surrounding the flower heads are strongly keeled, and bract tips are acute. **STEMS** are leafy, erect or prostrate, whitish to green, and covered with a dense layer of fine hair. Older stems have loose, fibrous bark.

HABITAT AND RANGE: Grows in many of California's mountains, throughout the intermountain West, and in the Great Plains. In California it is found from 750 m (2,500 ft) to 3,600 m (12,000 ft) in many kinds of shrublands and in open forests and woodlands.

REMARKS: Rubber rabbitbrush is a highly variable species with 22 subspecies recognized throughout its range; 12 grow in California.

Top left: Plate 1. Bristlecone fir, *Abies bracteata*.
Top right: Plate 2. Green ephedra, *Ephedra viridis*.
Bottom: Plate 3. Mountain juniper, *Juniperus occidentalis* var. *australis*.

Plate 4. Brewer spruce, *Picea breweriana*.

Plate 5. Foxtail pine, *Pinus balfouriana*.

Top left: Plate 6. Western white pine, *Pinus monticola*. *Top right:* Plate 7. Monterey pine, *Pinus radiata*. *Bottom:* Plate 8. Ghost pine, *Pinus sabiniana*.

Top: Plate 9. Redwood, *Sequoia sempervirens*.
Bottom: Plate 10. Giant sequoia, *Sequoiadendron giganteum*.

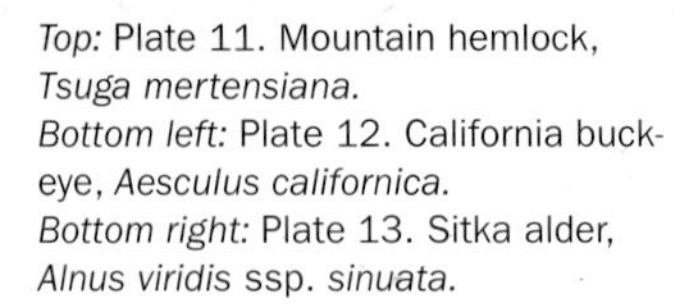

Top: Plate 11. Mountain hemlock, *Tsuga mertensiana.*
Bottom left: Plate 12. California buckeye, *Aesculus californica.*
Bottom right: Plate 13. Sitka alder, *Alnus viridis* ssp. *sinuata.*

Top left: Plate 14. Bearberry, *Arctostaphylos uva-ursi.*
Top right: Plate 15. Whiteleaf manzanita, *Arctostaphylos viscida.*
Left: Plate 16. Oregon-grape, *Berberis aquifolium.*
Bottom: Plate 17. Spice bush, *Calycanthus occidentalis.*

Left: Plate 18. Blue blossom, *Ceanothus thyrsiflorus.* *Bottom:* Plate 19. Western redbud, *Cercis occidentalis.*

Top: Plate 20. Curlleaf mountain-mahogany, *Cercocarpus ledifolius.* *Middle:* Plate 21. Mountain dogwood, *Cornus nuttallii.* *Bottom:* Plate 22. California hazel, *Corylus cornuta* var. *californica.*

Top left: Plate 23. Fremont silk tassel, *Garrya fremontii.*
Top right: Plate 24. California black walnut, *Juglans californica.*
Bottom left: Plate 25. Tanoak, *Lithocarpus densiflorus.*
Bottom right: Plate 26. Twinberry, *Lonicera involucrata.*

Plate 27. Black cottonwood, *Populus balsamifera* ssp. *trichocarpa.*

Top: Plate 28. Cliffrose, *Purshia mexicana* var. *stansburyana*.
Middle: Plate 29. Blue oak, *Quercus douglasii*.
Bottom: Plate 30. California black oak, *Quercus kelloggii*.

Plate 31. California black oak, *Quercus kelloggii*.

Top: Plate 32. Cascara, *Rhamnus purshiana.*
Left: Plate 33. Pacific rhododendron, *Rhododendron macrophyllum.*
Bottom: Plate 34. Skunkbush, *Rhus trilobata.*

Top left: Plate 35. Gummy gooseberry, *Ribes lobbii.*
Top right: Plate 36. Red flowering currant, *Ribes sanguineum.*
Bottom left: Plate 37. Black sage, *Salvia mellifera.*
Bottom right: Plate 38. Douglas spiraca, *Spiraea douglasii.*

Top: Plate 39. California bay, *Umbellularia californica*. *Bottom:* Plate 40. California fan palm, *Washingtonia filifera*.

PARRY RABBITBRUSH *Chrysothamnus parryi*

DESCRIPTION: A sparsely branched green shrub that may grow to be between 10 cm (4 in.) and 90 cm (36 in.) tall. **LEAVES** are drought-deciduous and green. Blades are threadlike to linear and 1 cm (.4 in.) to 7 cm (2.75 in.) long. **INFLORESCENCES** are rounded clusters of yellow flowers. Bracts surrounding the flowering heads are weakly keeled, and bract tips are long and tapering. **STEMS** are very leafy, erect or prostrate, white to green, and covered with a dense layer of fine hair.

HABITAT AND RANGE: Found in California's mountains and throughout the intermountain West, at elevations from 750 m (2,500 ft) to 3,600 m (12,000 ft). It populates treeless flats and forest openings and grows under trees in open forests and woodlands.

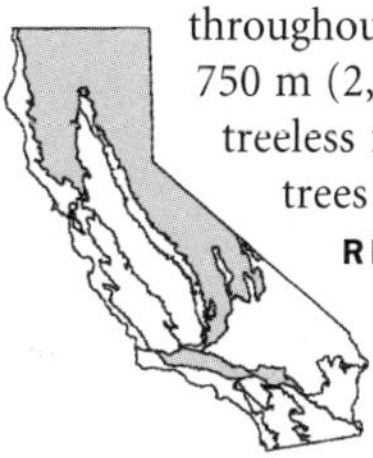

REMARKS: Parry rabbitbrush is a highly variable species with 12 subspecies recognized throughout its range; 6 grow in California.

YELLOW RABBITBRUSH *Chrysothamnus viscidiflorus*

DESCRIPTION: A many-branched, bright green shrub that may grow to be 1.5 m (5 ft) tall. **LEAVES** are drought-deciduous, green, and sticky. Blades are threadlike to oblong, and many are twisted or curly. **INFLORESCENCES** are tight, flat-topped clusters of yellow flower heads. **STEMS** are white, brittle, and hairless. Upper stems are green.

HABITAT AND RANGE: Found in the inner mountains of California and the intermountain West, at elevations from 900 m (3,000 ft) to 4,000 m (13,000 ft). It grows with big sagebrush and is associated with trees in an array of woodlands.

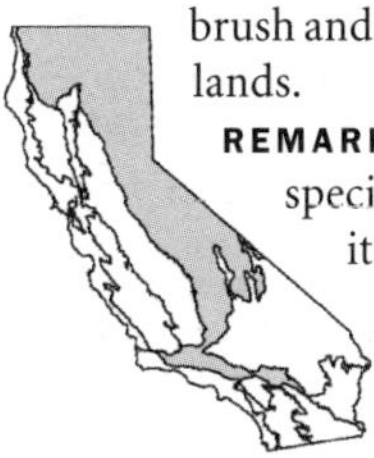

REMARKS: Yellow rabbitbrush is a highly variable species with 5 subspecies recognized throughout its range; 4 grow in California. Its value as browse for wildlife varies among subspecies. With its distinctive curly leaves, the species is easy to recognize.

Figure 96
Black bush,
Coleogyne ramosissima

COLEOGYNE (BLACK BUSH)

The genus *Coleogyne* has 1 species that grows in the intermountain West and the Mojave and Colorado Deserts.

BLACK BUSH (Fig. 96) ***Coleogyne ramosissima***

DESCRIPTION: A highly branched, aromatic shrub that may grow to be between 20 cm (8 in.) and 2 m (6.5 ft) tall. **LEAVES** are opposite, deciduous, and clustered on the stems. Blades are linear or club shaped and 1.2 cm (.5 in.) to 2 cm (.75 in.) long, with smooth margins. **INFLORESCENCES** consist of a single flower. **FLOWERS** are yellow or brown. Sepals provide the flower color, as there are no petals. **FRUITS** are achenes. **TWIGS** are spine tipped and ashy gray; they turn black with age. **HABITAT AND RANGE:** In California this species grows in the southern Great Basin, Mojave, Sonoran, and Colorado Deserts and their adjoining mountain slopes, between 600 m (2,000 ft) and 1,500 m (5,000 ft). It can mix with members of the interior coniferous woodlands or form striking, black-hued scrublands, where it is the dominant plant.

REMARKS: Black bush is highly susceptible to fire and does not sprout after being injured by fire. Individual shrubs may live for centuries. Outside of California, black bush is commonly called black brush.

CORNUS (DOGWOOD)

The genus *Cornus* has about 50 small tree and shrub species, almost all of which occur in the Northern Hemisphere. Sixteen of them are native to the United States, and of these, 1 tree and 4 shrub species are native to California.

Leaves are simple and usually deciduous and opposite. Margins are generally entire or fine-toothed. Veins are usually prominent. Lateral veins are evenly spaced near the mid vein, then curved toward the leaf tip near the margins. Flowers are small and clustered in multibranched or unbranched spherical inflorescences. Flower parts appear in 4s or 6s. Some species have large showy bracts surrounding the flowers, making inflorescences themselves look like flowers. Fruits are spherical to egg shaped, red, white, or blue drupes.

Some dogwood species are grown as ornamentals for spring flowers and/or autumn foliage colors. California dogwoods grow on moist lower slopes and in riparian zones.

1. Flowers are clustered in multibranched inflorescences that appear during or after leaf emergence 2
1. Flowers are clustered in unbranched inflorescences that appear before leaf emergence 3
   2. Lower leaf surfaces are mostly hairless. Twigs are gray or yellowish brown. Veins on the lower leaf surfaces are not prominent . . . . **brown dogwood (*C. glabrata*)**
   2. Lower leaf surfaces have a few stiff, sharp hairs. Twigs are reddish. Veins on the lower leaf surfaces are prominent . . . . **red osier (*C. sericea*)**

3. Twigs are green when young, turning reddish later. Flower clusters have 4 to 7 showy white to pinkish basal bracts . . . . **mountain dogwood** (*C. nuttallii*)
3. Twigs are gray or yellowish brown. Flower clusters have 4 ephemeral, nonshowy brownish basal bracts . . . . **blackfruit dogwood** (*C. sessilis*)

BROWN DOGWOOD *Cornus glabrata*

DESCRIPTION: An erect or spreading, multistemmed shrub that ranges in height from 1.25 m (4 ft) to 4.5 m (15 ft) and can form dense thickets. **LEAVES** are deciduous, opposite, simple, and 2.5 cm (1 in.) to 5 cm (2 in.) long, with a broad elliptical shape. Margins are entire and wavy. Upper and lower surfaces are light green and mostly hairless. Veins on lower surfaces are not prominent. **FLOWERS** are clustered in multibranched, spherical inflorescences that appear as or after the leaves emerge. **FRUITS** are spherical, white or blue, set on small red stalks, and about 6 mm (.25 in.) in diameter. **TWIGS** are gray or yellowish brown.

HABITAT AND RANGE: Grows in woodlands in Oregon and California, except for the Great Basin and warm deserts. It is found on moist flats and along streams below 1,500 m (5,000 ft); in southern California it is considered uncommon.

REMARKS: Brown dogwood is cultivated as an ornamental. It is most commonly found in fencerows.

MOUNTAIN DOGWOOD *Cornus nuttallii*

(Fig. 97; Pl. 21)

DESCRIPTION: An erect, small, single-stemmed tree. Mature plants are usually 3 m (10 ft) to 12 m (40 ft) tall and 15 cm (6 in.) to 30 cm (12 in.) in diameter. The largest grows in Clatskanie, Oregon, and is 18.3 m (60 ft) tall and 1.3 m (54 in.) in diameter. Crowns of forest-grown trees are narrow, and those of open-grown trees are rounded. **LEAVES** are deciduous, opposite, simple, and 7.5 cm (3 in.) to 12.5 cm (5 in.) long, with a broad elliptical shape. Margins are entire and wavy. Upper surfaces are green and have fine hairs, and lower surfaces are pale green and

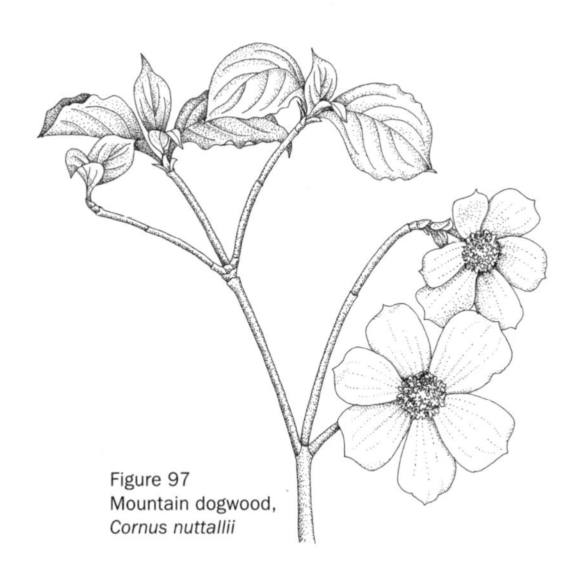
Figure 97
Mountain dogwood,
Cornus nuttallii

hairy. Veins are prominent, with lateral, evenly spaced veins that curve toward the leaf tip near the margins. **FLOWERS** are clustered in unbranched spherical heads and appear before the leaves emerge. Flower clusters have 4 to 7 showy white to pinkish bracts about 5 cm (2 in.) to 7.5 cm (3 in.) long. **FRUITS** are red or reddish orange drupes that measure about 12 mm (.5 in.) long. **TWIGS** are initially green, turning reddish later.

HABITAT AND RANGE: Grows in coniferous forests and woodlands from British Columbia to San Diego County. In California it is most abundant in the northern Coast Ranges, the Klamath Mountains, and the northern Sierra Nevada. Elsewhere, it is uncommon or rare. It is found below 2,000 m (6,500 ft) on moist lower slopes and in riparian zones.

REMARKS: Mountain dogwood is shade tolerant at all stages. It sprouts vigorously following injury from fire or cutting. The wood of mountain dogwood has been used to make piano keys and golf club heads. Native Americans used it to make harpoons. The species' red leaves in the fall add a show of color among the dark conifers.

Figure 98
Red osier,
Cornus sericea

RED OSIER (Fig. 98) ***Cornus sericea***

DESCRIPTION: A multistemmed shrub that often ranges in height from 1.5 m (5 ft) to 4.5 m (15 ft) and can form dense thickets. The largest grows in White Bird, Idaho, and is 7.9 m (26 ft) tall and 10 cm (4 in.) in diameter. **LEAVES** are deciduous, opposite, simple, elliptical, and 4 cm (1.5 in.) to 9 cm (3.5 in.) long. Margins are entire and wavy. Upper surfaces are green and seemingly hairless and lower surfaces are pale green and hairy. Veins are prominent; lateral, evenly spaced veins curve toward the leaf tip near the margins. **INFLORESCENCES** are clustered and multibranched. **FLOWERS** appear as or after the leaves emerge. **FRUITS** are spherical, white or blue, and about 6 mm (.25 in.) in diameter. **TWIGS** are reddish and generally hairless.

HABITAT AND RANGE: Grows in forests and woodlands throughout much of North America. In California it is found in wet meadows, swamps, riparian zones, and moist forests below 2,700 m (9,000 ft).

REMARKS: Red osier grows best under partially closed canopies. Two subspecies of *C. sericea* are recognized: *C. s.* ssp. *sericea* is distinguished by its hairy leaves, and the ridges on the faces of its seeds; *C. s.* ssp.

occidentalis has hairless leaves, and the faces of its seeds are smooth. *C. stolonifera* is one of many names that have been applied to *C. sericea* in California manuals.

BLACKFRUIT DOGWOOD *Cornus sessilis*

DESCRIPTION: A spreading, multistemmed shrub that can range in height from 1.5 m (5 ft) to 3 m (10 ft). **LEAVES** are deciduous, opposite, simple, and 4.5 cm (1.75 in.) to 10 cm (4 in.) long, with a broad elliptical shape. Margins are entire and wavy. Upper surfaces are light green and hairless and lower surfaces are pale green and hairy. Veins are prominent, with lateral, evenly spaced veins that curve toward the leaf tip near the margins. **INFLORESCENSES** are unbranched, spherical clusters. Flower clusters have 4 ephemeral, brownish yellow bracts that are about 1 cm (.4 in.) long. **FLOWERS** appear before the leaves emerge. **FRUITS** are black, egg shaped, shiny, and about 1.2 cm (.5 in.) long. Fruit stalks are about 2 cm (.75 in.) long. **TWIGS** are gray or yellowish brown and hairless.

HABITAT AND RANGE: Grows in forests and woodlands, usually along streams or in other moist habitats. It is found in the foothills and mountains of northern California, at elevations from 150 m (500 ft) to 1,200 m (4,000 ft).

REMARKS: Blackfruit dogwood, with its many small, brownish yellow flowers, is easily overlooked as it blooms in the early spring. It has the typical dogwood leaf, but the fruits are very different from those of other dogwoods.

CORYLUS (HAZEL)

The genus *Corylus* has about 15 species of trees and shrubs in the Northern Hemisphere; 2 are native to the United States, and 1 of these is native to California.

CALIFORNIA HAZEL *Corylus cornuta* var. *californica*

(Fig. 99; Pl. 22)

DESCRIPTION: An erect, multistemmed (occasionally single-stemmed) shrub or small tree. Mature plants are typically

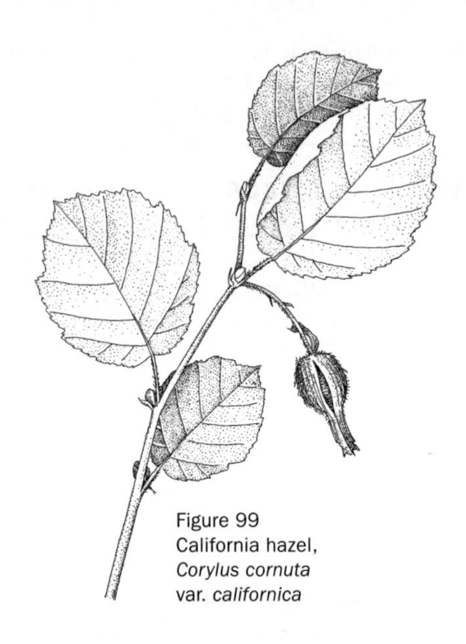

Figure 99
California hazel,
Corylus cornuta
var. *californica*

2 m (6.5 ft) to 3.5 m (12 ft) tall and 15 cm (6 in.) to 30 cm (12 in.) in diameter. The largest grows in Lincoln County, Oregon, and is 15.2 m (50 ft) tall and 53 cm (21 in.) in diameter. Crowns are rounded, with numerous, ascending branches. **LEAVES** are deciduous, simple, alternate, egg shaped to round, soft, hairy, 5 cm (2 in.) to 10 cm (4 in.) long, and 4 cm (1.5 in.) to 7.5 cm (3 in.) wide. Margins are doubly serrated. Upper surfaces are dark green and lower surfaces are light green. **INFLORESCENCES** are unisexual. Pollen-forming **FLOWERS** are droopy, hairy catkins measuring 4 cm (1.5 in.) to 7 cm (2.75 in.) long that appear before the leaves in the spring. Seed-forming flowers arise among a cluster of small, scaly buds and have red stigmas. **FRUITS** are spherical nuts surrounded by papery husks. **TWIGS** are hairy and zigzagged. **BARK** is smooth.

HABITAT AND RANGE: Grows in forests from Del Norte County along the coast to the Santa Cruz Mountains. Inland, it is found in the Klamath Mountains, Cascades, and western Sierra Nevada. Outside of California it occurs in British Columbia, Washington, and Oregon. The species typically inhabits cool,

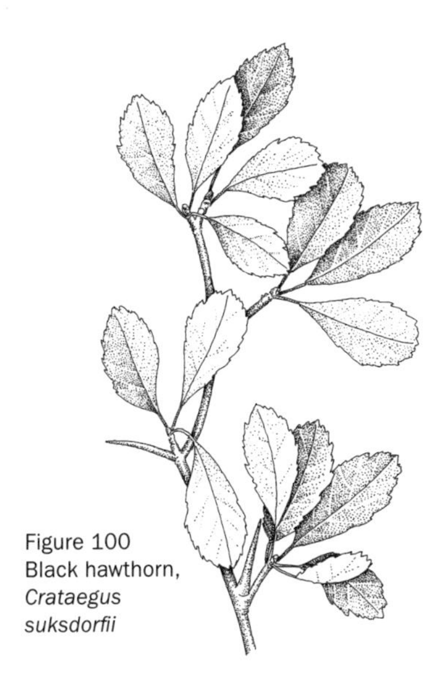

Figure 100
Black hawthorn,
Crataegus suksdorfii

moist, shady sites, often near streams. It ranges from sea level to 2,100 m (7,000 ft). The other hazel variety grows in the eastern United States.

REMARKS: This species can sprout from its root crown following fire or cutting. Wildlife species and livestock eat its foliage, and birds and small mammals devour the sweet-tasting nuts. It is a rare treat to find an intact nut in the fall. Native Americans used hazel twigs for basket making.

CRATAEGUS (HAWTHORN)

The genus *Crataegus* has about 280 species, which grow throughout the temperate Northern Hemisphere; of these, 2 are native to California.

BLACK HAWTHORN (Fig. 100) *Crataegus suksdorfii*

DESCRIPTION: A shrub or small tree that may grow to be 9 m (30 ft) tall. **LEAVES** are deciduous, simple, and alternate; they mostly occur on small lateral twigs. Leaf blades are round,

2.5 cm (1 in.) to 7.5 cm (3 in.) long, dark green, and hairless. Margins have teeth on the upper half of the blade only. Bases are wedge shaped. **INFLORESCENCES** consist of several flowers in flat clusters. **FLOWERS** are fragrant, white, and 12 mm (.5 in.) wide, with 5 petals. **FRUITS** are black pomes with 1 to 5 seeds. **TWIGS** may end in stout thorns.

HABITAT AND RANGE: Grows in forest edges from British Columbia south to the north coast of California, below 1,500 m (5,000 ft).

REMARKS: A second, less common hawthorn, which grows in Modoc County, is *C. douglasii;* it has leaves that are lobed above the middle and long thorns. Species of ornamental hawthorns (*C. erythropoda, C. monogyna*) have escaped from cultivation. The wood and thorns of hawthorns are known for their high strength. Many are planted as ornamentals for their flowers and fruits.

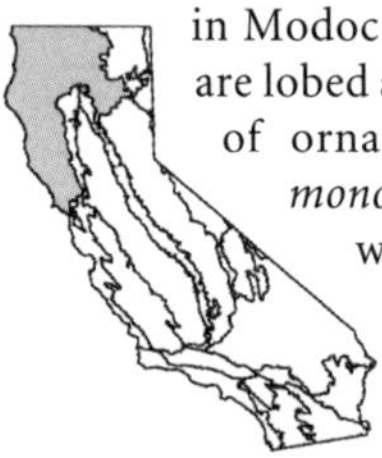

CYTISUS (BROOM)

The genus *Cytisus* has 33 species growing in Europe, western Asia, and northern Africa. Several of these species grown in California have escaped from cultivation and are noxious weeds.

SCOTCH BROOM (Fig. 101) ***Cytisus scoparius***

DESCRIPTION: A shrub with dark green, ribbed stems that may grow to be 3 m (10 ft) tall. **LEAVES** are alternate, deciduous, and inconspicuous. Blades on young stems are simple; those on older stems are compound, with 3 leaflets. Leaflets are 6 mm (.25 in.) to 20 mm (.75 in.) long. **INFLORESCENCES** are individual or a few flowers borne in the leaf axils. **FLOWERS** are about 6 mm (.25 in.) long and golden yellow, sometimes white, creamy yellow, or even yellow and maroon. **FRUITS** are brown or black legumes that are hairy along the margins and less than 25 mm (1 in.) long. **TWIGS** are 5-angled and dark green.

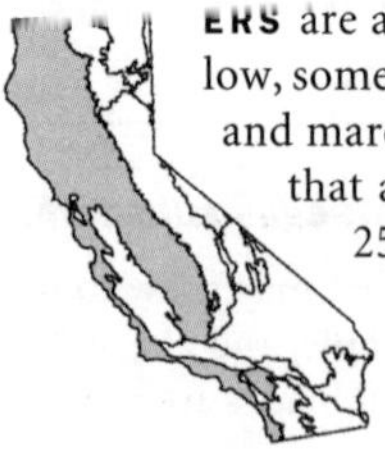

HABITAT AND RANGE: Common throughout California in disturbed areas

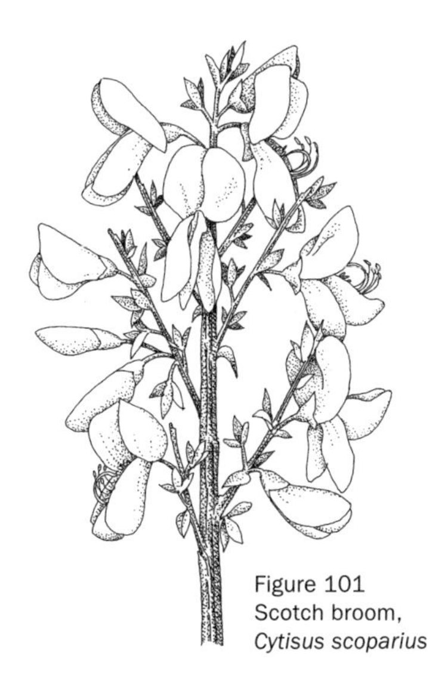

Figure 101
Scotch broom,
Cytisus scoparius

at elevations below 900 m (3,000 ft), except for the deserts. The species is native to the Mediterranean region of Europe.

REMARKS: Scotch broom is one of the most invasive of the introduced brooms; others include *C. multiflorus* and *C. striatus,* which are locally common along the central coast of California. See also descriptions of French broom (*Genista*), gorse (*Ulex*), and Spanish broom (*Spartium*). In time, Scotch broom creates a seed bank from which seedlings are recruited after a fire.

DENDROMECON (BUSH POPPY)

The genus *Dendromecon* has 2 species that grow in California and Baja California.

BUSH POPPY (Fig. 102) ***Dendromecon rigida***

DESCRIPTION: A shrub with showy yellow flowers. Shrubs may grow to be 3 m (10 ft) tall. **LEAVES** are alternate, simple, evergreen, and leathery. Blades are lance shaped, 2.5 cm (1 in.) to 10 cm (4 in.) long, and grayish green. Margins have fine teeth. **INFLORESCENCES** consist of individual flowers at the ends of

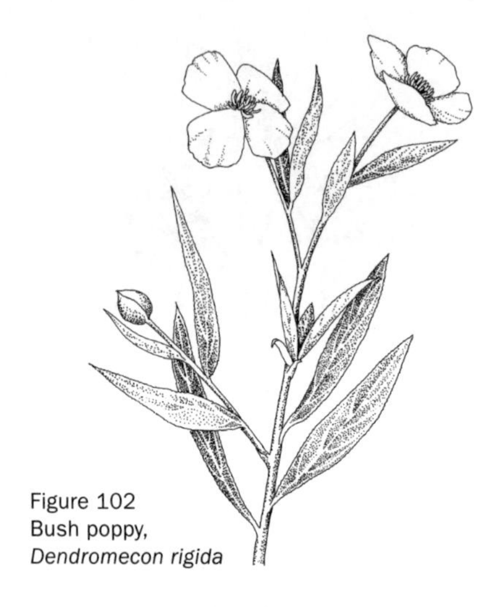

Figure 102
Bush poppy,
Dendromecon rigida

branches. **FLOWERS** are showy and measure 2 cm (.75 in.) to 6 cm (2.5 in.) broad. They resemble small California poppy flowers. The yellow petals surround many stamens. Flower stalks are 2.5 cm (1 in.) to 7.5 cm (3 in.) long. **FRUITS** are oblong capsules that open by 2 valves.

HABITAT AND RANGE: Grows in chaparrals and woodlands in the Coast Ranges and Sierra Nevada foothills, below 1,600 m (5,500 ft).

REMARKS: Another *Dendromecon* species, the Channel Islands tree poppy (*D. harfordii*), is an uncommon component of Channel Islands vegetation. It has leaves shorter and wider than those of *D. rigida*. Both have flowers that resemble California poppies.

EASTWOODIA (YELLOW-ASTER)

The monotypic genus *Eastwoodia* is a shrub with showy yellow heads of flowers. *Eastwoodia* is endemic to California.

YELLOW-ASTER *Eastwoodia elegans*

DESCRIPTION: An open-branched shrub that may grow to be 1m (3 ft). **LEAVES** are alternate, simple, drought-deciduous, and somewhat fleshy. Blades are linear and from 2 cm (.75 in.) to 4 cm (1.5 in.) long. **INFLORESCENCES** are open clusters of a few pincushion-like yellow flower heads. **STEMS** are marked with fine longitudinal lines. **BARK** is white and shreds with age. **FRUITS** are hairy achenes less than 2 mm (.1 in.) long.

HABITAT AND RANGE: Grows in the eastern foothills of the central Coast Ranges and the western foothills of the southern Sierra Nevada. It occurs below 1,250 m (4,000 ft). The species grows in grasslands and open woodlands.

REMARKS: Alice Eastwood, this plant's namesake, was curator and head of the Department of Botany at the California Academy of Sciences until her retirement in 1949, at the age of ninety. She was responsible for saving the academy's type collection (the plant specimens on which a binomial is based) after the San Francisco earthquake of 1906.

ENCELIA (ENCELIA)

The genus *Encelia* has 13 shrub species growing throughout the western parts of the Western Hemisphere. Five of these species grow in California.

Encelia has simple, drought-deciduous leaves that are alternate in arrangement. Flowers are set in showy yellow pincushion- or sunflower-like heads. Fruits are dry achenes.

1. Leaves are covered with curly hairs . . . . **brittlebush** (*E. farinosa*)
1. Leaves lack curly hairs . . . . **California brittlebush** (*E. californica*)

CALIFORNIA BRITTLEBUSH *Encelia californica*

DESCRIPTION: A mounded shrub with green leaves and showy yellow flower heads. Plants may grow to be 1.5 m (5 ft) tall. **LEAVES** are drought-deciduous, alternate, simple, and scattered along the stems. Blades are diamond shaped, from 3 cm (1.2 in.) to 6 cm (2.4 in.) long, and hairless. **INFLORESCENCES** are open sprays with sunflower-like heads with yellow flowers. **STEMS** are many, slender, and hairless. **FRUITS** are achenes about 6 mm (.25 in.) long.

HABITAT AND RANGE: Grows in the coastal parts of the cen-

Figure 103
Brittlebush,
Encelia farinosa

tral Coast Ranges, in southern California, and into Baja California. It occurs below 600 m (2,000 ft). The species grows in coastal scrubs, where it can dominate.

REMARKS: California brittlebush readily establishes in disturbed areas, making it a useful plant in ecological restoration. It is a popular ornamental in xeriscape gardening.

BRITTLEBUSH (Fig. 103) *Encelia farinosa*

DESCRIPTION: An open, mounded shrub with silver leaves and showy yellow flower heads. Plants may grow to be 1.5 m (5 ft) tall. **LEAVES** are drought-deciduous, alternate, simple, and crowded at the ends of stems. Blades have a broad lance shape, measure from 2 cm (.75 in.) to 8 cm (3 in.) long, and are covered with fine gray or white hairs. **INFLORESCENCES** are open

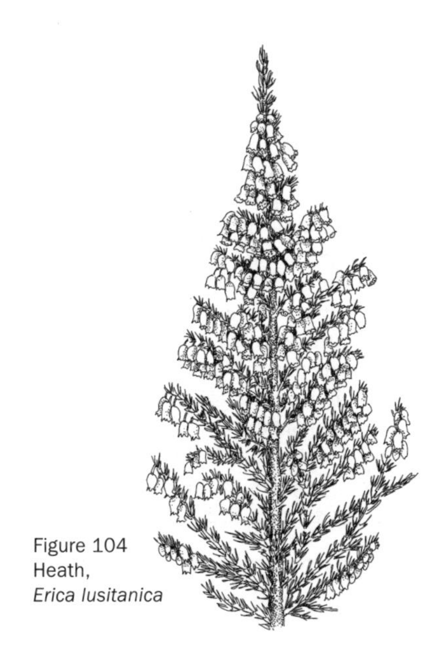

Figure 104
Heath,
Erica lusitanica

with yellow sunflower-like heads. The heads are elevated well above the leaves and are at the ends of golden stems. **STEMS** are many, fragrant, and covered with fine white hairs. **FRUITS** are achenes about 4 mm (.15 in.) long.

HABITAT AND RANGE: Grows in southern California, east to Utah, and south to northern Mexico. It occurs below 1,000 m (3,200 ft). The species grows in coastal and desert scrubs.

REMARKS: Brittlebush is also called incienso; resin that collects on the stems of the plants is burned as incense. The plant readily establishes itself in disturbed areas such as roadsides, which makes it useful in restoration. It is a popular ornamental in xeriscape gardening.

ERICA (HEATH)

Erica is a large African and European genus with more than 600 species.

HEATH (Fig. 104) *Erica lusitanica*

DESCRIPTION: An erect shrub with bright green, needlelike leaves. Shrubs may grow to be 2 m (6.5 ft) tall. **LEAVES** are alternate, simple, evergreen, and leathery. Blades are less than 12 mm (.5 in.) long. They appear needlelike because they are tightly rolled under, hiding the lower surfaces. **INFLORESCENCES** consist of single flowers arising from leaf axils. **FLOWERS** are 4-parted, urn shaped, and white to pink. **FRUITS** are spherical capsules.

HABITAT AND RANGE: An ornamental shrub that has escaped cultivation along the north coast of California. It is native to southwestern Europe.

REMARKS: Like the brooms, heath can create seed banks from which seedlings are recruited after disturbance.

ERICAMERIA (GOLDENBUSH)

The genus *Ericameria* has 23 species growing throughout western North America; 17 of these grow in California.

INTERIOR GOLDENBUSH (Fig. 105) *Ericameria linearifolia*

DESCRIPTION: A many-branched shrub that may grow to be 1.5 m (5 ft) tall. **LEAVES** are drought-deciduous and grayish green. Blades are linear and 1 cm (.4 in.) to 5 cm (2 in.) long. **INFLORESCENCES** consist of single daisylike heads. **FLOWERS** are showy, yellow, and 1 cm (.4 in.) long. **FRUITS** are achenes. **STEMS** are hairless or have small, fine hairs.

HABITAT AND RANGE: Grows throughout the Central Valley and surrounding foothills and deserts and the intermountain West. It occurs below 1,800 m (6,000 ft). The species grows in many kinds of chaparrals and scrubs and in forests and woodlands.

REMARKS: Interior goldenbush is one of several goldenbushes that can grow in woodlands, but many species are more restricted to specific habitats such as rock outcrops, dunes, and chaparrals. Some

Figure 105
Interior goldenbush,
Ericameria linearifolia

are abundant after fire. Turpentine-brush (*E. laricifolia*) is another wide-ranging species and is aromatic.

ERIODICTYON (YERBA SANTA)

The genus *Eriodictyon* is made up of 9 species of aromatic shrubs that grow in the southwestern United States and Mexico.

MOUNTAIN BALM *Eriodictyon californicum*

(Fig. 106)

DESCRIPTION: An open, coarse shrub with strongly scented, sticky leaves; it grows to be 3 m (10 ft) tall. **LEAVES** are alternate, simple, evergreen, and somewhat leathery. Blades are elliptical and measure from 5 cm (2 in.) to 10 cm (4 in.) long. They are green on the upper surfaces and may have fine teeth on the margins. **INFLORESCENCES** consist of open, ter-

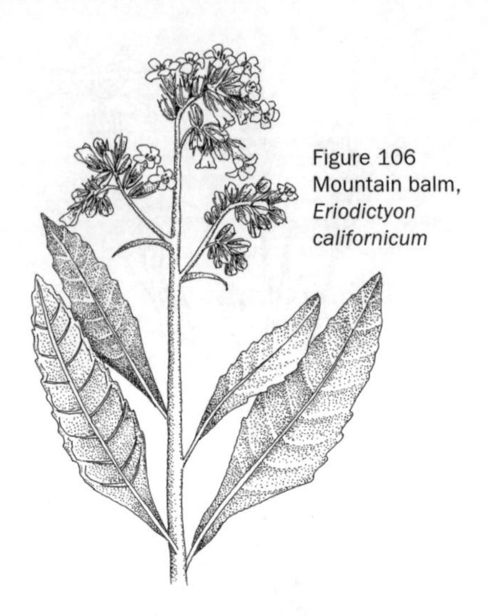

Figure 106
Mountain balm,
Eriodictyon californicum

minal clusters of flowers. **FLOWERS** are 5-parted, funnel shaped, and white to purple. **FRUITS** are capsules opening along 4 valves. **TWIGS** are sticky and hairless. Older stems have shredding bark.

HABITAT AND RANGE: Grows in woodlands and chaparrals of northern California mountains, below 1,800 m (6,000 ft). This species easily establishes itself on disturbed areas such as roadsides.

REMARKS: The common name, yerba santa, refers to all species of this genus. In southern California, 2 commonly encountered *Eriodictyon* species are the thickleaf yerba santa (*E. crassifolium*), with densely hairy upper blade surfaces, and hairy yerba santa (*E. trichocalyx*), with sparsely hairy upper blade surfaces.

ERIOGONUM (WILD BUCKWHEAT)

Eriogonum is the most diverse genus of plants in California. It is mostly composed of herbs and subshrubs; only 1 shrub species is common. The genus has about 250 species in North America, and half of these live in California.

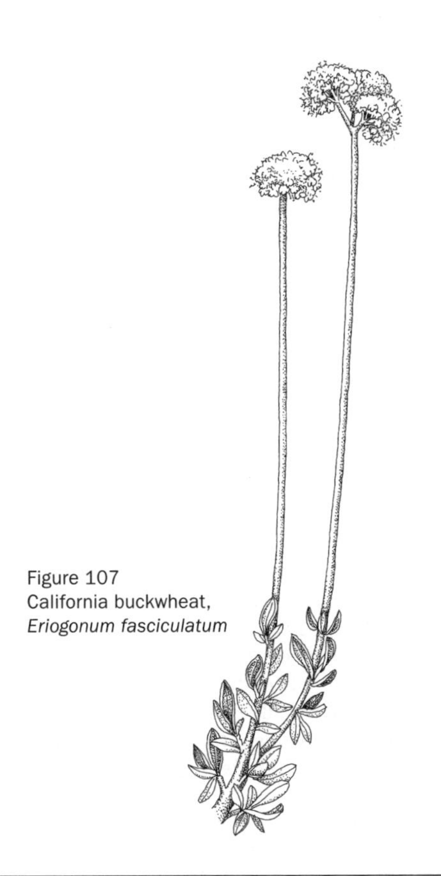

Figure 107
California buckwheat,
Eriogonum fasciculatum

CALIFORNIA BUCKWHEAT (Fig. 107) *Eriogonum fasciculatum*

DESCRIPTION: A many-stemmed, spreading shrub that can grow to be 1 m (3.3 ft) tall and 1.25 m (4 ft) broad. **LEAVES** are alternate, clustered at the nodes, and evergreen. Blades appear needlelike, are from 12 mm (.5 in.) to 25 mm (1 in.) long, and are tightly rolled under, hiding the lower surfaces. **INFLORESCENCES** consist of terminal clusters of flowers. **FLOWERS** are 6-parted, funnel shaped, and white to pink. **FRUITS** are achenes. **TWIGS** are covered with loose hairs. Older stems have shredding bark.

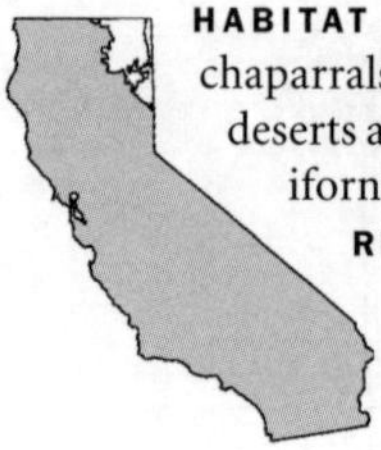

HABITAT AND RANGE: Grows in woodlands, chaparrals, and scrubs below 2,100 m (7,000 ft) in deserts and mountains of central and southern California.

REMARKS: Four varieties of California buckwheat are recognized; some are differentiated by range and habitat. This species is an important honey plant.

EUCALYPTUS (GUM TREE)

The genus *Eucalyptus* has about 500 species of trees and shrubs native to Australia. None are native to North America, but blue gum is a common component of California's landscape.

BLUE GUM (Fig. 108) ***Eucalyptus globulus***

DESCRIPTION: A non-native, erect, tall, single-stemmed tree. Mature trees are typically 30 m (100 ft) to 55 m (180 ft) tall and 60 cm (2 ft) to 120 cm (4 ft) in diameter. The largest in the United States lives in Fort Ross State Park and is 50.3 m (165 ft) tall and 3.4 m (135 in.) in diameter. Some California trees grow to be 80 m (260 ft) tall. Trees have long, straight trunks and narrow, rounded crowns. **LEAVES** are evergreen, simple, alternate, leathery, lance to sickle shaped, aromatic, and 10 cm (4 in.) to 33 cm (13 in.) long. Young leaves, and leaves from sprouts, are opposite and have a broad egg shape. **FLOWERS** are solitary in the axils of upper leaves. They are about 5 cm (2 in.) wide and have numerous showy, white to creamy stamens and 1 pistil. **FRUITS** are 4-parted, woody, warty capsules that are 2 cm (.75 in.) to 2.5 cm (1 in.) wide. **BARK** is deciduous and shreds in long strips, revealing smooth bluish gray, tan, and whitish inner bark.

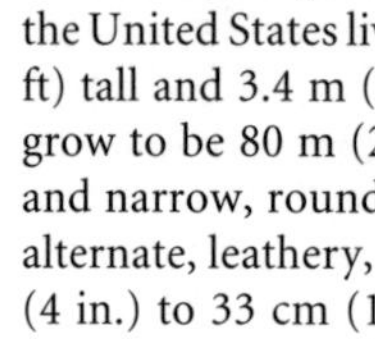

HABITAT AND RANGE: Native to southeastern Australia and Tasmania. The species is widely planted throughout the world. It was introduced to California in 1856, where it was planted along the coast, in coastal mountains, and in the Central Valley, below 300 m (1,000 ft) in elevation.

REMARKS: Blue gum is shade intolerant. In California it reproduces near es-

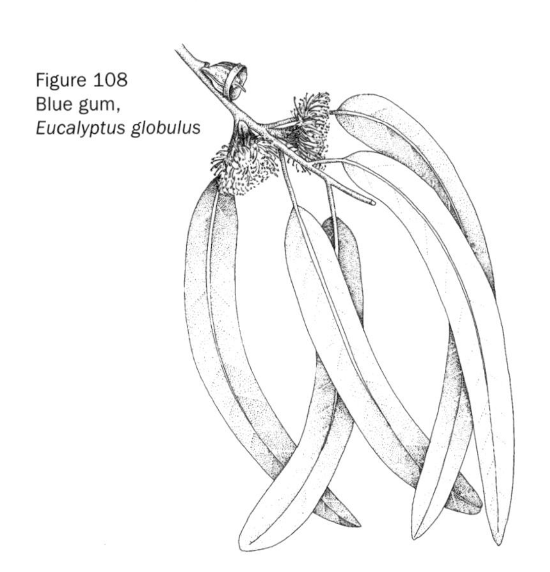
Figure 108
Blue gum,
Eucalyptus globulus

tablished stands but does not aggressively invade wildlands. Blue gum is fire adapted. Its bark, litter, and leaves all have volatile oils that exacerbate fire behavior. The species sprouts prolifically following injury from fire or cutting. It is used for firewood around the world, but it makes low-quality lumber because of warping and cracking during drying. In California, blue gum is planted as a windbreak, to block noise or views, or as an ornamental. Numerous pharmaceutical applications have been developed from blue gum oil, including cold and cough medicines. It is also used in disinfectants and ointments.

EUONYMUS (BURNING BUSH)

The genus *Euonymus* has about 180 species, with most occurring in Asia. One species is native to California.

WESTERN BURNING BUSH (Fig. 109) — ***Euonymus occidentalis***

DESCRIPTION: An open, often climbing shrub or a small tree. Shrubs may grow to be 6 m (20 ft) tall. **LEAVES** are op-

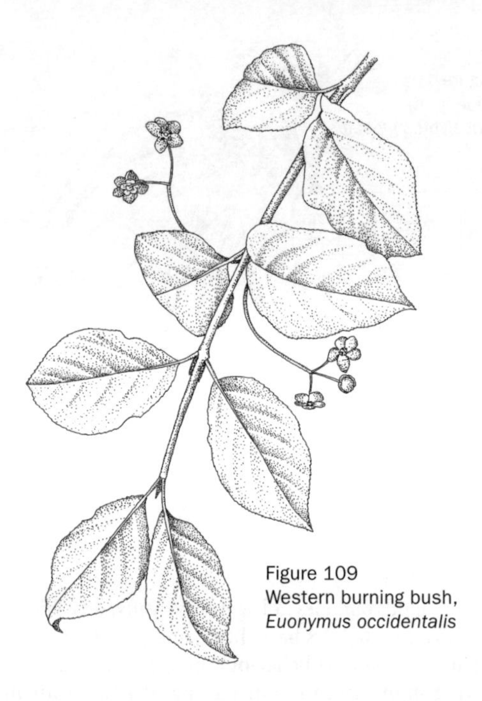

Figure 109
Western burning bush,
Euonymus occidentalis

posite, simple, deciduous, and thin. Blades are elliptical, from 3 cm (1.25 in.) to 10 cm (4 in.) long, and green, with fine teeth on the margins. **INFLORESCENCES** consist of open, axillary clusters of 1 to 5 flowers arising on slender stalks that measure 2 cm (.75 in.) to 7.5 cm (3 in.) long. **FLOWERS** are 5-parted, less than 12 mm (.5 in.) across, saucer shaped, and maroon with white margins **FRUITS** are 3-lobed capsules that open by valves. The seeds are surrounded by a red, fleshy aril, and they hang from the open capsule valves. **TWIGS** are slender and 4-angled.

HABITAT AND RANGE: Grows in shaded coastal coniferous forests of northern California and the Peninsular Ranges of southern California, below 2,100 m (7,000 ft). This shrub occurs as an individual or as populations of a few individuals.

REMARKS: The northern California populations have red twigs and leaves with pointed tips (*E. o.* var. *occidentalis*). The southern California populations have white twigs and leaves with round tips (*E. o.* var. *parishii*). Some *Euonymus* plants turn bright crimson-pink in the autumn before dropping their leaves, hence the common name burning bush. Members of the genus are commonly cultivated.

FRAXINUS (ASH)

The genus *Fraxinus* has about 70 species; 16 are native to North America, and 4 of these are native to California. Most species occur in temperate habitats in the Northern Hemisphere, although some are found in the tropics of the Southern Hemisphere. Most species are trees.

Leaves are deciduous, opposite, usually thin and hairless, and pinnately compound (rarely, they have only 1 leaflet). Leaflet margins are entire or serrated. Flowers are inconspicuous and clustered and may have petals. Pollen-bearing and seed-bearing flowers are found on separate plants. Fruits are single samaras with an enlarged papery wing. Twigs are slender to stout and flat at nodes. Bark is smooth to furrowed and generally gray.

Ashes typically grow in moist areas, especially along streams. Some species, though, grow in dry upland sites. Ash is famous for its hard, durable wood. It is used as lumber and for making veneer, tool handles, baseball bats, and novelty items. Seeds are eaten by wildlife.

1. Most leaves are simple; some may be compound, with 2 or 3 leaflets . . . . **singleleaf ash (*F. anomala*)**
1. Leaves are compound, with 3 to 9 leaflets 2
    2. Leaflets have either no stalks or very short ones. The mature plant is a medium-sized tree around 23 m (75 ft) tall . . . . **Oregon ash (*F. latifolia*)**
    2. Leaflets have evident stalks. The mature plant is a large shrub or small tree less than 11 m (35 ft) tall 3
3. Twigs are 4-angled in cross section . . . . **foothill ash (*F. dipetala*)**
3. Twigs are round in cross section . . . . **velvet ash (*F. velutina*)**

SINGLELEAF ASH *Fraxinus anomala*

DESCRIPTION: A single- or multistemmed large shrub or small tree. Mature plants are typically 2 m (6.5 ft) to 6 m (20 ft) tall and 5 cm (2 in.) to 10 cm (4 in.) in diameter. The largest grows in Mesa County, Colorado, and is 7.3 m (24 ft) tall and 15 cm (6 in.) in diameter. Crowns are rounded. **LEAVES** are deciduous, opposite, and 2.5 cm (1 in.) to 5 cm (2 in.) long. Most leaves are simple, but some are pinnately compound, with 2 or 3 leaflets. Leaflets are egg shaped to round. Leaflet margins are serrated, scalloped, or nearly entire. Leaflet tips have a broad, tapered to round shape, and bases are tapered to horizontal. Upper leaflet surfaces are dark green and hairless, while lower surfaces are light green and hairless. Leaflets are stalked. **FRUITS** are single samaras about 2 cm (.75 in.) long. **TWIGS** are 4-angled in cross section, hairy, and glandular. **BARK** is gray.

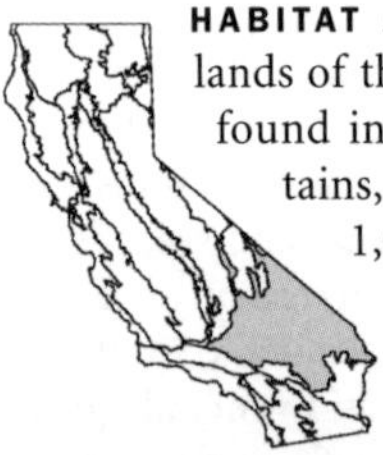

HABITAT AND RANGE: Grows in riparian woodlands of the American Southwest. In California it is found in the northern and eastern desert mountains, along streams in canyons and gulches from 1,000 m (3,500 ft) to 2,400 m (8,000 ft).

REMARKS: Singleleaf ash provides needed cover for wildlife.

FOOTHILL ASH (Fig. 110) *Fraxinus dipetala*

DESCRIPTION: An erect, single-stemmed large shrub or small tree. Mature plants are usually 2 m (6.5 ft) to 5.5 m (18 ft) tall and 10 cm (4 in.) to 20 cm (8 in.) in diameter. The largest grows in Lake County and is 10.4 m (34 ft) tall and 28 cm (11 in.) in diameter. **LEAVES** are deciduous, opposite, pinnately compound (usually 3 to 7 leaflets), and 5 cm (2 in.) to 15 cm (6 in.) long. Leaflets are egg shaped to round and 1.2 cm (.5 in.) to 4 cm (1.5 in.) long. Leaflet margins are generally serrated, sometimes entire. Leaflet tips and bases are round to tapered. Upper leaflet surfaces are dark green and hairless and lower surfaces are light green and hairless. Lateral leaflets have short stalks. The terminal leaflet has a longer stalk. **INFLORESCENCES** form large, showy white clusters that measure 3 cm (1.2 in.) to 12 cm (4.75 in.) long. **FLOWERS** have 2 petals, each about 6 mm (.25 in.) long. **FRUITS** are single samaras about 2.5 cm (1 in.) long. **TWIGS** are 4-angled in cross section (sometimes round). **BARK** is gray.

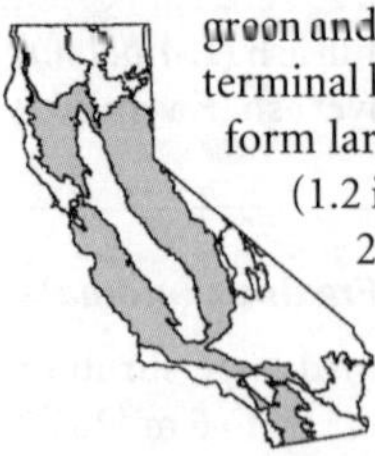

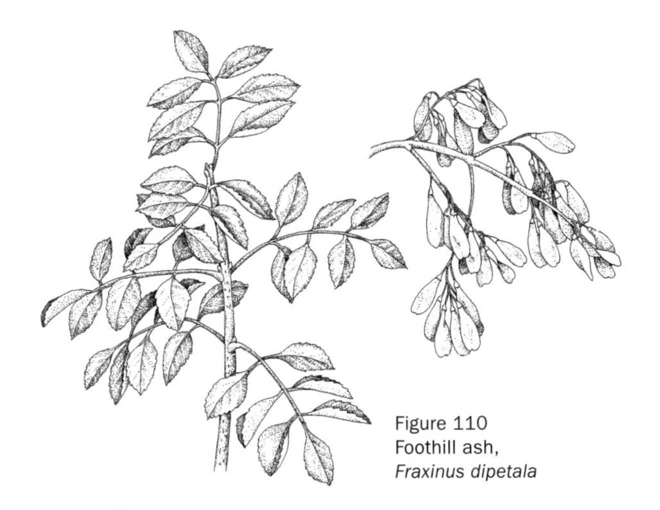
Figure 110
Foothill ash,
Fraxinus dipetala

HABITAT AND RANGE: Grows in chaparrals and woodlands. It occurs primarily in central and southwestern California, but it is also found in Baja California. It ranges from 90 m (300 ft) to 1,200 m (4,000 ft).
REMARKS: Foothill ash produces showy, ornamental flowers from March to June.

OREGON ASH *Fraxinus latifolia*

DESCRIPTION: A medium-sized, single-stemmed tree. Mature trees are typically 11 m (35 ft) to 24 m (80 ft) tall and 30 cm (1 ft) to 60 cm (2 ft) in diameter. The largest grows on Sauvie Island, Oregon, and is 18 m (59 ft) tall and 2.1 m (84 in.) in diameter. Crowns are narrow to broad, depending on tree density. Trees live to be about 250 years old. **LEAVES** are deciduous, opposite, pinnately compound (usually 5 to 7 leaflets), and 12.5 cm (5 in.) to 35 cm (14 in.) long. Leaflets are elliptical to egg shaped and 7.5 cm (3 in.) to 18 cm (7 in.) long. Leaflet margins are entire to serrated. Leaflet tips and bases are tapered. Upper leaflet surfaces are light green and generally hairless and lower surfaces are pale green and hairy. Lateral leaflets have either no stalk or a very short one. The terminal leaflet is stalked. **FRUITS** are single samaras about 4 cm (1.5 in) long. **TWIGS** are stout, and round in cross section between nodes but flat at the nodes. **BARK** is up to 4 cm (1.5 in.) thick, furrowed, and gray. **HABITAT AND RANGE:** Grows in broadleaved and riparian woodlands from British Columbia to central California. In California it usually grows along streams, seeps, and wet meadows at elevations below 1,700 m (5,500 ft).

REMARKS: Oregon ash is intermediate in its shade tolerance. It is used as firewood and to make tool handles, furniture, and barrels. In Oregon, Native Americans believed that poisonous snakes avoided Oregon ash stands. *F. velutina* and *F. latifolia* have similar characteristics where their ranges overlap south of the Kings River.

VELVET ASH (Fig. 111) *Fraxinus velutina*

DESCRIPTION: A small, single-stemmed tree. Mature trees are typically 4.5 m (15 ft) to 9 m (30 ft) tall and 15 cm (6 in.) to 30 cm (12 in.) in diameter. The largest grows in Santa Cruz City, Arizona, and is 23.2 m (76 ft) tall and 1.6 m (62 in.) in diameter. Crowns are rounded, with spreading branches. **LEAVES** are deciduous, opposite, pinnately compound (usually 3 to 5 leaflets), and 10 cm (4 in.) to 15 cm (6 in.) long. Leaflets are elliptical to egg shaped and 2.5 cm (1 in.) to 4 cm (1.5 in.) long. Leaflet margins are entire to scalloped. Leaflet tips are pointed and bases tapered. Upper leaflet surfaces are light green and hairless, while lower surfaces are pale green and densely hairy. Lateral leaflets have short stalks. The terminal leaflet has a longer stalk. **FRUITS** are single samaras. **TWIGS** are slender, and round in cross section between nodes but flat at the nodes. **BARK** is thin, gray, and furrowed, with furrows separated by flat ridges.

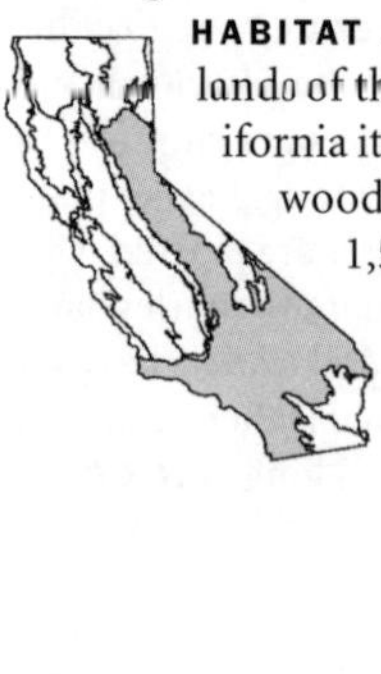

HABITAT AND RANGE: Grows in riparian woodlands of the American Southwest and Mexico. In California it occurs along streams in desert canyons and woodlands. It ranges from 200 m (650 ft) to 1,500 m (5,000 ft).

REMARKS: Velvet ash produces abundant shade, providing welcome relief on hot days. It is planted as a shade

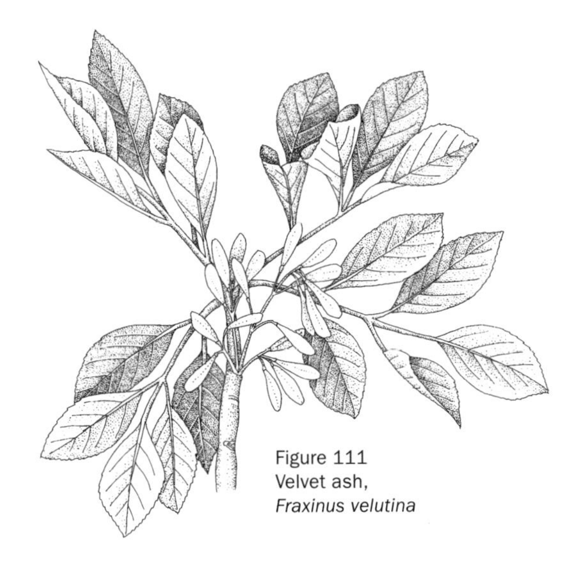

Figure 111
Velvet ash,
Fraxinus velutina

tree in Arizona, southern California, and Mexico. *F. velutina* and *F. latifolia* have similar characteristics where their ranges overlap south of the Kings River.

FREMONTODENDRON (FLANNELBUSH)

The genus *Fremontodendron* has 2 species of shrubs and small trees, which are limited to California, Arizona, and Mexico. Both species are native to California.

FLANNELBUSH *Fremontodendron californicum*

(Fig. 112)

DESCRIPTION: A low to erect shrub or small, single-stemmed tree. As a shrub it typically ranges in height from 30 cm (1 ft) to 3.5 m (12 ft), and as a tree it can grow to be 9 m (30 ft) tall. The largest lives in North Fork and is 8 m (26 ft) tall and 33 cm (13 in.) in diameter. Crowns are rounded, with long, open branches that remain close to the ground. **LEAVES** are evergreen, alternate, simple, leathery, round to egg shaped, and 1 cm

Figure 112
Flannelbush,
Fremontodendron californicum

(.4 in.) to 5 cm (2 in.) long. Margins are entire and usually pinnately or palmately 3-lobed (sometimes 5- or 7-lobed). Leaf tips are round to pointed. Upper surfaces are dark green and somewhat hairy, while lower surfaces are paler and densely hairy. Leaves often occur on spur shoots. **FLOWERS** are showy and usually solitary, and they have 5 petal-like yellow sepals measuring 2.5 cm (1 in.) to 4 cm (1.5 in.) across. Petals are absent. Flowers often occur on spur shoots. **FRUITS** are hairy, somewhat spherical capsules that measure 12 mm (.5 in.) to 40 mm (1.5 in.) long. **BRANCHES** are hairy. **BARK** is fissured. The inner bark is gelatinous.

HABITAT AND RANGE: Grows in chaparrals and woodlands throughout California, except in the deserts, and into Arizona and Baja California, at elevations from 400 m (1,300 ft) to 2,000 m (6,500 ft).

REMARKS: Flannelbush sprouts vigorously following fire or other injury. Deer find its foliage very palatable. The species is planted as an ornamental. Two subspecies of *F. californicum* are recognized: *F. c.* ssp. *californicum,* an erect shrub or tree that is widely distributed in California; and *F. c.* ssp. *decum-*

bens, characterized by its nearly prostrate habit, a rare subspecies found only on Pine Hill in El Dorado County. A second *Fremontodendron* species, *F. mexicanum,* also rare in California, is found in Orange, San Diego, and Imperial Counties, as well as Baja California.

GARRYA (SILK TASSEL)

The genus *Garrya* has 14 species of shrubs and small trees growing in the western United States, Central America, and the Caribbean. Six of these are native to California.

Leaves are evergreen, leathery, and opposite in arrangement. Plants have seed-bearing or pollen-bearing flowers on separate plants. Flowers are arranged in droopy or upturned catkins that may be showy. Male flowers contain 4 stamens that protrude from fused bracts arranged in dangling catkins. Female flowers contain 1 pistil with 2 styles that extend beyond fused bracts. Berries are initially fleshy; later the outer part may become dry and possibly separate from the juicy pulp, which surrounds 2 hard seeds. The bark, leaves, and fruits contain an alkaloid, garryine, which is used as a tonic.

In California, coastal silk tassel and Fremont silk tassel are the more common examples of this genus in forests and woodlands. *G. buxifolia* grows on serpentine soils in the western Klamath Mountains. *G. congdonii, G. flavescens,* and *G. veatchii* are characteristic of low-elevation chaparrals and desert scrubs.

1. Leaf margins are undulating, and lower blade surfaces have feltlike hairs . . . . **coastal silk tassel** (*G. elliptica*)
1. Leaf margins are not wavy, and lower blade surfaces lack feltlike hairs . . . . **Fremont silk tassel** (*G. fremontii*)

COASTAL SILK TASSEL (Fig. 113) ***Garrya elliptica***

DESCRIPTION: A shrub or small tree that can grow to be 9 m (30 ft) tall. The largest grows in Brookings, Oregon, and is 8.8 m (29 ft) tall and 23 cm (9 in.) in diameter. **LEAVES** are opposite, simple, evergreen, and leathery. Blades are elliptical to oval in shape and 4 cm (1.5 in.) to 6 cm (2.5 in.) long. Upper leaf surfaces are dark green and hairless, while lower leaf sur-

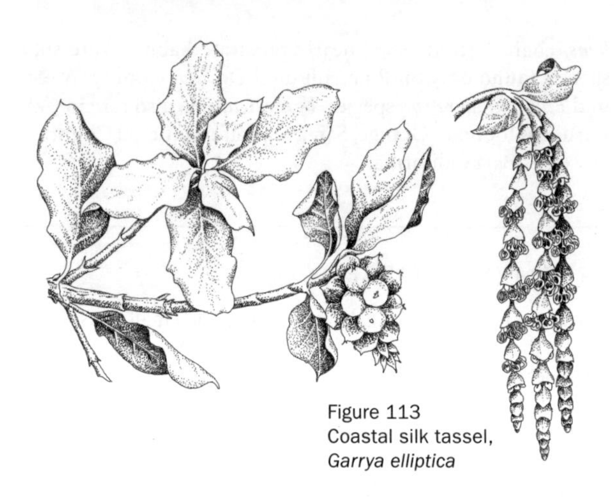

Figure 113
Coastal silk tassel,
Garrya elliptica

faces are paler and feltlike, with short curly or wavy hairs. Most leaf margins are turned under and undulating. **FLOWERS** are arranged in catkins. Pollen-bearing catkins are 7.5 cm (3 in.) to 20 cm (8 in.) long. Seed-bearing catkins are 5 cm (2 in.) to 9 cm (3.5 in.) long. Plants bloom from January to February. **FRUITS** are black or purple berries and are densely covered with white hairs. Pulp is very juicy when ripe.

HABITAT AND RANGE: Most commonly found scattered in open forests or coastal scrubs and chaparrals. On a few coastal bluffs it is the dominant plant. It occurs in coastal Oregon and California, where it grows below 600 m (2,000 ft).

REMARKS: Coastal silk tassel is the tallest of *Garrya* species, often developing into a small tree. It is cultivated for its winter-blooming, pollen-bearing catkins and dark grayish green leaves.

FREMONT SILK TASSEL (Pl. 23) *Garrya fremontii*

DESCRIPTION: A shrub that grows to be 3 m (10 ft) tall. **LEAVES** are opposite, simple, evergreen, and leathery. Blades are elliptical to oval in shape and 2 cm (.75 in.) to 6 cm (2.5 in.) long. Blades are hairless, are shiny above, and usually become yellowish green with age. Leaf margins are not wavy. **FLOWERS**

are arranged in catkins. Pollen-bearing catkins are 7.5 cm (3 in.) to 20 cm (8 in.) long. Seed-bearing catkins are 4 cm (1.5 in.) to 5 cm (2 in.) long. Plants bloom from January to April. **FRUITS** are hairless purple or black berries.

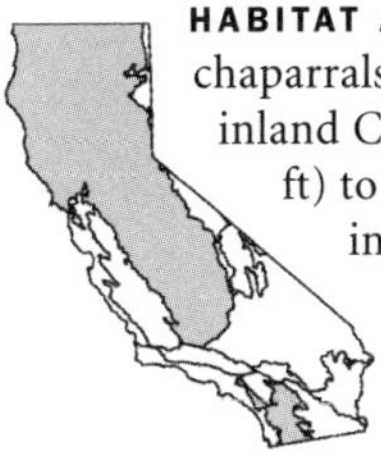

HABITAT AND RANGE: Most commonly grows in chaparrals and associated woodlands and forests in inland California, at elevations from 600 m (2,000 ft) to 2,100 m (7,000 ft). It also grows in Washington and Oregon.

REMARKS: Fremont silk tassel is the most desirable of the silk tassels for landscape use.

GAULTHERIA (WINTERGREEN)

The genus *Gaultheria* has about 200 shrub species growing in South America, North America, western Asia, and Australia. Three of these species are native to California, of which salal is commonly encountered.

SALAL (Fig. 114) ***Gaultheria shallon***

DESCRIPTION: A shrub with large showy leaves that grows to be 2 m (6.5 ft) tall. The species spreads from underground rhizomes. **LEAVES** are alternate, simple, evergreen, and leathery. Blades are 4 cm (1.5 in.) to 10 cm (4 in.) long and hairless; they have a broad egg shape, with a heart-shaped base and acute tip. **INFLORESCENCES** consist of simple flower clusters that measure 7.5 cm (3 in.) to 15 cm (6 in.) long. Conspicuous reddish bracts occur along the red stems. **FLOWERS** are 5-parted and about 1 cm (.4 in.) long. Sepals are red, and petals form a white to pink urn. **FRUITS** are spherical capsules covered with a fleshy calyx at maturity. The apparent "berry" is purple or black. **TWIGS** zigzag between the leaves.

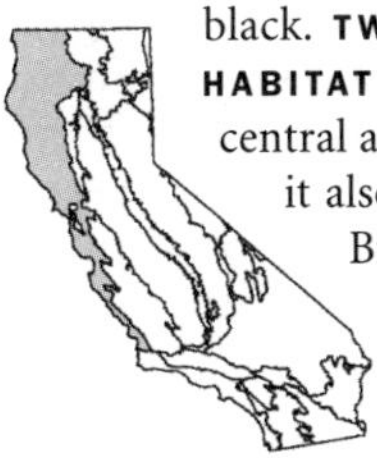

HABITAT AND RANGE: Often associated with the central and northern coastal coniferous forests, but it also grows in coastal scrubs and forests from British Columbia south to California.

REMARKS: The fruit when ripe is readily eaten by birds. The fleshy calyx and the capsule are the unit of disper-

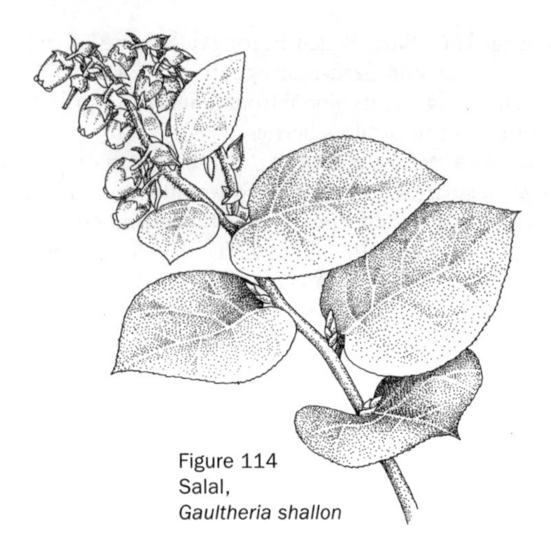

Figure 114
Salal,
Gaultheria shallon

sal. Two other members of *Gaultheria* grow in the mountains of northern California. The small, prostrate, alpine wintergreen (*G. humifusa*) can be found in wet sites near subalpine lakes in the Klamath Mountains and Sierra Nevada. The slightly larger Oregon wintergreen (*G. ovatifolia*) grows in the Klamath Mountains at montane elevations.

GENISTA (FRENCH BROOM)

The genus *Genista* is made up of 87 species from Europe, western Asia, and northern Africa. In California some species have escaped from cultivation and are noxious weeds.

FRENCH BROOM (Fig. 115) *Genista monspessulana*

DESCRIPTION: A shrub with dark green, ribbed or angled stems that may grow to be 3 m (10 ft) tall. **LEAVES** are alternate, deciduous, and conspicuous. Blades are compound and composed of 3 leaflets. Each leaflet is 12 mm (.5 in.) long. **INFLORESCENCES** are 2 to 9 flowers in tight clusters that are borne in leaf axils. **FLOWERS** are bright yellow. **FRUITS** are hairy brown legumes less than 2 cm (.75 in.) long. **TWIGS** are roundish, 5-angled, and dark green. Mature twigs have silky, silvery hairs.

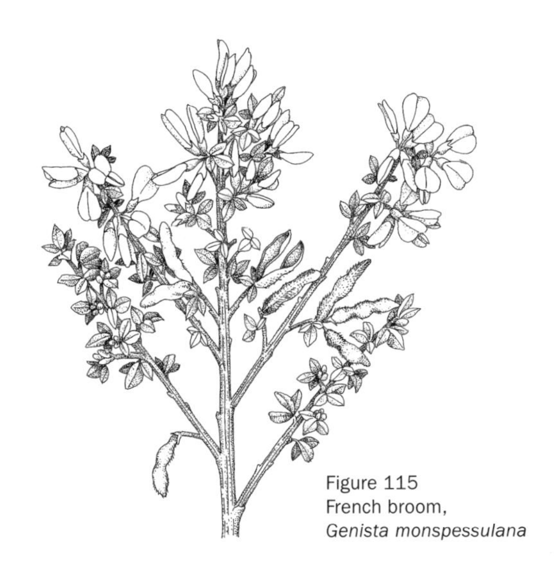

Figure 115
French broom,
Genista monspessulana

HABITAT AND RANGE: Common in disturbed areas below 450 m (1,500 ft) along California's coastline and the Sierra Nevada. It is native to the Mediterranean region in Europe.

REMARKS: French broom is one of the most invasive of the introduced brooms; others include *G. canariensis, G. linifolia, G. maderensis,* and *G. stenopetala,* which are locally common in the state. Also, see descriptions of gorse, Scotch broom, and Spanish broom. Through time French broom creates a seed bank from which seedlings are recruited after fire.

GRAYIA (HOP-SAGE)

Grayia is a genus with only 1 species that occurs in California's deserts and the intermountain West.

HOP-SAGE (Fig. 116) ***Grayia spinosa***

DESCRIPTION: A rigidly branched shrub that may grow to be 1 m (3.3 ft) tall. **LEAVES** are alternate, evergreen, fleshy, and 6 mm (.25 in.) to 40 mm (1.75 in.) long. Blades are elliptical,

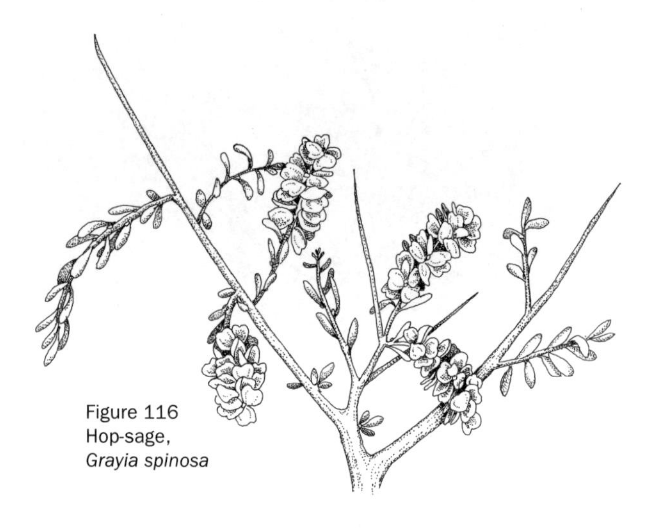

Figure 116
Hop-sage,
Grayia spinosa

flat, and entire; they may have soft, branching hairs. **INFLORESCENCES** are 2 separate pollen-bearing and seed-bearing terminal spikes that measure 6 mm (.25 in.) to 20 mm (.75 in.) long and are most often found on separate plants. **FLOWERS** are small, drab, and yellow or brown. **FRUITS** are utricles. Fused white or red-tinged, leaflike bracts surround the fruits. **TWIGS** are straw colored to gray, stiff, and spinelike on older plants.

HABITAT AND RANGE: In California the species grows in the interior foothills of the San Joaquin Valley and the state's deserts, at elevations between 500 m (1,600 ft) and 2,700 m (9,000 ft). It can mix with members of the interior coniferous woodlands and desert scrubs.

REMARKS: Hop-sage is browsed by sheep and cattle. The generic name honors Asa Gray, a Harvard University professor who described many California plants.

HETEROMELES (TOYON)

The genus *Heteromeles* has 1 species that grows in California and Baja California.

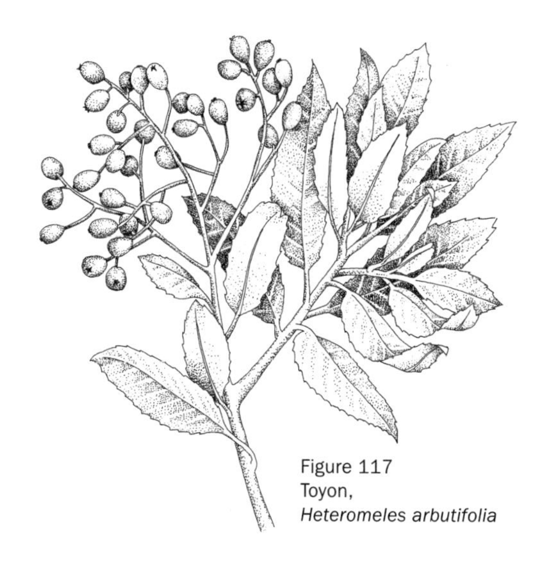

Figure 117
Toyon,
Heteromeles arbutifolia

TOYON (Fig. 117) ***Heteromeles arbutifolia***

DESCRIPTION: A shrub or small tree that typically grows to be 4.5 m (15 ft) tall. The largest lives in Santa Barbara County and is 9.7 m (32 ft) tall and 45 cm (18 in.) in diameter. **LEAVES** are alternate, evergreen, and leathery. Blades are glossy green, elliptical, and 5 cm (2 in.) to 10 cm (4 in.) long, with regularly toothed margins. Petioles are 12 mm (.5 in.) to 25 mm (1 in.) long. **INFLORESCENCES** are terminal, flat-topped clusters of many flowers. **FLOWERS** are white. **FRUITS** are pomes that are red or sometimes yellow. Fruits are around 6 mm (.25 in.) in diameter. **TWIGS** are gray with fine hairs.

HABITAT AND RANGE: A common shrub in the state's chaparrals and woodlands. It occurs in the foothills surrounding the Central Valley and throughout California, except for the deserts, at elevations below 1,200 m (4,000 ft).

REMARKS: Toyon is one of California's exceptionally attractive shrubs. The dark green foliage and red fruits are resplendent in the winter.

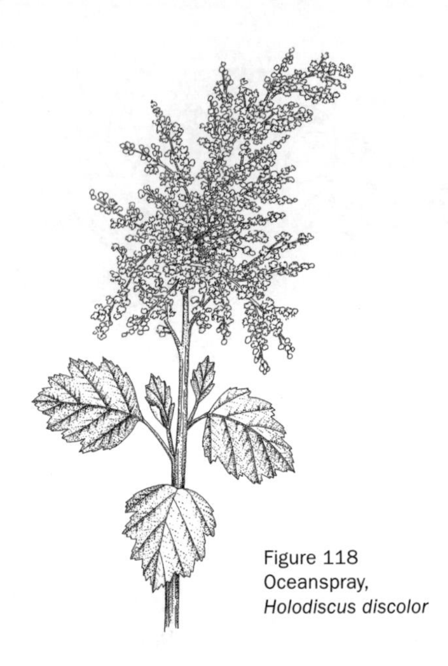

Figure 118
Oceanspray,
Holodiscus discolor

HOLODISCUS (CREAM BUSH)

Holodiscus is a genus of about 5 highly variable species that grow in western North, Central, and South America. Two of these are native to California.

1. Leaf blades have teeth on the side and top margins. Leaf bases are broad . . . . **oceanspray** (*H. discolor*)
1. Leaf blades have teeth restricted to upper margins. Leaf bases are wedge shaped . . . **rock-spiraea** (*H. microphyllus*)

OCEANSPRAY (Fig. 118) ***Holodiscus discolor***

DESCRIPTION: A shrub that may grow to be between 1.5 m (5 ft) and 6 m (20 ft) tall. **LEAVES** are alternate, deciduous, and thin, and they occur on short lateral twigs of older stems. Blades are dull green, hairy, egg shaped, and 12 mm (.5 in.) to 7.5 cm (3 in.) long. Teeth are found on both the side and top margins. Prominent parallel veins resemble chevrons. Bases are broad to flat. The amount and color of hair, especially on the lower blade surfaces, varies greatly by region. **INFLORESCENCES** are terminal and conical and located well beyond the foliage. They measure 10 cm (4 in.) to 25 cm (10 in.) long and are very showy when flowers are in bloom. Infructescences are persistent. **FLOWERS** are white. **FRUITS** are small brown achenes. **TWIGS** are brown and hairy. Older **BRANCHES** have thin sheets of bark peeling off.

HABITAT AND RANGE: Grows on cliffs overlooking the ocean and throughout northern California's forests and woodlands below 1,800 m (6,000 ft), and in southern California's mountains from 300 to 1,300 m (1,000 to 4,200 ft). It is also found throughout the western United States.

REMARKS: Oceanspray's leaf characteristics are quite variable. Some botanists consider oceanspray and rock-spiraea to be members of a single species. Others not only differentiate between the 2 species but also distinguish several varieties of oceanspray. The species' straight branches have been used to make arrow shafts.

ROCK-SPIRAEA *Holodiscus microphyllus*

DESCRIPTION: A shrub that may grow to be 1 m (3.3 ft) tall. **LEAVES** are alternate, deciduous, and thin. They occur on long lateral twigs on older branches. Blades are dull, hairy, round, and 12 mm (.5 in.) to 4 cm (1.5 in.) long. Bases are wedge shaped. Margins above the middle of the blade are toothed. **INFLORESCENCES** are terminal and conical and measure 2 cm (.75 in.) to 8 cm (3 in.) long. They are often mixed with leaves and are very showy when flowers are in bloom. Infructescences are persistent. **FLOWERS** are white or creamy. **FRUITS** are small brown achenes. **TWIGS** are gray and hairy. Older **BRANCHES** have thin sheets of bark peeling off.

HABITAT AND RANGE: Grows in California's open areas in forests and woodlands, and in dry, rocky alpine habitats, ranging from 600 m (2,000 ft) to 4,000 m (13,000 ft). It is found throughout the western United States.

REMARKS: Rock-spiraea has especially variable leaf characteristics. *H. m.*

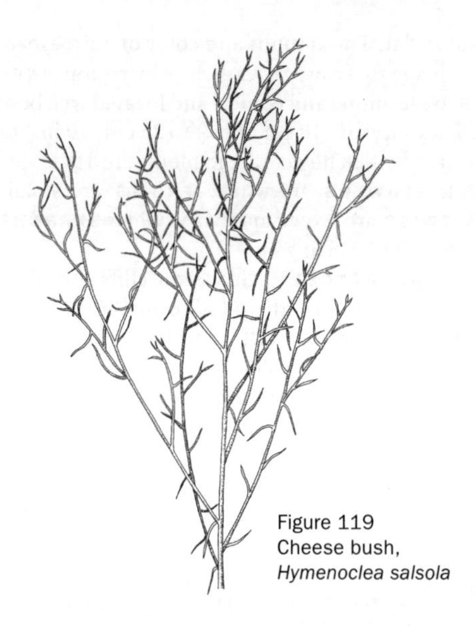

Figure 119
Cheese bush,
Hymenoclea salsola

var. *glabrescens,* a plant that is almost hairless or has fine hairs, occurs in northern coastal California mountains. *H. m.* var. *microphyllus,* found in the more inland mountains, has distinctively hairy leaves. Some botanists consider rock-spiraea to be a form of *H. dumosus,* which grows in the intermountain West.

HYMENOCLEA (CHEESE BUSH)

The genus *Hymenoclea* has 2 shrub species that grow in the southwestern United States and northern Mexico.

CHEESE BUSH (Fig. 119) *Hymenoclea salsola*

DESCRIPTION: A diffusely branched, spreading shrub that may grow to be 2 m (6.5 ft) tall. **LEAVES** are drought-deciduous, alternate, simple, and often resinous. Blades are threadlike and from 2 cm (.75 in.) to 5 cm (2 in.) long. Their margins curl upward. **INFLORESCENCES** are of 2 kinds. The pollen-bearing ones are mixed with the seed-bearing ones. A seed-bearing head has 1 pistil surrounded by a set of papery bracts, giving the head a pearly glow in the sun. **STEMS** are thin and pale-straw colored. **FRUITS** are achenes surrounded by winged burs.

HABITAT AND RANGE: Grows in the San Joaquin Valley, the eastern foothills of the Coast Ranges, southern California, and the deserts, at elevations below 1,800 m (6,800 ft). It can grow as an individual plant or in extensive thickets around, along, and in washes. Outside of California it grows in the southwestern United States.

REMARKS: The name cheese bush comes from the odor of its crushed leaves. Other common names are burrobrush, pearl bush, and white burrobush. The common name white cheese bush is sometimes used to distinguish this species from the singlewhorl burrobrush (*H. monogyra*), which has a similar range in California. Cheese bush is closely related to ragweed and causes hay fever.

ISOMERIS (BLADDERPOD)

The genus *Isomeris* has 1 species that grows in southern California and Baja California.

BLADDERPOD (Fig. 120) ***Isomeris arborea***

DESCRIPTION: An ill-scented, profusely branched shrub that often grows to be 2.75 m (9 ft) tall. **LEAVES** are alternate, evergreen, and simple or compound. When compound, they have 3 leaflets. Leaflets are elliptical, from 12 mm (.5 in.) to 4 cm (1.5 in.) long, and greenish to gray. **INFLORESCENCES** consist of terminal flower clusters measuring from 1 cm (.4 in.) to 30 cm (12 in.) long. **FLOWERS** are 4-parted and yellow. **FRUITS** are inflated elliptical capsules that measure 12 mm (.5 in.) to 38 mm (1.5 in.) long and extend beyond the petals on stalks that are 2.5 cm (1 in.) to 5 cm (2 in.) long. **TWIGS** are covered with fine hairs.

HABITAT AND RANGE: Commonly grows in woodlands and scrubs in central and southern California's deserts and mountains, below 1,200 m (4,000 ft). It is also found in Baja California.

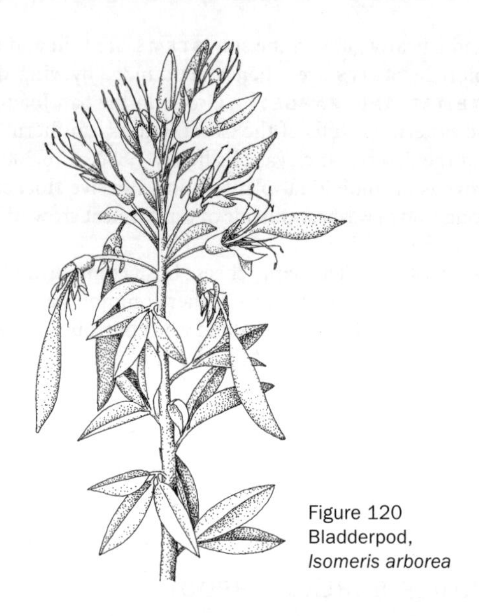

Figure 120
Bladderpod,
Isomeris arborea

REMARKS: Bladderpod is a winter- and early-spring-blooming shrub often seen along highways and roads. The unusual stalked fruit and flowers make this shrub easy to remember.

JUGLANS (WALNUT)

The genus *Juglans* has about 20 tree species: 6 are native to North America, and 1 of these is native to California.

CALIFORNIA BLACK WALNUT *Juglans californica* (Fig. 121; Pl. 24)

DESCRIPTION: A small, single- or multistemmed tree. Mature trees are typically 3 m (10 ft) to 12 m (40 ft) tall and 30 cm (1 ft) to 90 cm (3 ft) in diameter. The largest grows in Napa County and is 35 m (115 ft) tall and 1.9 m (77 in.) in diameter. Early European settlers noted larger individuals up to 1.8 m (6 ft) in diameter. Crowns are broad-spreading. Trunks are often forked from the base, with branches originating low on the stems. **LEAVES** are deciduous, alternate, pinnately compound

Figure 121
California black walnut,
Juglans californica

(usually 11 to 19 leaflets), and 15 cm (6 in.) to 30 cm (12 in.) long. Leaflets are lance shaped to egg shaped and 2.5 cm (1 in.) to 10 cm (4 in.) long. Leaflet margins are serrated. Leaflet tips are sharp pointed and bases are round or heart shaped. Upper leaflet surfaces are yellowish green and hairless, while lower surfaces are pale yellowish green with hairs at the main veins. **INFLORESCENCES** are of 2 kinds. The pollen-bearing flowers occur in showy catkins; on the same plant, the seed-bearing flowers occur as small clusters of 1 to 3 flowers. **FRUITS** are spherical nuts surrounded by a fleshy husk measuring 2.5 cm (1 in.) to 4 cm (1.5 in.) in diameter. Nuts have hard shells with vertical grooves. **TWIGS** are brown and hairy and have chambered piths. **BARK** is thick, dark brown to black, and deeply furrowed; furrows are separated by broad irregular ridges.

HABITAT AND RANGE: Grows in broadleaved and riparian woodlands in central and southern California, at elevations from 50 m (150 ft) to 900 m (3,000 ft). It is typically found near streams and on lower slopes.

REMARKS: Two varieties of *J. californica* are recognized: the southern California black walnut (*J. c.* var. *californica*), and the northern California black walnut (*J. c.* var. *hindsii*). The southern California variety is more widespread and characteristically has multiple stems and smaller fruits. The northern California variety usually is single-stemmed and has larger fruits. Most, if not all, natural stands of the northern variety are associated with former Native American camps. Settlers cut many of the largest specimens. Today, the northern variety has become widely naturalized in the interior Coast Ranges and the Central Valley. California black walnut is used as a shade tree and as rootstock for grafted English walnuts. Most of the plants in the species' northern range are apparently the result of hybridization with the eastern black walnut (*J. nigra*). There are only 3 native stands of the northern variety.

KALMIA (LAUREL)

The genus *Kalmia* has 10 species of shrubs that are associated with boggy environments at high latitudes or high-elevation mountains in the Northern Hemisphere.

LAUREL (Fig. 122) ***Kalmia polifolia***

DESCRIPTION: A short, spreading shrub with dark green leaves and showy flowers. Shrubs often grow to be 45 cm (1.5 ft) tall. **LEAVES** are mostly opposite, simple, evergreen, and leathery. Blades are less than 25 mm (1 in.) long, and margins are loosely or tightly rolled under. The upper surface is hairless and the lower surface is densely hairy. **INFLORESCENCES** consist of single flowers arising from leaf axils at the tops of stems. **FLOWERS** are 5-parted, with petals fused only at the base. They have a wide cup shape and are pink to rose-purple. **FRUITS** are spherical capsules. **TWIGS** are brown and have swollen nodes associated with leaves.

HABITAT AND RANGE: Grows in the Klamath, Cascade, and Warner Mountains and the Sierra Nevada, in boggy areas associated with subalpine forests and alpine meadows at elevations from 2,100 m (7,000 ft) to 3,600 m (12,000 ft). It ranges north to Alaska and east to New England.

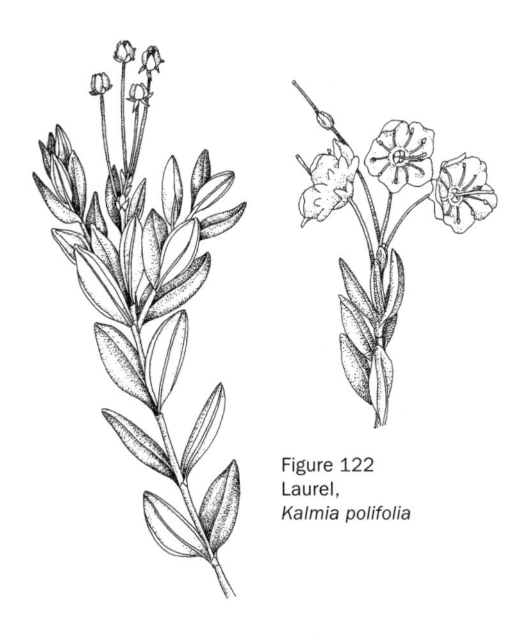

Figure 122
Laurel,
Kalmia polifolia

REMARKS: Two varieties of *K. polifolia* are recognized. *K. p.* var. *polifolia* has tightly rolled-under margins and *K. p.* var. *microphylla* has loosely rolled-under margins. This shrub is very poisonous to sheep and cattle.

KECKIELLA (KECKIELLA)

The genus *Keckiella* is made up of 7 species and many subspecies. The genus occurs throughout California and in surrounding states. Prior to 1970, *Keckiella* was included in the genus *Penstemon,* a group of mainly perennial herbs with showy flowers.

Keckiella is a collection of drought-deciduous shrubs and subshrubs with wandlike stems and simple leaves that are generally opposite in arrangement. Flowers are brown, red, white, or yellow 2-lipped tubes. Fruits are dry capsules.

1. Leaf blades are linear to elliptical, and flowers are typically yellow . . . . **yellow keckiella (*K. antirrhinoides*)**
1. Leaf blades are egg shaped to round, and flowers are typically red 2

2. Leaf bases are heart shaped . . . . straggly keckiella (*K. cordifolia*)

2. Leaf bases are wedge shaped . . . . red keckiella (*K. corymbosa*)

YELLOW KECKIELLA *Keckiella antirrhinoides*

DESCRIPTION: A shrub with stems from 60 cm (2 ft) to 2.5 m (8 ft) tall. The showy flowers change from yellow to black with age. **LEAVES** are opposite, evergreen, elliptical or linear, and from 10 mm (.4 in.) to 20 mm (.75 in.) long. The bases taper directly to the stem, and the margins are entire. **INFLORESCENCES** consist of long, terminal clusters of flowers. **FLOWERS** are yellow, with petals forming a tube with 2 lips. **FRUITS** are elliptical capsules. **TWIGS** are covered with fine, sometimes gray, hairs.

HABITAT AND RANGE: Grows in chaparrals and woodlands at elevations from 100 m (350 ft) to 1,600 m (5,000 ft). It ranges from the San Bernardino Mountains through the Peninsular Ranges to Baja California. It is also found in the Colorado and Mojave Deserts.

REMARKS: In the mountains the plants have stems with fine hairs (var. *antirrhinoides*), whereas those in the deserts have gray hairs (var. *microphylla*). Yellow keckiella is cultivated for its long-lasting, showy flowers.

STRAGGLY KECKIELLA *Keckiella cordifolia*

DESCRIPTION: A bushy shrub that is often seen climbing over other shrubs. Its stems can grow to be 3 m (10 ft) long. The showy flowers can be found throughout the year. **LEAVES** are opposite, dark green, heart shaped to round, and from 20 mm (.75 in.) to 65 mm (2.5 in.) long. The margins are entire or toothed. **INFLORESCENCES** consist of long, terminal clusters of flowers. **FLOWERS** are red or scarlet with petals forming a tube with 2 lips. **FRUITS** are elliptical capsules. **TWIGS** are covered with fine hairs or can be hairless.

HABITAT AND RANGE: Grows in chaparrals and woodlands below 1,200 m (4,000 ft) in the coastal mountains south of San

Figure 123
Red keckiella,
Keckiella corymbosa

Luis Obispo County and throughout the mountains of southern California.

REMARKS: Individual straggly keckiella can have yellow rather than red flowers. The similar red keckiella (*K. corymbosa*) has wedge-shaped rather than heart-shaped leaves and grows in northern California mountains. Both color forms of scraggly keckiella are cultivated for their showy flowers.

RED KECKIELLA (Fig. 123) *Keckiella corymbosa*

DESCRIPTION: A small, mat-forming shrub that often grows to be 60 cm (2 ft) tall, with showy flowers that bloom in midsummer. **LEAVES** are opposite, drought-deciduous, dark green, egg shaped or wider above the middle, and from 12 mm (.5 in.) to 4 cm (1.5 in.) long. The bases are wedge shaped, and the margins are entire or have 3 to 5 teeth. **INFLORESCENCES** consist of long, terminal clusters of flowers. **FLOWERS** are brick red, with petals forming a tube with 2 lips. **FRUITS** are elliptical capsules. **TWIGS** are hairless or may be covered with fine hairs.

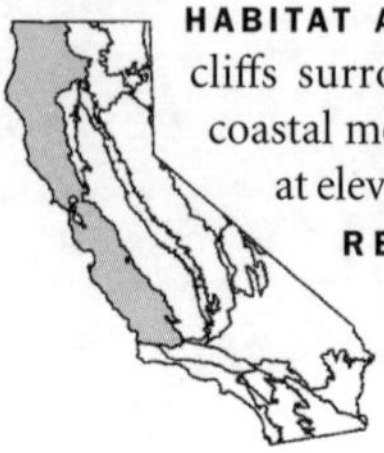

HABITAT AND RANGE: Grows on rocky slopes and cliffs surrounded by forests and woodlands in the coastal mountains of northern California. It is found at elevations of 100 m (350 ft) to 1,600 m (5,000 ft).

REMARKS: The similar red-colored straggly keckiella (*K. cordifolia*) has heart-shaped rather than wedge-shaped leaves and grows in southern California mountains.

KRASCHENINNIKOVIA (WINTER FAT)

Grows in California's deserts and the intermountain West. It is the only New World member of this genus found in arid environments throughout the Northern Hemisphere. In older books the names *Ceratoides lanata* and *Eurotia lanata* are used for winter fat.

WINTER FAT (Fig. 124) ***Krascheninnikovia lanata***

DESCRIPTION: An erect or spreading shrub that may grow to be 1 m (3 ft) tall. **LEAVES** are simple, alternate, and clustered and either are drought-deciduous or stay on the plant until the next spring. Blades are linear and from 6 mm (.25 in.) to 30 mm (1.2 in.) long, with tightly rolled-under margins. **INFLORESCENCES** are tight clusters of flowers found in the leaf axils at the upper portions of the branches. **FLOWERS** are small and either pollen bearing or seed bearing. **FRUITS** are utricles surrounded by 2 bracts that are densely covered with long, silvery hairs. **TWIGS** have a dense covering of star-shaped or unbranched hair. The young hairs are white, becoming rusty with age.

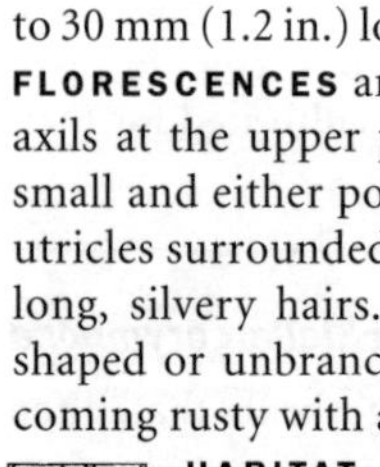

HABITAT AND RANGE: In California the species grows in the Great Basin and Mojave Deserts and in the southern San Joaquin Valley and adjacent foothills, at elevations from 100 m (350 ft) to 2,700 m (8,900 ft). It occurs in interior coniferous woodlands and desert scrubs. It ranges throughout the intermountain West.

REMARKS: Winter fat is an important

Figure 124
Winter fat,
Krascheninnikovia lanata

forage plant for livestock and wildlife, especially during winter, when forage is scarce. It is a useful shrub for reclamation of disturbed sites in arid climates.

LARREA (CREOSOTE BUSH)

The genus *Larrea* has 5 shrub species growing in the Americas, 1 of which is native to California.

CREOSOTE BUSH (Fig. 125) *Larrea tridentata*

DESCRIPTION: An erect, multistemmed shrub that can grow to be 60 cm (2 ft) to 4 m (13 ft) tall. Branches are brittle and they spread wide. The species is very long-lived, with many plants surviving more than 1,000 years. A clone in Johnson Valley has been estimated to be 9,400 years old. **LEAVES** are evergreen, opposite, and compound, with 2 leaflets. Leaflets have a broad lance shape to sickle shape and are about 1 cm (.4 in.) long, silky haired, and 3-veined from the base. Margins are entire. Leaflet tips are pointed and bases have a broad round shape. Leaflets have a strong resinous odor. **FLOWERS** are yellow, solitary at the ends of twigs, 5-petaled, and about 6 mm (.25 in.) long. **FRUITS** are

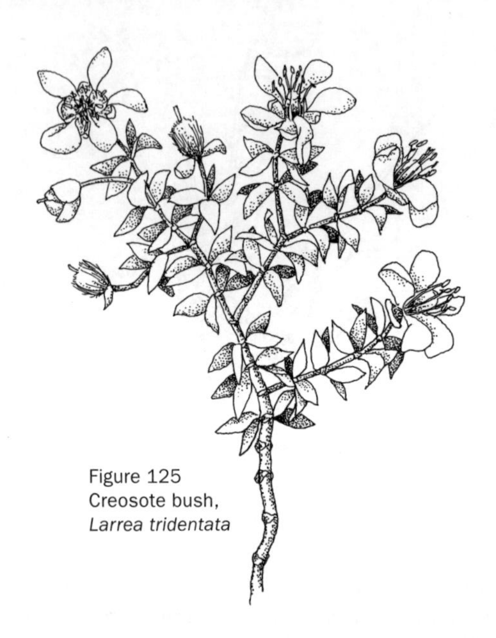

Figure 125
Creosote bush,
Larrea tridentata

spherical capsules about 6 mm (.25 in.) long and covered with whitish or rusty hairs. They have evergreen styles. **TWIGS** are reddish initially, becoming gray with dark rings at nodes.

HABITAT AND RANGE: Widely distributed in the deserts of the southwestern United States and Mexico. In California it is quite common in the Mojave, Colorado, and Sonoran Deserts. It is usually found growing below 1,500 m (5,000 ft).

REMARKS: Creosote bush provides vital habitat for many desert wildlife species. Small mammals eat its foliage and seeds, but it is generally unpalatable to larger browsing species. Native Americans had a variety of medicinal uses for creosote bush. The same or a closely related species, *L. divaricata,* grows in southern South America.

LEDUM (LABRADOR-TEA)

Ledum is composed of 2 or 3 species of shrubs associated with swampy or boggy environments in northern North America, Europe, and Asia.

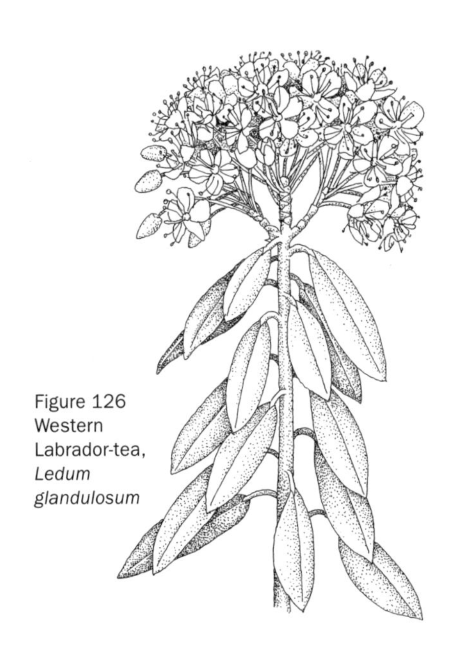

Figure 126 Western Labrador-tea, *Ledum glandulosum*

WESTERN LABRADOR-TEA *Ledum glandulosum*

(Fig. 126)

DESCRIPTION: An erect or spreading shrub with fragrant, droopy leaves clustered at the tops of its stems. Shrubs often grow to be 1.5 m (5 ft) tall. **LEAVES** are alternate, simple, evergreen, and leathery. Blades are elliptical and from 4 cm (1.5 in.) to 6 cm (2.5 in.) long. Margins may be tightly rolled under. Upper surfaces are hairless and dark green, while lower surfaces have fine hairs. **INFLORESCENCES** are terminal, flat-headed, open clusters of flowers on stalks about 2.5 cm (1 in.) long. **FLOWERS** are 5-parted, with petals fused only at their bases so they appear separate. Flowers have a wide cup shape and are creamy white. **FRUITS** are elliptical capsules whose valves open from the base. **TWIGS** are yellowish green and covered with hairs and glands.

HABITAT AND RANGE: Grows in marshy or boggy areas from sea level to 3,600 m (12,000 ft) in the mountains of northern California. It ranges from western Canada and the Rocky Mountains to California.

REMARKS: Leaf characteristics such as the degree to which margins are rolled under, color of the upper surfaces, and hairiness on the lower surfaces vary greatly among plants. Hybridization with eastern Labrador-tea (*L. groenlandicum*) may explain *L. ×columbianum*, coastal populations with tightly rolled-under leaf margins (the "×" indicates that hybridization probably created this form).

LEPIDOSPARTUM (SCALEBROOM)

The genus *Lepidospartum* is made up of 3 species that grow throughout western North America. Two of these are found in California.

SCALEBROOM *Lepidospartum squamatum*

DESCRIPTION: A broomlike shrub that often grows to be 2 m (6.5 ft) tall. **LEAVES** are of 2 kinds. On new branches, the leaves are egg shaped, 2 cm (.75 in.) to 3 cm (1.25 in.) long, and covered with white, woolly hairs. As branches age, the wool is lost and the stems become green and the leaves scaly. **INFLORESCENCES** are dense clusters of heads, with 3 to 5 yellow flowers. **FRUITS** are achenes. **STEMS** are covered with white, woolly hairs when young; as they age they become green. The stems are erect; plants spread, forming thickets.

HABITAT AND RANGE: Grows in southern California, in the warm deserts, and on the central coast, below 1,800 m (6,000 ft). It also grows in coastal scrubs, chaparrals, and woodlands. The species is often associated with washes and stream terraces.

REMARKS: The closely related greenbroom (*L. latisquamum*) grows only locally and is associated with limestone.

LEUCOTHOE (BLACK-LAUREL)

The genus *Leucothoe* has about 8 species of shrubs and trees found in the United States, Central America, China, and Japan. One of these is native to California.

Figure 127
Black-laurel,
Leucothoe davisiae

BLACK-LAUREL (Fig. 127) ***Leucothoe davisiae***

DESCRIPTION: An erect, spreading shrub that develops extensive colonies on forest floors. Shrubs often grow to be 1.5 m (5 ft) tall. **LEAVES** are alternate, simple, evergreen, dark green, and somewhat leathery. Blades are elliptical, from 12 mm (.5 in.) to 6 cm (2.5 in.) long, and hairless. **INFLORESCENCES** are erect, simple, terminal clusters of flowers less than 15 cm (6 in.) long. **FLOWERS** are 5-parted, urn shaped, and white. **FRUITS** are spherical capsules. **TWIGS** are brown and hairless.

HABITAT AND RANGE: Occurs in the Klamath Mountains, Sierra Nevada, and Cascades at elevations from 1,200 m (4,000 ft) to 2,500 m (8,500 ft). It grows at the edges of meadows and under forest trees in wet sites. The species is also found in Oregon.

REMARKS: Black-laurel is rarely seen as a single-stemmed shrub; instead it is found in small to extensive thickets.

LITHOCARPUS (TANOAK)

The genus *Lithocarpus* has about 300 species of trees and shrubs, 1 of which is native to California. All others are found in southeastern Asia and Indonesia.

TANOAK (Fig. 128; Pl. 25) ***Lithocarpus densiflorus***

DESCRIPTION: Usually an erect, medium-sized, single-stemmed tree. On harsh sites or in forest understories it can be multistemmed and/or a shrub. Mature trees are typically 15 m (50 ft) to 27 m (90 ft) tall and 30 cm (1 ft) to 90 cm (3 ft) in diameter. The largest grows in Curry County, Oregon, and is 43.9 m (144 ft) tall and 2.2 m (87 in.) in diameter. Crowns are rounded at the tops of long, clear trunks. Multistemmed and shrub forms have broad, spreading crowns. The trees live to be 300 or 400 years old. **LEAVES** are evergreen, simple, alternate, leathery, egg shaped to oblong, and 2.5 cm (1 in.) to 14 cm (5.5 in.) long. Margins are serrated to entire. Upper surfaces are dark green and somewhat hairy, while lower surfaces are light green and densely hairy. Some plants have leaves that lack or nearly lack hairs. **FLOWERS** are unisexual. Pollen-bearing flowers are upright catkins measuring 7.5 cm (3 in.) to 10 cm (4 in.) long. Seed-bearing flowers are located at the bases of the pollen-bearing catkins. **FRUITS** are egg shaped to spherical, yellowish brown acorns that measure 2.5 cm (1 in.) to 4 cm (1.5 in.) long. They occur singly or in pairs. Acorn cups are shallow, with bristly scales. **TWIGS** are hairy. **BARK** is 2.5 cm (1 in.) to 7.5 cm (3 in.) thick, fissured, and dark grayish brown.

HABITAT AND RANGE: Grows in forests, woodlands, and chaparrals from southwestern Oregon to Ventura County. It occurs below 2,000 m (6,500 ft), mostly on fertile soils with sufficient summer moisture.

REMARKS: Tanoak is very shade tolerant and is often found as an understory species in deep shade. It is susceptible to fire and sprouts vigorously following injury. Since the mid 1990s, a bark-invading fungus (an unnamed species in the genus *Phytophthora*) has killed thousands of tanoak trees from Big Sur to Sonoma County. Two varieties of tanoak are rec-

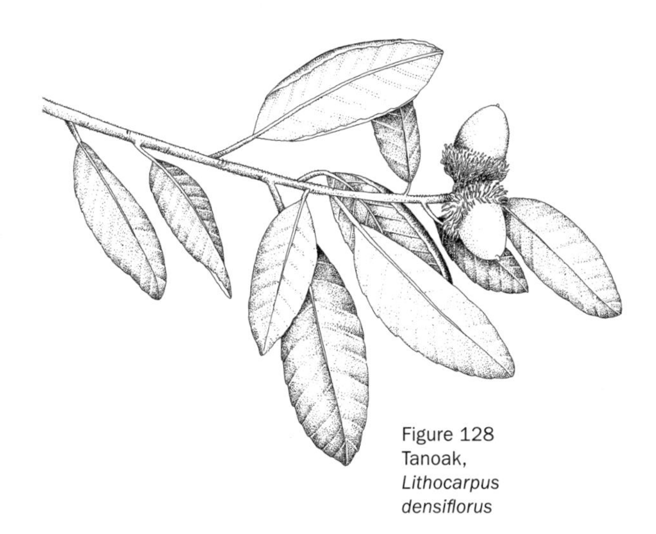
Figure 128
Tanoak,
Lithocarpus densiflorus

ognized: the usual tree-sized form, *L. d.* var. *densiflorus,* and a shrub form, *L. d.* var. *echinoides,* whose margins are entire, wavy, or few-toothed and whose blades have obscure veins on their lower surfaces. Tanoak acorns have been a staple in the diets of Native Americans and a variety of wildlife. Tannin extracted from the species' bark has been used to tan hides, and its wood serves as firewood and lumber.

LONICERA (TWINBERRY)

The genus *Lonicera* is made up of over 200 species of shrubs and vines. They are found in temperate and subtropical regions of the Northern Hemisphere.

Lonicera species have opposite, simple leaves and paired, tubular flowers. The fruits are paired berries. Species typically have fragrant flowers, and many are used as ornamentals. A number of the shrubs are called twinberries, and the vines are often called honeysuckles. This book does not include vines.

1. Flowers are purple and the 2 pistils are fused . . . . **purple-flowered honeysuckle** (*L. conjugialis*)
1. Flowers are yellow and the 2 pistils are separate . . . . **twinberry** (*L. involucrata*)

PURPLE-FLOWERED HONEYSUCKLE *Lonicera conjugialis*

DESCRIPTION: An erect, spreading shrub with dark-colored flowers and red fruits. Shrubs can grow to be 2 m (7 ft) tall. **LEAVES** are simple, opposite, deciduous, thin, elliptical, and from 2 cm (.74 in.) to 8 cm (3 in.) long. Plants may have scattered hairs. **INFLORESCENCES** consist of a pair of flowers that may be surrounded by a pair of minute bracts. **FLOWERS** are 2 united, dull purple, 2-lipped floral tubes. **FRUITS** are a pair of fused, bright red berries. **TWIGS** are slender.

HABITAT AND RANGE: Grows in northern California mountains at elevations from 1,200 m (4,000 ft) to 3,200 m (10,500 ft) and is commonly associated with moist places in forests.

REMARKS: In California there are 5 *Lonicera* species that are vines with the common name honeysuckle. Members of the genus typically have fragrant flowers, and many are used as ornamentals.

TWINBERRY (Fig. 129; Pl. 26) *Lonicera involucrata*

DESCRIPTION: An erect, spreading shrub with colorful flowers and fruits; it often grows to be 3 m (10 ft) tall. **LEAVES** are opposite, simple, deciduous, dark green, elliptical, and from 5 cm (2 in.) to 12.5 cm (5 in.) long. Small hairs can often be seen on the margins. **INFLORESCENCES** consist of a pair of flowers surrounded by 2 pairs of leafy bracts in the leaf axils. The bracts become red to purple and fleshy with age. **FLOWERS** are 2 upright, yellow or reddish-tinged, 2-lipped floral tubes protruding from the bracts. **FRUITS** are a pair of fused black berries surrounded by bracts. **TWIGS** are angled when young.

HABITAT AND RANGE. Grows in northern California coastal and montane forests below 3,000 m (9,500 ft) and is commonly associated with moist places. This species occurs from Alaska to Mexico.

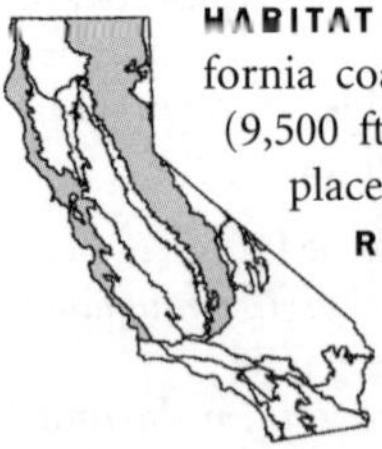

REMARKS: Twinberries from the mountains, *L. i.* var. *involucrata,* have yellow floral tubes, while coastal plants, *L. i.* var. *ledebouri,* have red-tinged floral tubes.

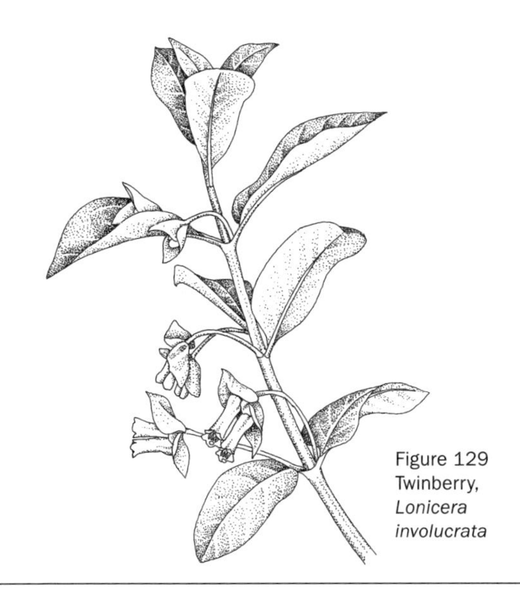

Figure 129
Twinberry,
Lonicera involucrata

LUPINUS (LUPINE)

The genus *Lupinus* has about 200 species native to North and South America as well as Mediterranean Europe.

The genus consists of annual and perennial herbs and shrubs, typically with palmately compound leaves and distinctive, showy flowers arranged in dense spikes.

The leaves and seeds of some species are eaten and some plants are used ornamentally. The leaves and seeds are toxic in some species. The name *Lupinus* refers to *lupus,* the wolf, from the antiquated belief that lupines steal from the soil.

YELLOW BUSH LUPINE *Lupinus arboreus*

(Fig. 130)

DESCRIPTION: A shrub with round stems that may grow to be 3 m (10 ft) tall. **LEAVES** are alternate and evergreen and may be covered with silky gray hairs. Blades are compound, with 5 to 12 linear leaflets radiating from the base. Each leaflet is 2 cm (.75 in.) to 6 cm (2.4 in.) long. **INFLORESCENCES** are terminal spikes. **FLOWERS** are lilac, purple, or yellow. **FRUITS** are

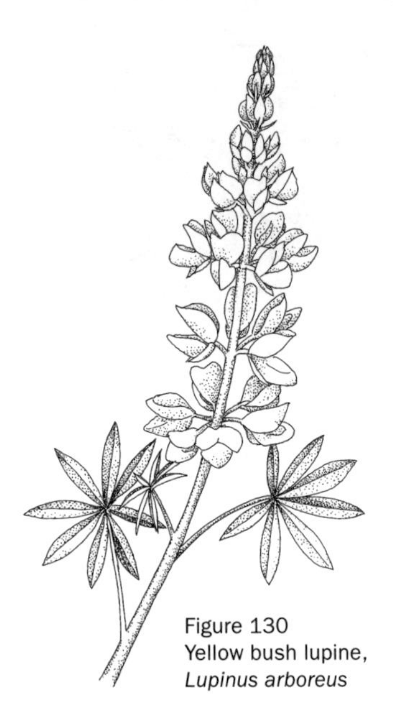

Figure 130
Yellow bush lupine,
Lupinus arboreus

pea-pod-like. They break open along 2 sides as the pod dries. **TWIGS** are dark green to gray.

HABITAT AND RANGE: Locally common on dunes or disturbed areas below 100 m (300 ft) along the coast from Washington south to Santa Barbara County.

REMARKS: Yellow bush lupine is considered native from Sonoma County south, and non-native to the north, where the plant has been introduced. There is a large degree of variation in flower color, and some flowers are even bicolored (blue and yellow). Much of the variation may be due to hybridization of yellow bush lupine with other lupines, such as *L. rivularis*.

LYCIUM (BOXTHORN)

The genus *Lycium* has more than 100 species, which grow in deserts worldwide. California has 10 native and 1 introduced species.

Figure 131
Anderson boxthorn,
Lycium andersonii

Lycium is a collection of shrubs with leafy thorns and clustered leaves that are alternate in arrangement. Flowers are bell shaped or tubular and are white, tinged with green or purple. Fruits are berries.

The genus goes by many names, including boxthorn, buckthorn, bullberry, rabbit thorn, tomatilla, water jacket, and wolfberry. The name *Lycium* refers to the ancient country of Lycia in Asia Minor. Some Old World species have medicinal uses.

1. Twigs are hairless. Flower and calyx tubes are of different lengths; the flowers are much longer than the calyxes .... **Anderson boxthorn** (*L. andersonii*)
1. Twigs are covered with fine, sticky hairs. Flower and calyx tubes are about the same length; the flowers are barely longer than the calyxes........ **desert-thorn** (***L. brevipes***)

ANDERSON BOXTHORN *Lycium andersonii*

(Fig. 131)

DESCRIPTION: An intricately branched, thorny shrub that may grow to be 2.75 m (9 ft) tall. **LEAVES** are alternate, clustered, entire, drought-deciduous, and fleshy. Leaf blades are pear shaped, 3 mm (.12 in.) to 13 mm (.5 in.) long, and grayish green. **INFLORESCENCES** are solitary flowers in the forks of branches. **FLOWERS** are lavender tinged and funnel shaped. **FRUITS** are orange or red berries that look like small tomatoes and measure 3 mm to 8 mm in diameter. **TWIGS** are gray and hairless.

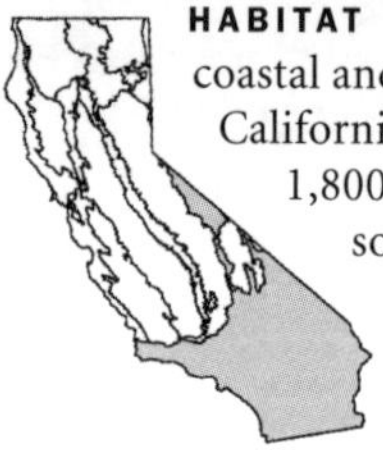

HABITAT AND RANGE: Grows in woodlands, coastal and desert scrubs, and chaparrals of southern California and the warm deserts, at elevations below 1,800 m (6,000 ft). It ranges throughout the southwestern United States and northern Mexico.

REMARKS: Birds and mammals eat Anderson boxthorn's berries.

DESERT-THORN *Lycium brevipes*

DESCRIPTION: An intricately branched, thorny shrub that grows to be 3 m (9 ft) tall. **LEAVES** are alternate, clustered, drought-deciduous, and fleshy. Leaf blades are pear shaped, from 5 mm (.2 in.) to 15 mm (.6 in.) long, and grayish green. **INFLORESCENCES** are solitary flowers in the forks of branches. **FLOWERS** are white to lavender or pink and bell shaped. **FRUITS** are red berries that look like small tomatoes. **TWIGS** have fine hairs.

HABITAT AND RANGE: Grows in coastal scrubs of southern California and scrubs in the Colorado and Sonoran Deserts, at elevations below 600 m (1,900 ft). It ranges into northern Mexico.

REMARKS: Birds and mammals eat desert-thorn's berries.

MALOSMA (LAUREL SUMAC)

The genus *Malosma* has 1 shrub species that grows in southern California and Baja California. In older books, laurel sumac was included in the genus *Rhus*.

LAUREL SUMAC (Fig. 132) *Malosma laurina*

DESCRIPTION: An erect shrub or small tree that can grow to be 2 m (6.5 ft) to 5 m (16 ft) tall. Crowns are dense and

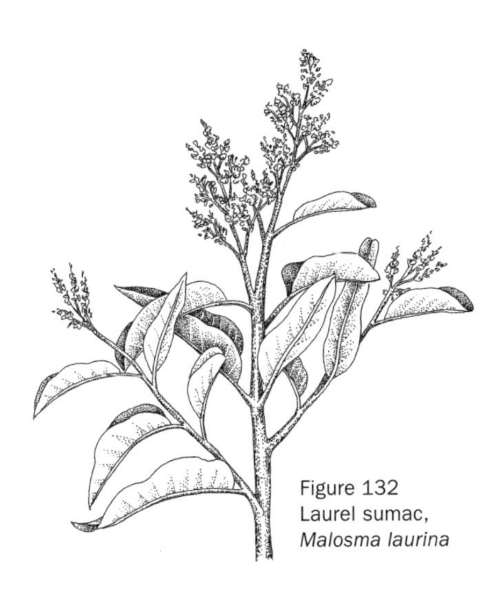

Figure 132
Laurel sumac,
Malosma laurina

rounded. Plants can live to be at least 50 years old. **LEAVES** are alternate, simple, evergreen, leathery, and aromatic. Blades have an elliptical to narrow egg shape, are more or less folded along the midrib, and measure from 5 cm (2 in.) to 10 cm (4 in.) long. Tips are tapered, ending with a short, abrupt point. Margins are entire and often reddish. Upper and lower surfaces are smooth and have reddish veins. Petioles are 15 mm (.6 in.) to 25 mm (1 in.) long. **INFLORESCENCES** have dense, branched, terminal clusters of small white flowers. **FRUITS** are hairless whitish drupes about 3 mm (.1 in.) in diameter. **TWIGS** are slender and reddish. **BARK** is smooth and brown to reddish.
HABITAT AND RANGE: Grows in chaparrals, woodlands, and scrubs from Santa Barbara County to northern Baja California. It can be found on foothills, mesas, and coastal slopes and in valleys in southern California, at elevations below 900 m (3,000 ft).
REMARKS: Following fire, laurel sumac can sprout from massive underground burls. Its seeds can germinate once the heat of a fire cracks the hard seed coats. The attractiveness of the plant makes it a favorite landscape species in southern California. Its leaves are unpalatable to wildlife and livestock, although bees collect pollen from its flowers and various mammals and birds eat its fruits.

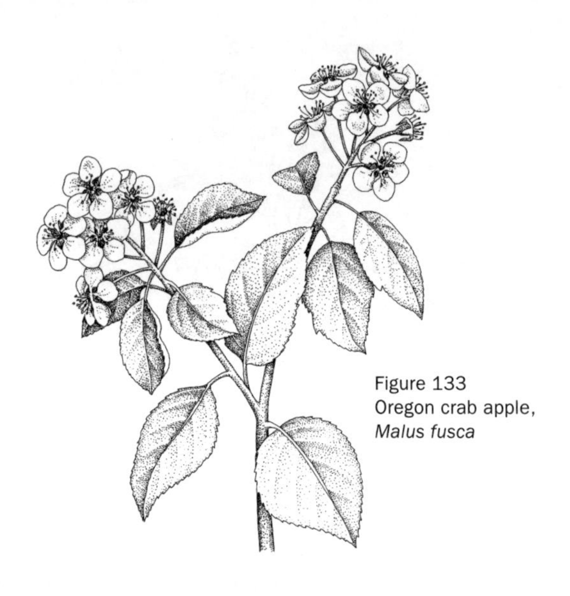

Figure 133
Oregon crab apple,
Malus fusca

MALUS (APPLE)

The genus *Malus* has about 25 species of shrubs and trees that grow in the Northern Hemisphere.

OREGON CRAB APPLE (Fig. 133) ***Malus fusca***

DESCRIPTION: A large shrub or small tree that often grows in thickets. Mature plants are typically 3 m (10 ft) to 9 m (30 ft) tall and 15 cm (6 in.) to 30 cm (12 in.) in diameter. The largest grows in Nisqually National Wildlife Refuge, Washington, and is 24 m (79 ft) tall and 53 cm (21 in.) in diameter. Crowns are rounded and spreading. **LEAVES** are deciduous, alternate, simple, egg shaped to elliptical, and 2.5 cm (1 in.) to 9 cm (3.5 in.) long. Margins are serrated. Leaf tips are pointed and bases are round to tapered. Upper surfaces are dark green and hairless, while lower surfaces are green and slightly hairy. Petioles are 12 mm (.5 in.) to 5 cm (2 in.) long. Leaves occur on spur shoots. **FLOWERS** are white to pinkish white and 12 mm (.5 in.) in diameter. Flowers are in clusters of 4 to 10. **FRUITS** are apples that measure 12 mm (.5 in.) to 25 mm (1 in.) long and are yellowish to reddish. **TWIGS** are hairy the first year, have small lenticels, and are sometimes thorny. **BARK** is reddish brown, with large flat scales.

HABITAT AND RANGE: Grows in woodlands and forests from Alaska to California. In California it occurs on moist lower slopes and in riparian zones, at elevations below 750 m (2,500 ft).

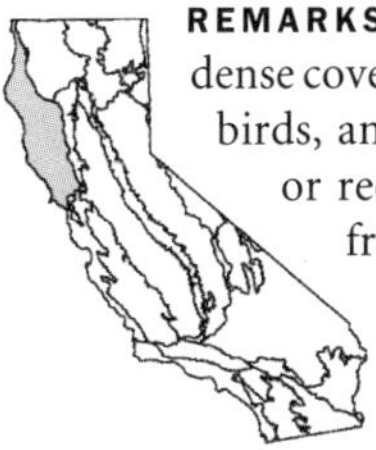

REMARKS: Oregon crab apple provides excellent, dense cover for many wildlife species. Small mammals, birds, and deer eat its fruits. Its leaves turn orange or red in the fall. Jams and jellies can be made from the sour apples. Seeds from discarded cores of the apple (*M. sylvestris*) often become established and develop into trees in northern California.

MENZIESIA (MOCK-AZALEA)

The genus *Menziesia* has 7 species of shrubs found in western North America and eastern Asia. One species is native to California.

MOCK-AZALEA (Fig. 134) ***Menziesia ferruginea***

DESCRIPTION: An erect, straggly shrub that occurs individually in coastal forests. Shrubs may grow to be 4.5 m (15 ft) tall. **LEAVES** are alternate, simple, deciduous, and thin. Blades are elliptical, measure from 4 cm (1.5 in.) to 7 cm (2.75 in.) long, and have a scattering of rusty hairs. Margins are also hairy. **INFLORESCENCES** are axillary clusters of flowers. **FLOWERS** are 5-parted, urn shaped, and yellowish green with red-tinged petals. **FRUITS** are elongated capsules. **TWIGS** are brown, with a scattering of hairs.

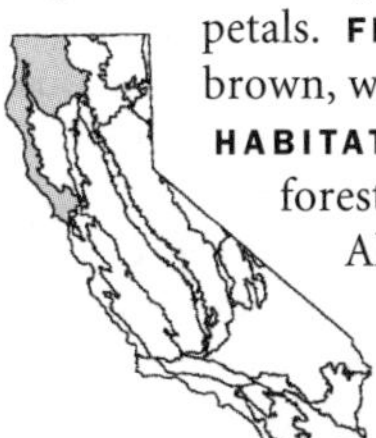

HABITAT AND RANGE: Grows in coniferous forests at elevations under 300 m (1,000 ft), from Alaska south to the north coast of California.

REMARKS: The genus name honors Archibald Menzies, botanist and surgeon on the Vancouver Expedition, who col-

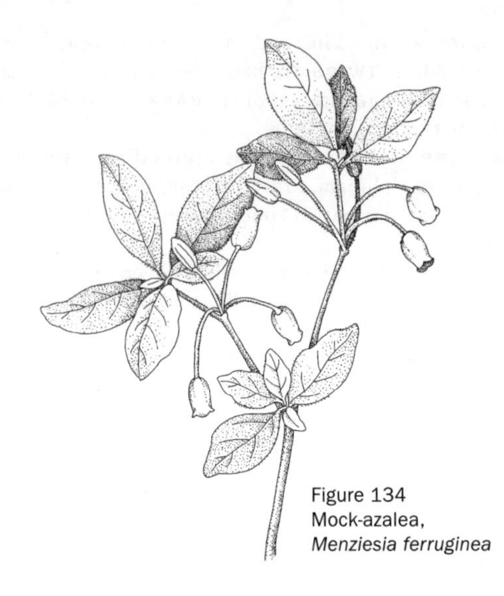

Figure 134
Mock-azalea,
Menziesia ferruginea

lected many specimens used by other botanists to name new genera and species in the late eighteenth century.

MIMULUS (MONKEYFLOWER)

The genus *Mimulus* has about 100 species of shrubs and herbs that are found in western North America, eastern Asia, Chile, southern Africa, Australia, and New Zealand.

BUSH MONKEYFLOWER *Mimulus aurantiacus* (Fig. 135)

DESCRIPTION: An erect, spreading shrub that can grow to be 1.25 m (4 ft) tall. **LEAVES** are opposite, drought-deciduous, dark green, oblong, linear to wider above the middle, and from 2.5 cm (1 in.) to 5 cm (2 in.) long, with margins rolled under. **INFLORESCENCES** consist of 1 or several flowers in the uppermost stem axils. **FLOWERS** are white, yellow, orange, or red, with petals forming a 2-lipped tube. **FRUITS** are elliptical capsules. **TWIGS** are hairless or may be covered with fine hairs. **HABITAT AND RANGE:** Grows on rocky slopes and cliffs sur-

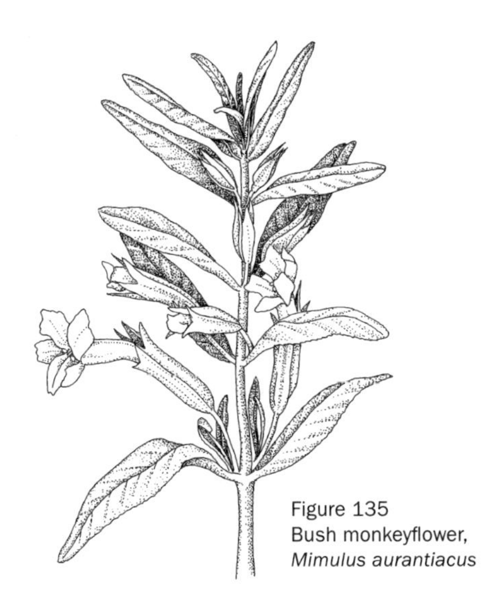

Figure 135
Bush monkeyflower,
Mimulus aurantiacus

rounded by chaparrals, scrubs, woodlands, and forests in much of California. It is found at elevations below 900 m (3,000 ft).

REMARKS: Bush monkeyflower has many regional forms based on flower color and other characteristics, so expect a great deal of variety in it throughout the state. Many people see the grinning face of a mime when they look at a monkeyflower head-on. Some botanists recognize another genus, *Diplacus,* as the genus of the shrubby members of *Mimulus.*

MYRICA (WAX-MYRTLE)

The genus *Myrica* has about 48 species of shrubs and small trees, 2 of which are native to California. *Myrica* species are found in temperate and tropical regions.

Leaves are simple, alternate, and either evergreen or deciduous. Margins are entire or serrated. The flowers, which are incomplete, are clustered in catkins. Pollen-bearing and seed-bearing catkins can be on either the same or separate plants. Fruits are egg-shaped or spherical drupelike nuts or nutlets.

1. Leaves are evergreen and hairless. Plants grow near the coast . . . . **Pacific wax-myrtle** (*M. californica*)
1. Leaves are deciduous and sparsely hairy. Plants grow in the western Sierra Nevada . . . . **Sierra sweet-bay** (*M. hartwegii*)

PACIFIC WAX-MYRTLE *Myrica californica*

(Fig. 136)

DESCRIPTION: A shrub or small tree that usually ranges in height from 3 m (10 ft) to 11 m (35 ft). The largest grows in Siuslaw National Forest, Oregon, and is 11.6 m (38 ft) tall and 41 cm (16 in.) in diameter. Crowns are rounded and full on open-grown trees, and straggly and thin on plants growing in partial shade. **LEAVES** are evergreen, alternate, simple, and oblong to lance shaped, with the widest part above the middle; they measure 5 cm (2 in.) to 11 cm (4.5 in.) long. Margins are coarse toothed from above the middle and entire near the base. Leaf tips are sharp pointed and bases are wedge shaped. Upper surfaces are dark green and shiny, while lower surfaces are pale green. **FRUITS** are clusters of spherical nuts about 6 mm (.25 in.) in diameter, which are covered by purplish resin and a waxy white coating. **BARK** is thin, smooth, and gray, with mottled white patches.

HABITAT AND RANGE: Grows in coastal forests and scrubs from Washington to southern California. In California it is found on moist, well-drained soils in partial shade or full sun, at elevations below 150 m (500 ft).

REMARKS: Pacific wax-myrtle has root nodules that fix atmospheric nitrogen into a biologically usable form. The species is planted as an ornamental. Pioneering settlers made candles out of its waxy fruits.

SIERRA SWEET-BAY *Myrica hartwegii*

DESCRIPTION: A shrub that ranges in height from 1 m (3.3 ft) to 2 m (6.5 ft). **LEAVES** are deciduous, alternate, simple, aromatic, and lance shaped, with the widest part above the middle; they are 4 cm (1.5 in.) to 7.5 cm (3 in.) long. Margins are serrated above the middle and entire near the base. Leaf tips are sharp pointed and bases are wedge shaped. Upper surfaces are

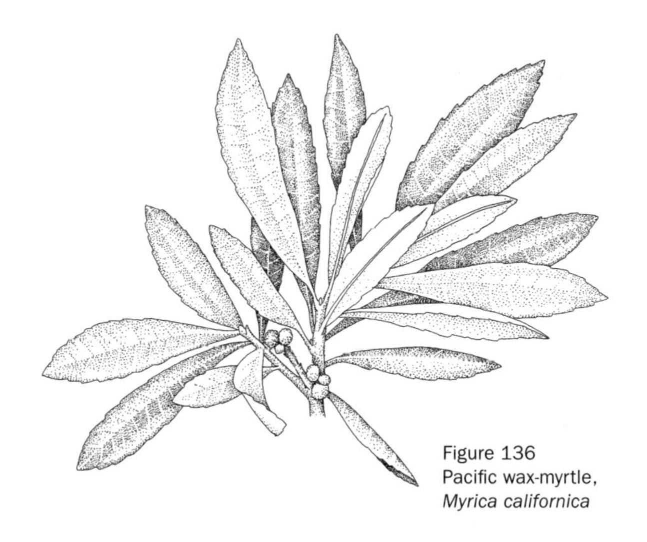

Figure 136
Pacific wax-myrtle,
Myrica californica

dull green and sparsely hairy, while lower surfaces are pale green and hairy. **FRUITS** are clusters of flat nutlets about 6 mm (.25 in.) long, with longer, enclosing, pointed bracts. **TWIGS** are slender and hairy. **BRANCHES** are reddish brown and hairless.

HABITAT AND RANGE: Grows in montane and riparian forests in the northern and central Sierra Nevada, at elevations from 300 m (1,000 ft) to 1,500 m (5,000 ft). It is found along stream banks and in other moist sites.

REMARKS: Sierra sweet-bay has root nodules that fix atmospheric nitrogen into a biologically usable form.

NEVIUSIA (SNOW WREATH)

The genus *Neviusia* has 2 species native to the United States. One of these is native to California.

SNOW WREATH *Neviusia cliftonii*

DESCRIPTION: A shrub with slender branches that may grow to be 30 cm (1 ft) tall. **LEAVES** are alternate, deciduous, and individually positioned along the stems. Blades are heart shaped and 2 cm (.75 in.) to 6 cm (2.5 in.) long, with long, sparse hairs. **INFLORESCENCES** consist of a few flowers in terminal, flat clusters that measure 1 cm (.4 in.) to 3 cm (1.2 in.) long. **FLOWERS** are showy and white. The showiness results from the many white stamens, as the petals are small. Sepals are toothed, egg shaped, and about 6 mm (.25 in.) long. **FRUITS** consist of several achenes per flower.

HABITAT AND RANGE: Grows in the eastern Klamath Mountains and Cascades near Lake Shasta but is considered uncommon. It is found in forest understories on limestone soils, at elevations from 300 m (1,000 ft) to 600 m (2,000 ft).

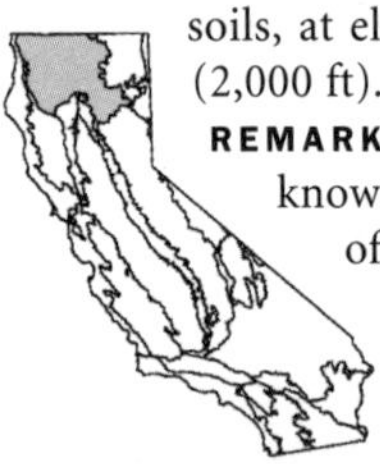

REMARKS: Until 1992, snow wreath was not known to occur in California. The other member of the genus grows in the hills of Alabama. The genus is also known from leaf fossils in the Miocene floras of Oregon. Snow wreath generally resembles ninebark.

NOLINA (NOLINA)

The genus *Nolina* has about 25 species native to the United States and Mexico. Three of these are native to California.

PARRY NOLINA *Nolina parryi*

DESCRIPTION: An erect, treelike plant with a dense rosette of long, swordlike leaves on top of a short, thick, woody stem. Stems have few or no branches and range in height from 1 m (3.3 ft) to 2 m (6.5 ft). **LEAVES** are evergreen, stiff, thick, swordlike, concave, green, and 60 cm (2 ft) to 1 m (3.3 ft) long. Margins have small teeth. Leaves are broad near their whitish, fleshy bases, tapering to sharp tips. **INFLORESCENCES** are clustered, erect, and branched and reach about 2 m (6.5 ft) long. They form on a stalk measuring between 1 m (3.3 ft) and 2 m (6.5 ft) long. Stalks have large, papery, evergreen bracts. **FLOWERS** are oblong and whitish, and each measures about 6 mm (.25 in.) long. **FRUITS** are papery capsules, notched at tips and bases, that measure about 12 mm (.5 in.) long and wide.

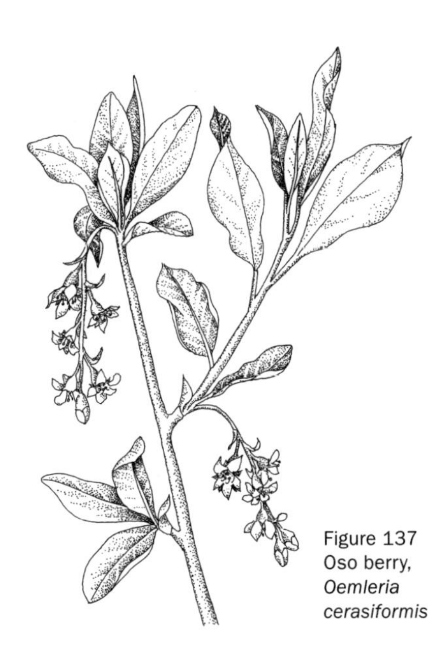

Figure 137
Oso berry,
Oemleria cerasiformis

HABITAT AND RANGE: Grows in woodlands, chaparrals, and coastal scrubs in southern California, Arizona, and Baja California, at elevations from 900 m (3,000 ft) to 1,200 m (4,000 ft).
REMARKS: Two other *Nolina* species grow in California: *N. bigelovii,* which inhabits parts of the Mojave and Colorado Deserts; and *N. interrata,* an endangered species found on gabbro soils in San Diego County and Baja California.

OEMLERIA (OSO BERRY)

The genus *Oemleria* has 1 species that grows from California north to British Columbia.

OSO BERRY (Fig. 137) ***Oemleria cerasiformis***

DESCRIPTION: A shrub or small tree that may grow to be between 1 m (3.3 ft) and 4 m (13 ft) tall. **LEAVES** are alternate, deciduous, and thin. Blades are light green, oblong, and 4 cm (1.5 in.) to 6 cm (2.5 in.) long, with smooth margins. **INFLO-**

RESCENCES have 2 kinds of flowers, with the seed-bearing and pollen-bearing flowers found on separate plants. The inflorescences are terminal and are composed of droopy clusters of flowers that measure 2.5 cm (1 in.) to 10 cm (4 in.) long. **FLOWERS** are white and fragrant. **FRUITS** consist of 1 to 5 drupes per flower. Fruits are about 1 cm (.4 in.) long and change from green to yellow to red and to black with age. **TWIGS** are gray and smooth. Piths are chambered.

HABITAT AND RANGE: Grows from British Columbia south to California's northern and coastal mountains and the Central Valley. It is associated with woodlands, forest edges, and chaparrals below 1,500 m (5,000 ft).

REMARKS: Oso berry is most noticeable in the early spring, when the fragrant white flowers stand out, especially among plants growing in fencerows.

OLNEYA (IRONWOOD)

The genus *Olneya* has 1 species that grows in the deserts of the southwestern United States and northwestern Mexico.

IRONWOOD (Fig. 138) ***Olneya tesota***

DESCRIPTION: A grayish tree with scaly bark that typically grows to be 7.5 m (25 ft) tall. The largest lives in Maricopa County, Arizona, and is 13.7 m (45 ft) tall and 1.4 m (54 in.) in diameter. **LEAVES** are alternate and deciduous, with a pair of spines at the bases of leaves. Blades are compound, composed of 4 to 9 pairs of gray-green leaflets arranged along main blade axes that extend beyond the leaflets. **INFLORESCENCES** are rounded clusters of 2 to 4 flowers borne in the leaf axils. **FLOWERS** are pale rose-purple blossoms that resemble pea flowers. **FRUITS** are legumes with constrictions between the seeds. **TWIGS** are gray, with stipular spines.

HABITAT AND RANGE: Occurs along washes and in floodplains below 600 m (2,000 ft) in the Colorado and Sonoran Deserts in southeastern California and in Arizona and northern Mexico.

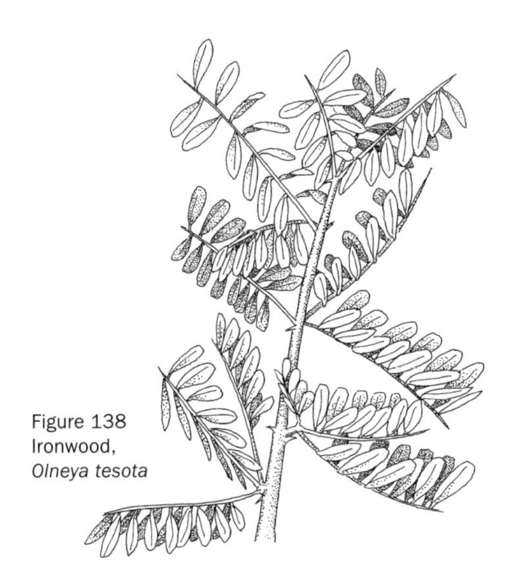

Figure 138
Ironwood,
Olneya tesota

REMARKS: Ironwood is one of the taller and more easily identified trees of the Colorado and Sonoran Deserts. It can be distinguished by its gray, leafy appearance.

PAXISTIMA (BOXWOOD)

The genus *Paxistima* has 2 species found in North America, 1 on each coast.

OREGON BOXWOOD (Fig. 139) *Paxistima myrsinites*

DESCRIPTION: A compact shrub that typically grows to be 1 m (3 ft) tall. **LEAVES** are opposite, simple, leathery, evergreen, and thick. Blades are elliptical, from 6 mm (.25 in.) to 25 mm (1 in.) long, with round tips and fine-toothed margins. **INFLORESCENCES** consist of small axillary clusters of 1 to 3 flowers. **FLOWERS** are 4-parted, saucer shaped, and green or maroon. **FRUITS** are capsules that open by 2 valves. **TWIGS** are corky and 4-angled.
HABITAT AND RANGE: Grows commonly in the forests of northern California mountains, at elevations from 600 m

Figure 139
Oregon boxwood,
Paxistima myrsinites

(2,000 ft) to 2,100 m (7,000 ft). It ranges throughout the western mountains and into Mexico.

REMARKS: Oregon boxwood is a common shrub in the understories of montane coniferous forests. It shares with *Euonymus* the strange characteristic of being a shrub with maroon flowers. In some books the genus name is spelled *Pachistima.*

PHILADELPHUS (MOCK-ORANGE)

The genus *Philadelphus* has about 65 shrub species that grow in temperate and subtropical latitudes of the Northern Hemisphere.

WILD MOCK-ORANGE *Philadelphus lewisii*

(Fig. 140)

DESCRIPTION: A showy shrub that grows to be 3 m (10 ft) tall. **LEAVES** are opposite, deciduous, and simple. Leaf blades are lance shaped, from 3 cm (1.25 in.) to 9 cm (3.5 in.) long, and green, with 3 or 5 veins originating from the base. Margins are entire or toothed. **INFLORESCENCES** are showy and consist of

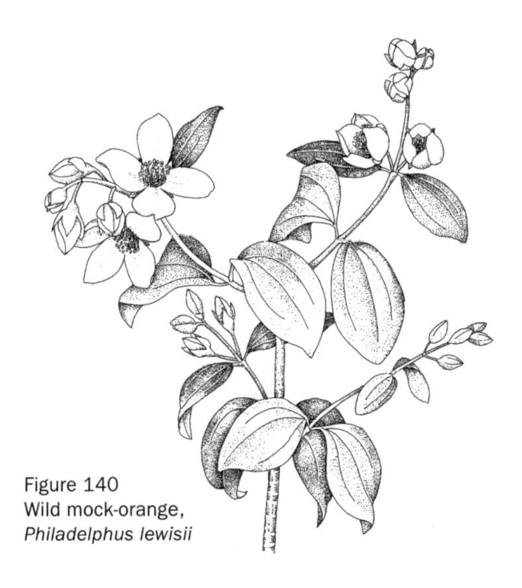
Figure 140
Wild mock-orange,
Philadelphus lewisii

terminal clusters of about 6 flowers. **FLOWERS** are white, with many stamens; flowers measure about 2.5 cm (1 in.) wide. **FRUITS** are many-seeded capsules. **TWIGS** are brown, aging to gray.

HABITAT AND RANGE: Grows from western Canada south to the forests and woodlands of northern California's hills and mountains, at elevations below 1,400 m (4,500 ft).

REMARKS: Wild mock-orange is a conspicuous summer-blooming shrub that occurs as individual shrubs in forests. There is great variation in its leaf characteristics, but the species is distinctive from the southern deserts' little mock-orange (*P. microphyllus*) with its smaller leaves, inflorescences, and flowers.

PHYLLODOCE (MOUNTAIN HEATHER)

The genus *Phyllodoce* has up to 7 shrub species growing at high latitudes (circumboreal) and in western mountains at high elevations. Two of these species grow in California.

Phyllodoce is a collection of low shrubs that have erect or prostrate stems and can form small to extensive thickets. Shrubs are typically less than 50 cm (20 in.) tall. Leaves are simple, alternate, and evergreen. Blades are less than 15 mm (.6 in.) long and appear to be needlelike, as they are tightly rolled under, which hides their lower surfaces. California's *Phyllodoce* species have 5-lobed red flowers with a broad tubular shape. Fruits are spherical capsules.

1. Flowers are bell shaped, with stamens extending beyond the corolla . . . . **Brewer heather** (*P. breweri*)
1. Flowers are urn shaped, with stamens included in the corolla . . . . **Cascade heather** (*P. empetriformis*)

BREWER HEATHER *Phyllodoce breweri*

DESCRIPTION: A shrub with erect to prostrate stems that forms small to extensive clusters. Shrubs typically are less than 30 cm (1 ft) tall. **LEAVES** are alternate, simple, and evergreen. Blades are less than 12 mm (.5 in.) long and appear to be needlelike, as they are tightly rolled under, which hides their lower surfaces. **INFLORESCENCES** are short, terminal clusters of flowers. **FLOWERS** are 5-parted, bell shaped, and rose-purple, with stamens exserted beyond the corolla. The corolla lobes are as long as, or longer than, the corolla tube. **FRUITS** are spherical capsules.

HABITAT AND RANGE: Grows in the southern Cascades (only near Mount Lassen), through the Sierra Nevada, and in the San Bernardino Mountains in open subalpine forests and alpine meadows, at elevations from 2,000 m (6,500 ft) to 3,600 m (12,000 ft).

REMARKS: Brewer heather is one of 2 mountain heathers growing at high elevations in California. They can be differentiated by range alone. Brewer heather occurs south of Mount Lassen.

CASCADE HEATHER *Phyllodoce empetriformis*

(Fig. 141)

DESCRIPTION: A shrub with erect to prostrate stems that forms small to extensive thickets. Shrubs typically are less than 30 cm (1 ft) tall. **LEAVES** are alternate, simple, and evergreen.

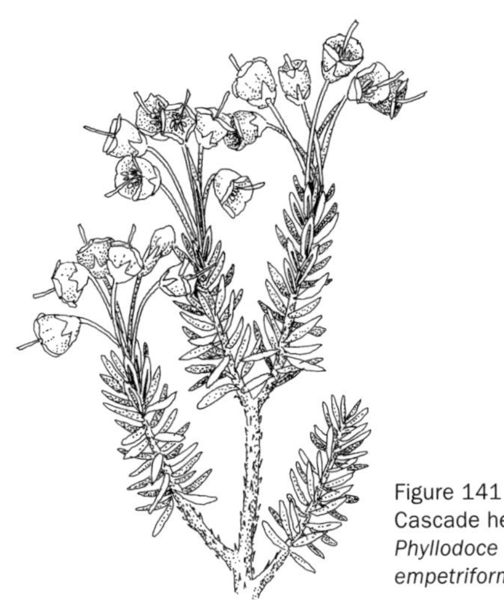

Figure 141
Cascade heather,
Phyllodoce empetriformis

Blades are less than 12 mm (.5 in.) long and appear to be needle-like, as they are tightly rolled under, which hides their lower surfaces. **INFLORESCENCES** are short, terminal clusters of flowers. **FLOWERS** are 5-parted, urn shaped, and rose-purple, with stamens included in the corolla. The corolla lobes are shorter than the corolla tube. **FRUITS** are spherical capsules. **HABITAT AND RANGE:** In California it grows in the Klamath Mountains and on Mount Shasta, in open subalpine forests and alpine meadows at elevations from 1,500 m (5,000 ft) to 2,700 m (9,000 ft). It ranges north through the northern Rocky Mountains to Alaska.

REMARKS: Cascade heather is one of 2 red-colored mountain heathers growing at high elevations in California. They can be differentiated by range alone. Cascade heather is found north of Mount Lassen.

PHYSOCARPUS (NINEBARK)

Physocarpus is a genus with about 10 species in North America and Asia. Two of these are native to California.

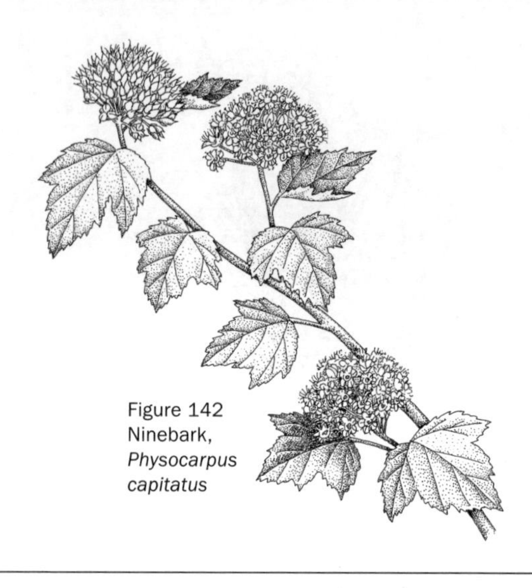

Figure 142
Ninebark,
Physocarpus capitatus

NINEBARK (Fig. 142) ***Physocarpus capitatus***

DESCRIPTION: A shrub that may grow to be between 1 m (3.3 ft) and 2 m (6.5 ft) tall. **LEAVES** are alternate, deciduous, and thin. Blades are 4 cm (1.5 in.) to 6 cm (2.5 in.) long and maplelike, with 3 or 5 lobes. The upper surface is hairless and the lower one may have star-shaped hairs. **INFLORESCENCES** are terminal, rounded clusters that extend from leafy twigs and measure 5 cm (2 in.) to 7.5 cm (3 in.) across. **FLOWERS** are white. **FRUITS** are a set of 1 to 5 follicles that open on 1 or both sides and are covered with starlike hairs. **TWIGS** are brown; as they age, long, thin sheets of brown bark peel off the branches.

HABITAT AND RANGE: Found in low-elevation and montane forests and woodlands in northern California mountains, at elevations below 1,400 m (4,500 ft).

REMARKS: Ninebark looks like just another rose family shrub with white flowers until you have seen an older plant. Its many layers of shreddy, brown bark make it quite distinctive. Use a hand lens to see the starlike hairs on the leaves and fruits. Another species, *P. alternans*, grows in mountains east of the Sierra Nevada. Its leaves are smaller than those of ninebark.

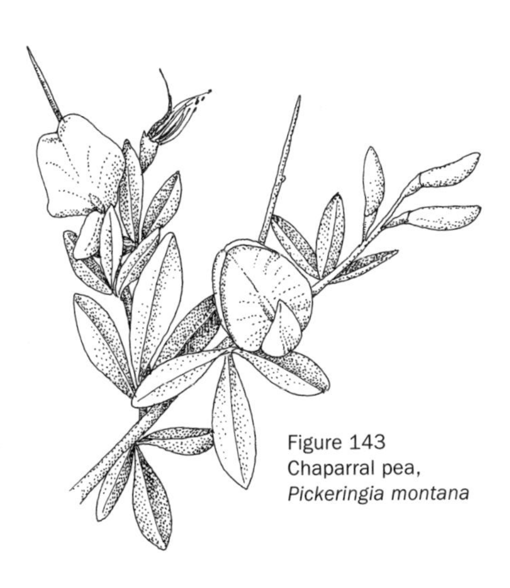
Figure 143
Chaparral pea,
Pickeringia montana

PICKERINGIA (CHAPARRAL PEA)

The genus *Pickeringia* has 1 shrub species that grows in California and Baja California.

CHAPARRAL PEA (Fig. 143) ***Pickeringia montana***

DESCRIPTION: A medium-sized, thorny shrub that can grow to be 60 cm (2 ft) to 2 m (6.5 ft) tall. **LEAVES** are palmately compound, with 3 leaflets (but are occasionally simple), alternate, and evergreen. Leaflets are egg shaped, with the broader part above the middle, and from 6 mm (.25 in.) to 13 mm (.5 in.) long. Leaf tips are pointed. Margins are entire. Surfaces are hairless or have minute hairs. Petioles are absent or nearly so. **INFLORESCENCES** are solitary flowers in unbranched clusters. **FLOWERS** are peas flowers with striking magenta-colored petals that measure about 13 mm (.5 in.) to 19 mm (.75 in.) long. **FRUITS** are seldom seen, but when produced they are flat legumes about 3 cm (1.25 in.) to 5 cm (2 in.) long. **TWIGS** are spiny and stiff.

HABITAT AND RANGE: Grows in foothill and lower montane chaparrals and woodlands from Mendocino County to northern Baja California, at elevations below 1,500 m (5,000 ft). It can also be found in the foothills of the Sierra Nevada from Butte County to Mariposa County and on Santa Cruz Island.

REMARKS: Chaparral pea regenerates primarily from sprouting root systems and rarely from seed. It is adapted to fire, able to sprout from its roots and root crown. Plants can occur singly or in impenetrable thickets. Chaparral pea provides important browse for deer and helps to replenish nitrogen in soils. The variety found in southern California, *P. m.* var. *tomentosa,* has hairier leaves and twigs than *P. m.* var. *montana,* found in northern California.

PLATANUS (SYCAMORE)

The genus *Platanus* has 6 or 7 tree species; 3 are native to North America, and 1 of these is native to California. *Platanus* is also found in Mexico, Europe, and Asia Minor.

CALIFORNIA SYCAMORE *Platanus racemosa*

(Fig. 144)

DESCRIPTION: An erect, leaning, or forked tree. It is medium sized and can be single- or multistemmed. Trunks are typically short and forked, with ascending, crooked, stout branches. Mature trees are commonly 9 m (30 ft) to 24 m (80 ft) tall and 30 cm (1 ft) to 90 cm (3 ft) in diameter. The largest grows in Goleta and is 27.4 m (90 ft) tall and 2.4 m (8 ft) in diameter. Crowns are irregularly rounded. Trees live more than 200 years. **LEAVES** are deciduous, round, simple, alternate, and 10 cm (4 in.) to 25 cm (10 in.) in diameter. They are palmately lobed with 3 to 5 lobes, which are toothed, entire, and somewhat narrow. The sinuses between lobes are about 50% of the lobe length. Upper surfaces are light green and lower surfaces are paler, with rusty hairs. Petioles are 12 mm (.5 in.) to 7.5 cm (3 in.) long, with conspicuous, leaflike stipules about 2.5 cm (1 in.) long. **FLOWERS** are unisexual and occur in spherical clusters that are around 6 mm (.25 in.) in diameter. There are usually 4 or 5 pollen-bearing flower clusters on new twigs. Seed-bearing flower clusters are found on older branches and number between 2 and 7. **FRUITS** are spherical clusters of hairy achenes about 2.5 cm

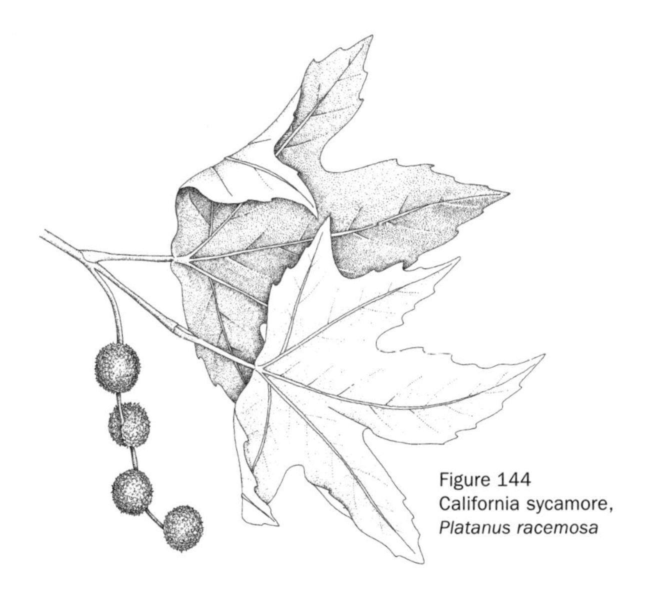

Figure 144
California sycamore,
Platanus racemosa

(1 in.) in diameter. **BARK** is 2.5 cm (1 in.) to 7.5 cm (3 in.) thick, furrowed, and dark brown at the bases of mature trees. Young trees, branches, and upper trunks have smooth, mottled, scaly bark that is white, tan, green, or gray.

HABITAT AND RANGE: Grows in riparian woodlands below 1,400 m (4,500 ft) in the warmer parts of California, except for the deserts. It also occurs in Baja California.

REMARKS: California sycamore is fast-growing and is the tallest tree in its native habitat. Its abundant shade helps to keep streams and riparian zones cool. Wildlife species use it for nesting and roosting and as hiding cover. Birds eat California sycamore seeds. The fungus anthracnose infects California sycamore in the spring and may occasionally completely defoliate trees. Few trees die, however, since a second set of leaves is often produced.

PLUCHEA (ARROW-WEED)

Pluchea is a genus of some 40 leafy herb and shrub species found in the warmer parts of the world. Two of these species grow in California.

ARROW-WEED *Pluchea sericea*

DESCRIPTION: A thicket-forming shrub that may grow to be 5 m (16 ft) tall. **LEAVES** are drought-deciduous, alternate, crowded, and simple. Blades have a narrow lance shape, measure from 2 cm (.75 in.) to 4 cm (1.5 in.) long, and are covered with silky hairs. **INFLORESCENCES** are flat-topped sets of purplish flower heads. **STEMS** are straight and covered with fine silvery hairs. **FRUITS** are achenes about 1 mm (.04 in.) long.
HABITAT AND RANGE: Grows in the San Joaquin Valley, the eastern foothills of the Coast Ranges, southern California, and the hot deserts. It occurs at elevations below 600 m (2,000 ft). The species grows as extensive thickets around springs, in stream bottoms, and in washes. The habitat may be saline.
REMARKS: The most-photographed stand of arrow-weed is near Stovepipe Wells in Death Valley and is called the Devil's Cornfield.

POPULUS (COTTONWOOD)

The genus *Populus* is restricted to the Northern Hemisphere and has about 40 species; about 10 are native to North America, and 4 of these are native to California (3 cottonwoods and 1 aspen). The genus is made up of small to tall trees.

Leaves are deciduous; alternate; simple; generally hairless; egg shaped, triangular, or round; and 7.5 cm (3 in.) to 12.5 cm (5 in.) long. Petioles are usually long and may be round or flat in cross section. Leaf tips are usually tapered. Small pollen-bearing and seed-bearing flowers are aggregated in droopy, slender catkins. The 2 kinds of flowers are found on separate trees. Mature seed-bearing catkins are 10 cm (4 in.) to 15 cm (6 in.) long and have 2- to 4-parted capsules housing several seeds. Seeds are small with long, silky hairs that help in wind dispersal. Some species disperse copious, cottony seed masses that can collect in depressions and street gutters. Twigs are slender to stout, hairy or hairless, and olive brown to reddish brown. They have evident lenticels. Buds have overlapping scales and may be resinous or nonresinous. Leaf scars are prominent and are usually triangular. Bark is gray, grayish brown, whitish gray, or white.

Cottonwoods and aspens typically grow in moist woodlands and forests, often in riparian areas. A considerable amount of breeding has been done with this genus, resulting in extremely fast-growing hybrid poplars and cottonwoods that are being used in energy plantations.

1. Leaves are lance shaped. Trees grow in the eastern Sierra Nevada and in canyons of the White Mountains .... **narrowleaf cottonwood** (*P. angustifolia*)
1. Leaves are triangular, egg shaped, or round. Trees are widely distributed in California 2
   2. Leaves are triangular. Leaf margins have coarse scallops .... **Fremont cottonwood** (*P. fremontii*)
   2. Leaves are egg shaped or round. Leaf margins have fine scallops 3
3. Leaves are egg shaped. Petioles are round in cross section .... **black cottonwood** (*P. balsamifera*)
3. Leaves have a broad egg shape to round shape. Petioles are flat in cross section .... **quaking aspen** (*P. tremuloides*)

NARROWLEAF COTTONWOOD *Populus angustifolia*

DESCRIPTION: A small- to medium-sized, single-stemmed tree. Mature trees are usually 9 m (30 ft) to 15 m (50 ft) tall and 30 cm (1 ft) to 45 cm (1.5 ft) in diameter. The largest grows in Malheur County, Oregon, and is 24 m (79 ft) tall and 2.5 m (100 in.) in diameter. Crowns are narrow and conical. Branches are upright, slender, and willowlike. The species is short-lived. **LEAVES** are deciduous, alternate, simple, lance shaped, 5 cm (2 in.) to 12.5 cm (5 in.) long, and 12 mm (.5 in.) to 25 mm (1 in.) wide. Margins are minutely serrated. Leaf tips are elongated and sharp pointed and bases are round. Upper surfaces are shiny yellowish green and hairless, while lower surfaces are paler. Petioles are flat at their bases. **FRUITS** hang in catkins that measure 10 cm (4 in.) to 15 cm (6 in.) long and are composed of many 2-parted capsules. Seeds have silky hairs. **TWIGS** are initially yellowish brown to orange, becoming whitish gray in the second year. **BUDS** are sharp pointed and have overlapping, sticky scales. **BARK** on mature trees is characterized by flat-topped ridges separated by deep furrows.

HABITAT AND RANGE: Grows in riparian woodlands

throughout the Rocky Mountains from Canada to Mexico. In California it is rare and is found in the eastern Sierra Nevada and the White Mountains. It ranges from 1,200 m (4,000 ft) to 1,800 m (5,900 ft).

REMARKS: Narrowleaf cottonwood is shade intolerant and sprouts from roots following injury from fire or cutting. It provides valuable wildlife habitat. Native Americans used it to make baskets. Narrowleaf cottonwood is planted as an ornamental.

BLACK COTTONWOOD *Populus balsamifera* ssp. *trichocarpa*

(Fig. 145; Pl. 27)

DESCRIPTION: A medium-sized to tall, single-stemmed tree. Mature trees are typically 30 m (100 ft) to 38 m (125 ft) tall and 45 cm (1.5 ft) to 90 cm (3 ft) in diameter. The largest grows in Willamette Mission State Park, Oregon, and is 48 m (158 ft) tall and 2.6 m (102 in.) in diameter. Crowns are open and broad and the trees have long, clear trunks. **LEAVES** are deciduous, alternate, simple, 7.5 cm (3 in.) to 15 cm (6 in.) long, and 5 cm (2 in.) to 10 cm (4 in.) wide. They have an egg shape to narrow egg shape; margins are finely scalloped. Leaf tips are elongated and sharp pointed and bases are round to heart shaped. Upper surfaces are shiny, dark green, and hairless, while lower surfaces are pale green with rusty streaks. Petioles are round in cross section. **FRUITS** hang in catkins that measure 10 cm (4 in.) to 15 cm (6 in.) long and are composed of many 3-parted capsules. Seeds have silky hairs. **TWIGS** are orange-brown. **BUDS** are large, sharp pointed, and fragrant. Scales are overlapping and sticky. **BARK** on mature trees has grayish, flat-topped ridges separated by deep furrows.

HABITAT AND RANGE: Grows in riparian woodlands from Alaska, the Pacific Northwest, and northern Rocky Mountains to Baja California. In California it is widely distributed, except in the southeastern deserts. It ranges from sea level to 3,000 m (10,000 ft).

REMARKS: Black cottonwood is shade intolerant and is crowded out by conifers or other shade-tolerant species when there are no recurring disturbances such as

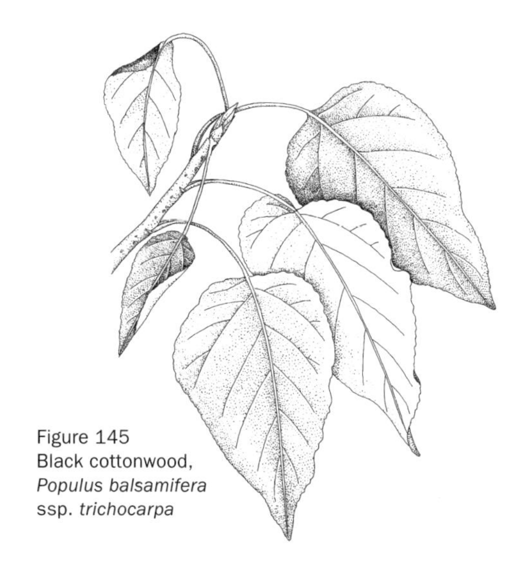
Figure 145
Black cottonwood,
Populus balsamifera
ssp. *trichocarpa*

flooding. The species is used as lumber and for pulp and windbreaks. Its rapid growth has sparked considerable interest in plantations in the Pacific Northwest. The species provides valuable wildlife habitat and streamside protection. Balsam poplar (*P. b.* var. *balsamifera*) grows in Alaska, Canada, and the eastern United States. Some botanists distinguish black cottonwood as a separate species (*P. trichocarpa*). It hybridizes with Fremont cottonwood.

FREMONT COTTONWOOD *Populus fremontii* ssp. *fremontii*

(Fig. 146)

DESCRIPTION: A medium-sized, single-stemmed tree. Mature trees are usually 12 m (40 ft) to 30 m (100 ft) tall and 60 cm (2 ft) to 90 cm (3 ft) in diameter. The largest grows in Santa Cruz County, Arizona, and is 28 m (92 ft) tall and 4 m (160 in.) in diameter. Crowns are broadly rounded to cylindrical. **LEAVES** are deciduous, alternate, simple, triangular, and 5 cm (2 in.) to 7.5 cm (3 in.) long and wide. Margins are coarsely scalloped. Leaf tips are sharp pointed and bases are truncated to more or less heart shaped. Upper and lower surfaces are yel-

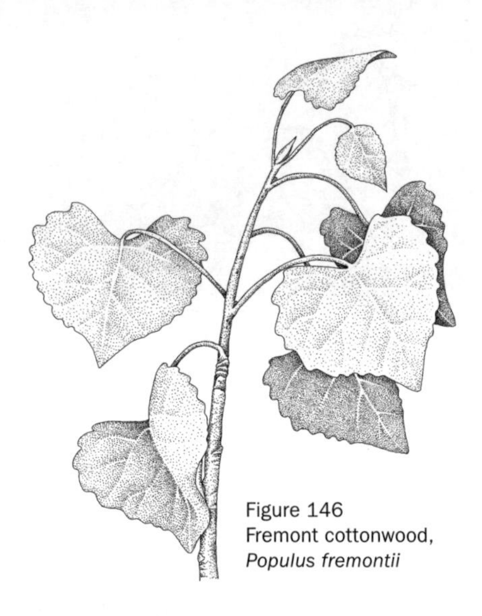

Figure 146
Fremont cottonwood,
Populus fremontii

lowish green, hairless to hairy, and often stained with milky resin. Petioles are yellow and flat in cross section. **FRUITS** hang in catkins that measure about 10 cm (4 in.) long and are composed of many 3-parted capsules. Seeds have silky hairs. **BUDS** at the tips of branches are about 1 cm (.4 in.) long, are sharp pointed, and have overlapping scales. **BARK** on mature trees has whitish, flat-topped ridges separated by furrows.

HABITAT AND RANGE: Grows in riparian woodlands in the southern and middle Rocky Mountains, the Great Basin, Mexico, and California. In California it is widely distributed at elevations below 2,000 m (6,500 ft), except on the Modoc Plateau. This species is replaced by black cottonwood near the coast and at higher elevations.

REMARKS: Fremont cottonwood is shade intolerant. It becomes established on disturbed sites near streams that have bare, moist soil. The species is used as firewood and to make fence posts and pallets. It provides critical habitat for wildlife throughout the southwest in "gallery" forests along streams. Cottonwoods in general, and especially Fremont cottonwood, have served as indicators of surface water. Widespread planting of Fre-

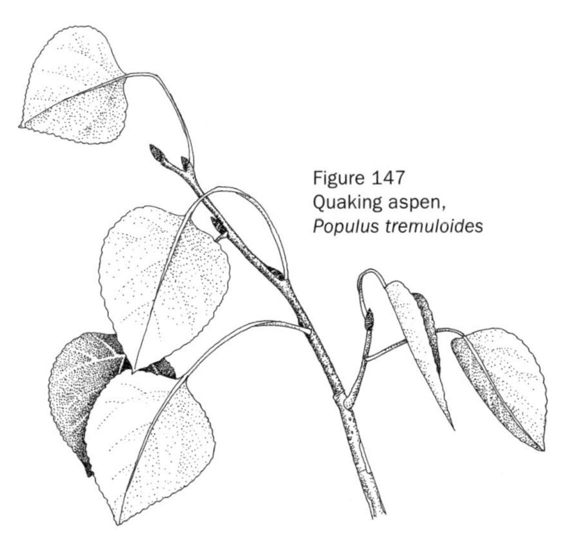
Figure 147
Quaking aspen,
Populus tremuloides

mont cottonwood has provided needed habitat and shade in the Southwest. Two subspecies are recognized: *P. f.* ssp. *fremontii* and the Arizona cottonwood (*P. f.* ssp. *mesetae*). Fremont cottonwood hybridizes with narrowleaf cottonwood and black cottonwood, but apparently not in California.

QUAKING ASPEN (Fig. 147) ***Populus tremuloides***

DESCRIPTION: A medium-sized, single-stemmed tree. Mature trees are often 6 m (20 ft) to 24 m (80 ft) tall and 10 cm (4 in.) to 60 cm (24 in.) in diameter. The largest grows in Ontonagon County, Michigan, and is 33.2 m (109 ft) tall and 1 m (39 in.) in diameter. Quaking aspen groves are usually composed of a mosaic of clones, with individual clones ranging in size from a few trees to tens of acres. Tree crowns are rounded on long, clear trunks. The species' individual stems are short-lived, with the oldest trees surviving between 100 and 200 years. Clones in the Great Basin, however, are estimated to be over 8,000 years old. **LEAVES** are deciduous, alternate, simple, and 4 cm (1.5 in.) to 7.5 cm (3 in.) in diameter. They have a broad egg shape to round shape. Margins are finely scalloped. Leaf tips are broadly tapered and bases are round to heart shaped. Upper surfaces are shiny and yellowish green, while lower surfaces are pale green.

Long, flat petioles cause aspen leaves to "tremble" in the wind. **FRUITS** hang in catkins that measure about 10 cm (4 in.) long and are composed of many 2-parted capsules. Seeds have silky hairs. **TWIGS** are shiny and brownish gray, with abundant lenticels. Buds are conical and nonresinous. **BARK** is initially greenish gray, gray, or white. Bark at the bases of large trees has dark brown, flat-topped ridges separated by dark furrows.
HABITAT AND RANGE: Grows in forests from the Arctic to Mexico and from the Atlantic to the Pacific Ocean. It occurs in a great diversity of habitat types. In California it is often found near montane and upper montane meadows and creeks, at elevations from 900 m (3,000 ft) to 3,000 m (10,000 ft).
REMARKS: Quaking aspen is shade intolerant. In the West, it reproduces almost exclusively from root suckers following fire, flooding, grazing, or cutting. The species is used to make pulp, manufactured boards, furniture, and novelty items. It provides valuable wildlife habitat and forage for a wide variety of species, notably deer, elk, and beaver. Quaking aspen clones provide the most extensive fall foliage color in the West.

POTENTILLA (CINQUEFOIL)

The genus *Potentilla* has about 500 species in the Northern Hemisphere. Most are small perennial herbs.

SHRUB CINQUEFOIL (Fig. 148) *Potentilla fruticosa*

DESCRIPTION: A multistemmed shrub that may grow to be 1.2 m (4 ft) tall. **LEAVES** are alternate, deciduous, and compound. Blades are pinnately compound or lobed and 12 mm (.5 in.) to 3 cm (1.25 in.) long, with 2 or 3 pairs of linear leaflets. The upper surface is green and hairy and the margin is rolled under. **INFLORESCENCES** are terminal clusters of 1 to several flowers. **FLOWERS** are yellow or white. **FRUITS** are many achenes in a dry, strawberry-like receptacle. **TWIGS** have silky hairs.
HABITAT AND RANGE: Grows throughout the northern parts of North America and Eurasia. In California it occurs in subalpine forests and woodlands and in alpine habitats, at elevations from 2,000 m (6,500 ft) to 4,000 m (13,000 ft).

Figure 148
Shrub cinquefoil,
Potentilla fruticosa

REMARKS: Unlike many other wide-ranging species, shrub cinquefoil shows little regional variation. Some botanists, however, segregate shrub cinquefoil from other *Potentilla* species by naming it *Petaphylloides floribunda*. By whatever name, the plant is hardy and sufficiently handsome to be a commonly planted ornamental.

PROSOPIS (MESQUITE)

The genus *Prosopis* has about 44 species in America, Asia, and Africa. Four of these grow in California, 2 of them native and 2 naturalized.

HONEY MESQUITE — *Prosopis glandulosa* var. *torreyana*

(Fig. 149)

DESCRIPTION: An erect, multistemmed shrub or small tree that grows to be between 2 m (6.5 ft) and 9 m (30 ft) tall. Stems are short, with many arched, crooked branches. Crowns are

Figure 149
Honey mesquite,
Prosopis glandulosa var. *torreyana*

broader than tall. **LEAVES** are deciduous, alternate, hairless, and pinnately compound. Primary leaflets form in pairs and are 5 cm (2 in.) to 10 cm (4 in.) long. Secondary leaflets have 7 to 17 pairs of leaflets that are each 1 cm (.4 in.) to 2.5 cm (1 in.) long, linear to oblong, and entire. **FLOWERS** are clustered in greenish yellow spikes 5 cm (2 in.) to 9 cm (3.5 in.) long. **FRUITS** are legumes in droopy clusters that measure 7.5 cm (3 in.) to 20 cm (8 in.) long. **TWIGS** have paired spines measuring from 6 mm (.25 in.) to 3 cm (1.25 in.) long that can be found at nodes and the bases of petioles.

HABITAT AND RANGE: Widely distributed in the deserts of the southwestern United States and Mexico. In California it is common in southern California's Mojave, Colorado, and Sonoran Deserts. Some populations occur in the San Joaquin Valley and are thought to have been planted about 120 years ago. The species is usually found growing in washes and on lower slopes, at elevations below 1,600 m (5,500 ft).

REMARKS: In addition to honey mesquite, 3 other mesquites grow in California: *P. pubescens* (screw bean), *P. strombulifera,* and *P. velutina.* All 3 are uncommon, and the last 2 are non-native.

Honey mesquite is used as firewood and lumber and for making fence posts and charcoal. Domestic livestock and wildlife species eat the protein-rich, sugary legumes. Many wildlife species utilize honey mesquite for cover. Native Americans exploited honey mesquite for food, shelter, medicines, clothing, and many other domestic uses. *P. glandulosa* var. *glandulosa* grows in upland settings and is known as an invader in historic grasslands of the southwestern United States.

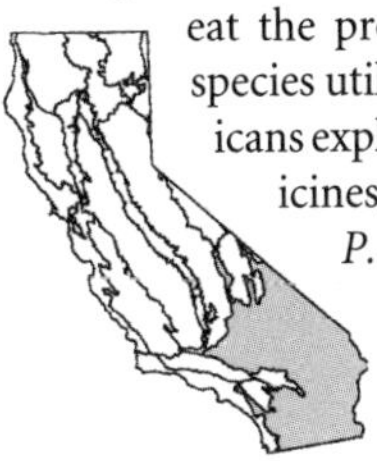

PRUNUS (CHERRY)

The genus *Prunus* has 400 species growing throughout the Northern Hemisphere. Eight of these are native to California.

Prunus is a collection of shrubs and trees with simple leaves that are alternate in arrangement. Many species have a pair of warty glands present at the tops of the petioles or at the bases of the leaf blades. Branches and trunks are typically smooth and covered with lenticels. Flowers are showy, pink to white blossoms that appear in the spring. Each flower has 5 petals and many stamens attached to the top of a cup surrounding a single pistil. On the same plant, flowers may be complete or seed-bearing or stamen-bearing. Fruits are drupes with 1 stone and typically embedded in fleshy pulp.

This genus includes many species that are cultivated for their edible fruit, for lumber, or for making novelty items. It is easy to find almond (*P. dulcis*), apricot (*P. armeniaca*), peach (*P. persica*), plum (*P. domestica*), sour cherry (*P. cerasus*), and sweet cherry (*P. avium*) plants growing wild as well as in yards and orchards. Ornamentals such as the evergreen cherry-laurels (*P. caroliniana, P. laurocerasus, P. lusitanica*) and the deciduous cherry plum (*P. cerasifera*) are spread by birds beyond their initial locations. Seeds and leaves are toxic, as they contain hydrocyanic acid.

Only the native California species are presented here.

1. Leaves are leathery and evergreen .... **hollyleaf cherry (*P. ilicifolia*)**
1. Leaves are thin and deciduous 2

2. Twigs are rigid and often thorny 3
2. Twigs are not rigid or thorny 6

3. Leaf blades have a linear or narrow oblong shape 4
3. Leaf blades are elliptical or round 5

4. Margins of the leaf blades are smooth . . . . **desert almond** (*P. fasciculata*)
4. Margins of the leaf blades have fine teeth . . . . **desert peach** (*P. andersonii*)

5. Leaves are clustered and blades are 12 mm (.5 in.) to 20 mm (.75 in.) long . . . . **desert apricot** (*P. fremontii*)
5. Leaves are not clustered, and blades are 20 mm (.75 in.) to 5 cm (2 in.) long . . . . **Klamath plum** (*P. subcordata*)

6. A pair of warty glands is usually evident at the bases of leaf blades . . . . **bitter cherry** (*P. emarginata*)
6. A pair of warty glands is usually evident on the petioles near the bases of leaf blades 7

7. Leaf tips taper to a point. Flowers and fruits form elongated, many-flowered clusters . . . . . . **choke cherry** (*P. virginiana*)
7. Leaf tips are round. Flowers and fruits form roundish, few-flowered clusters . . . . **Klamath plum** (*P. subcordata*)

DESERT PEACH (Fig. 150) *Prunus andersonii*

DESCRIPTION: A multistemmed, thorny shrub that may grow to be 4.5 m (15 ft) tall. **LEAVES** are deciduous and closely clustered on short branches. Leaf blades are narrow and oblong, 12 mm (.5 in.) to 25 mm (1 in.) long, with very fine-toothed margins. Petioles are absent to 6 mm (.25 in.) long. **INFLORESCENCES** consist of 1 to 3 flowers. **FLOWERS** are rose colored, white, or yellow; complete; and about 12 mm (.5 in.) broad. **FRUITS** are dark red and have a dense covering of fine hair. The pulp covering the stone is dry. **TWIGS** are rigid, and most end in sharp thorns.

HABITAT AND RANGE: Grows in open scrubs and woodlands in mountains of the Great Basin and northern Mojave Deserts, at elevations from 1,000 m (3,300 ft) to 2,300 m (7,500 ft).

REMARKS: Desert peach and desert almond may be confused. Desert peach is a Great Basin species, while desert almond's range is farther south.

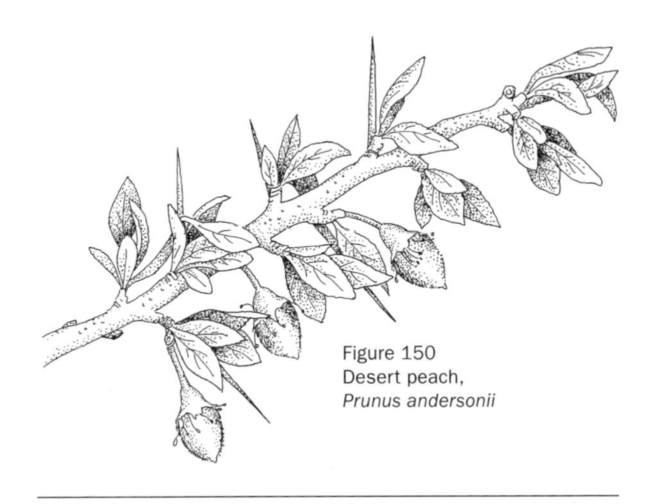
Figure 150
Desert peach,
Prunus andersonii

BITTER CHERRY (Fig. 151) ***Prunus emarginata***

DESCRIPTION: A shrub or small tree that usually grows to be between 1 m (3.3 ft) and 9 m (30 ft) tall. The largest lives in Seward Park, Seattle, and is 30.5 m (100 ft) tall and 45 cm (18 in.) in diameter. **LEAVES** are deciduous and clustered on the older branches. Leaf blades are narrow and oval or widest above the middle. They measure 2 cm (.75 in.) to 5 cm (2 in.) long and have fine-toothed margins. Petioles are 3 mm (.12 in.) to 12 mm (.5 in.) long. Leaf bases are wedge shaped. Upper blade surfaces are dark green and hairless and lower surfaces are hairless or hairy. A pair of warty glands is found on the basal margin of the leaf blade or on the petiole near the base of the blade. **INFLORESCENCES** consist of 5 to 12 flowers in short, flat-topped clusters. **FLOWERS** are white, complete, and about 12 mm (.5 in.) wide. **FRUITS** are 6 mm (.25 in.) to 12 mm (.5 in.) long, oval, and bright red, turning black with age. The pulp covering the stone is fleshy and very bitter. **TWIGS** are slender, red or gray, and hairy or hairless. Lenticels are evident.

HABITAT AND RANGE: Grows in montane chaparrals, open forests, woodlands, and other open habitats throughout western Canada and the United States. In California it grows below 2,800 m (9,200 ft).

REMARKS: Most Californians know

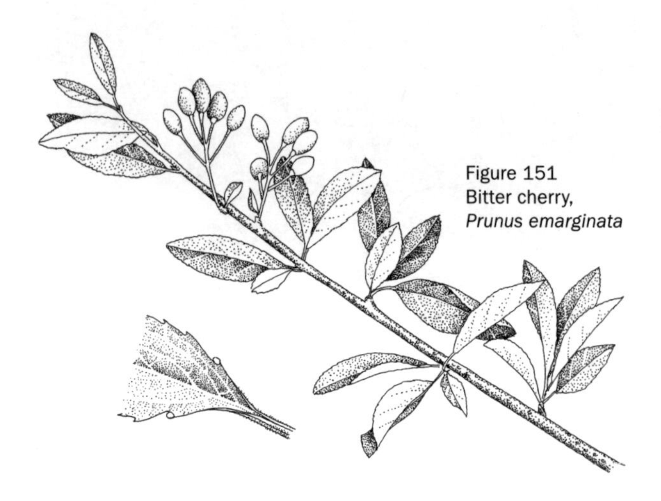
Figure 151
Bitter cherry,
Prunus emarginata

the shrub form of bitter cherry (*P. e.* var. *emarginata*), which has hairless leaves; but rather different-looking, tree-sized bitter cherry plants with hairy leaves (*P. e.* var. *mollis*) grow in coastal Del Norte County, Oregon, and Washington. Both forms provide important forage for wildlife. The species' bark is used in basketry and its seedlings are used in habitat restoration.

DESERT ALMOND *Prunus fasciculata*

DESCRIPTION: A multistemmed, thorny shrub that may grow to be between 1 m (3.3 ft) and 2 m (6.5 ft) tall. **LEAVES** are deciduous and densely clustered on short branches. Leaf blades are narrow and spoon shaped; they measure 6 mm (.25 in.) to 20 mm (.75 in.) long and have generally smooth margins. **INFLORESCENCES** consist of 1 to 3 flowers. **FLOWERS** are white or yellow, complete, and 4 mm (.15 in.) to 8 mm (.3 in.) wide. Only seed-bearing or stamen-bearing flowers are found on an individual plant. **FRUITS** are gray or red-brown, with dry pulp covering the stones. **TWIGS** are rigid, and most end in sharp thorns.

HABITAT AND RANGE: Grows in open chaparrals, scrubs, and woodlands of central and southern California, at elevations from 750 m (2,500 ft) to 2,000 m (6,500 ft). It ranges throughout the intermountain West and south into Baja California.

REMARKS: There are 2 subspecies of desert almond: *P. f.* var. *punctata,* which is characterized by hairless leaf blades and grows in the southern central Coast and Transverse Ranges; and *P. f.* var. *fasciculata,* which is found in deserts. Both forms are browsed by wildlife. Desert almond may be confused with dcsert peach, but it has a more southerly range than does desert peach.

DESERT APRICOT *Prunus fremontii*

DESCRIPTION: A multistemmed, often thorny shrub that may grow to be between 1 m (3.3 ft) and 4.5 m (15 ft) tall. **LEAVES** are deciduous and clustered on short branches. Leaf blades are round or egg shaped and 12 mm (.5 in.) to 20 mm (.75 in.) long, with fine teeth. Petioles are 2 mm (.08 in.) to 7 mm (.25 in.) long. Leaves arise from short spur shoots. Upper and lower surfaces are hairless. **INFLORESCENCES** consist of 1 to 10 flowers. **FLOWERS** are white, complete, and about 12 mm (.5 in.) broad. **FRUITS** are yellow and have fine hairs, with dry pulp covering the stones. **TWIGS** are rigid, and most end in thorns.

HABITAT AND RANGE: Grows in open chaparrals, scrubs, and woodlands of the eastern Peninsular Ranges, Baja California, and the western Colorado Desert, at elevations from 750 m (2,500 ft) to 2,000 m (6,500 ft).

REMARKS: The dry fruits of desert apricot and desert almond are southern California's spiny desert "cherries." Desert apricot's broad leaves, however, help to differentiate it from the narrow-leaved desert almond.

HOLLYLEAF CHERRY (Fig. 152) *Prunus ilicifolia*

DESCRIPTION: A shrub or tree that typically grows to be 15 m (50 ft) tall. The largest lives in Jolon, California, and is 17 m (56 ft) tall and 76 cm (30 in.) in diameter. **LEAVES** are evergreen, dark green, and leathery. Leaf blades are oval or elliptical, 2 cm (.75 in.) to 5 cm (2 in.) long, dark green, and shiny, with spiny or smooth margins. Petioles are 4 mm (.15 in.) to 25

Figure 152
Hollyleaf cherry,
Prunus ilicifolia

mm (1 in.) long. **INFLORESCENCES** are long clusters up to 5 cm (2 in.) in length. **FLOWERS** are white and about 6 mm (.25 in.) broad. **FRUITS** are dark purple or black, with a very thin covering of sweet pulp over the stone. **TWIGS** are gray or reddish brown.

HABITAT AND RANGE: Grows in chaparrals and woodlands of central and southern California, from sea level to 1,500 m (5,000 ft). It ranges south into Baja California.

REMARKS: There are 2 subspecies of hollyleaf cherry. On the Channel Islands, the Catalina cherry (*P. i.* var. *lyonii*) has leaf blades with smooth margins. On the mainland and on San Clemente and Santa Catalina Islands, hollyleaf cherry (*P. i.* var. *ilicifolia*) has leaf blades with spiny margins. Both varieties are planted as ornamentals on the mainland, where they hybridize. Some books treat these forms as separate species. Both varieties can become trees, but in most areas on the mainland the intervals between fires are sufficiently short that the plants remain shrubs.

KLAMATH PLUM *Prunus subcordata*

DESCRIPTION: A stiff-branched shrub or small tree that may grow to be between 2 m (6.5 ft) and 7.5 m (25 ft) tall. **LEAVES** are deciduous and open-clustered on branches. Leaf blades are round or elliptical, 2 cm (.75 in.) to 5 cm (2 in.) long, and 12 mm (.5 in.) to 4 cm (1.5 in.) wide, with very fine-toothed margins. Petioles are 4 mm (.15 in.) to 15 mm (.6 in.) long. A pair of warty glands is found on the petioles near the bases of blades. **INFLORESCENCES** consist of 1 to 7 flowers. **FLOWERS** are white or pink, complete, and about 12 mm (.5 in.) broad. **FRUITS** are oblong, dark red or yellow, hairless, and from 2 cm (.75 in.) to 2.5 cm (1 in.) long. The pulp covering stones may be fleshy or dry. **TWIGS** are rigid and crooked and may end in thornlike points.

HABITAT AND RANGE: Grows in open forests from southern Oregon to Santa Cruz County on the coast and in Kern County inland, at elevations below 2,000 m (6,500 ft). The species is often encountered along fencerows.

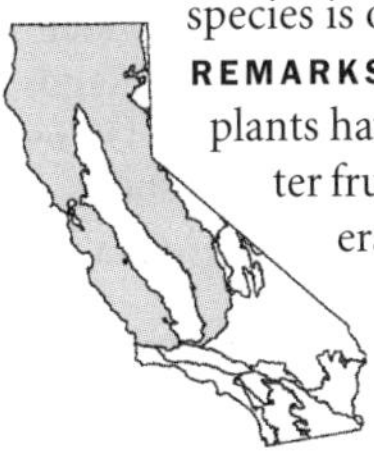

REMARKS: Klamath plum varies regionally. Some plants have yellow, sweet fruits. Others have red, bitter fruits. Some plants have white flowers and others are pink. Subspecific names exist to help describe this variation, but botanists do not agree on them, so they will not be included here. Many plants have tasty, edible fruits.

CHOKE CHERRY (Fig. 153) *Prunus virginiana*

DESCRIPTION: A flexible-branched shrub or small tree that often grows to be between 1 m (3.3 ft) and 6 m (20 ft) tall. The largest lives in Kootenai County, Idaho, and is 22.2 m (73 ft) tall and 43 cm (17 in.) in diameter. **LEAVES** are deciduous. Leaf blades have a broad oval shape or are widest above the middle, measure 6 cm (2.5 in.) to 9 cm (3.5 in.) long, and have fine-toothed margins. Leaf tips are sharp pointed. Petioles are 10 mm (.4 in.) to 25 mm (1 in.) long. A pair of warty glands is found on the petiole near the base of the blade. **INFLORESCENCES** consist of many flowers in long, cylindrical clusters that measure 5 cm (2 in.) to 12.5 cm (5 in.) long. **FLOWERS** are white, complete, and 6 mm (.25 in.) to 12 mm (.5 in.) broad. **FRUITS** are hairless and bright red, dark red, black, or yellow. The pulp

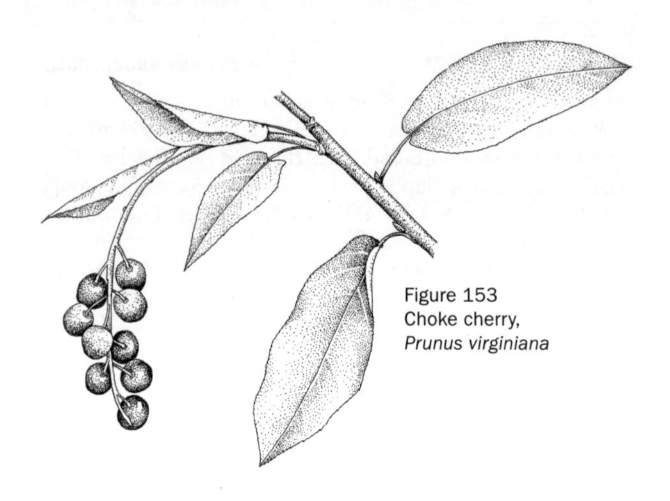

Figure 153
Choke cherry,
Prunus virginiana

covering the stone is fleshy and astringent. **TWIGS** are slender with fine gray hair that is lost with age. Lenticels are evident.

HABITAT AND RANGE: Grows in open forests, woodlands, and open habitats throughout southern Canada and the United States. In California it lives at elevations from 100 m (300 ft) to 2,900 m (9,500 ft), where it is often encountered along fencerows. Its flowers and fruits are showy.

REMARKS: There are 2 varieties of choke cherry. Western choke cherry (*P. v.* var. *demissa*) grows in California; its lower leaf blades are covered with downy hairs. Eastern choke cherry (*P. v.* var. *virginiana*) grows to the east, is tree-sized, and has hairless lower leaf blades. The fruits of both are edible, although astringent. Wildlife species eat choke cherry foliage.

PSOROTHAMNUS (SMOKE TREE)

The genus *Psorothamnus* has 9 species occurring in the deserts of the southwestern United States and northwestern Mexico.

SMOKE TREE (Fig. 154) ***Psorothamnus spinosus***

DESCRIPTION: A small tree or thorny shrub with gray, nearly leafless stems. The tree may grow to be 9 m (30 ft) tall.

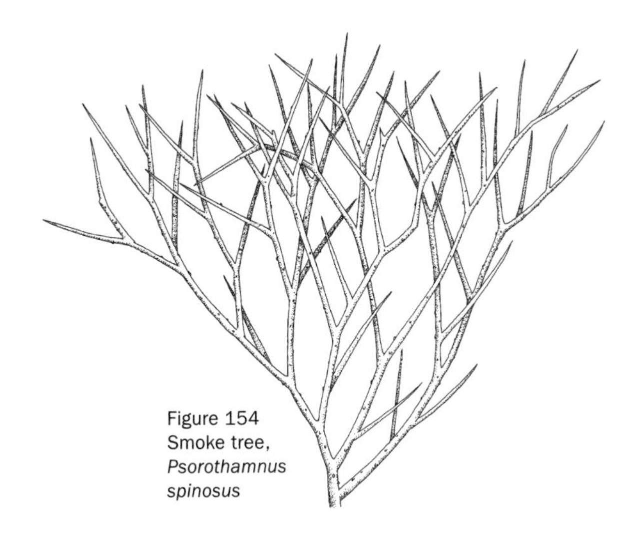
Figure 154
Smoke tree,
Psorothamnus spinosus

LEAVES are alternate, deciduous, and short-lived. Blades are linear, gland dotted, and 6 mm (.25 in.) to 20 mm (.75 in.) long. **INFLORESCENCES** are small clusters of 5 to 15 flowers borne along the stems. **FLOWERS** are indigo blue, very fragrant, and about 6 mm (.25 in.) long. The calyx has reddish dots. **FRUITS** are legumes about 6 mm (.25 in.) long with reddish dots. **TWIGS** are ashy gray thorns.

HABITAT AND RANGE: Grows along washes and in floodplains below 450 m (1,500 ft), in the Mojave, Colorado, and Sonoran Deserts in southeastern California, and the Sonoran Desert of Arizona and northwestern Mexico.

REMARKS: Smoke tree is easily identified as the smoke gray shrub in washes and floodplains in California's deserts. Older books called smoke tree *Dalea spinosus.* Like seeds of other wash species, smoke tree seeds must be prepared for germination by abrasion during floods.

PTELEA (HOP TREE)

The genus *Ptelea* has 3 species of shrubs and small trees limited to the United States and Mexico. One of these is native to California.

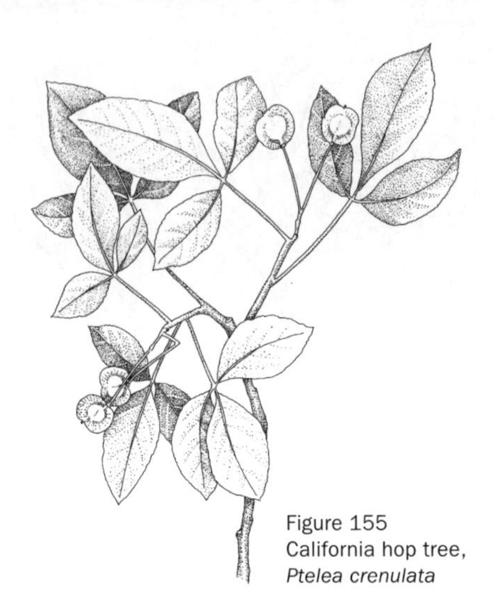

Figure 155
California hop tree,
Ptelea crenulata

CALIFORNIA HOP TREE ***Ptelea crenulata***
(Fig. 155)

DESCRIPTION: A shrub or small, single-stemmed tree that typically ranges in height from 2 m (6.5 ft) to 4.5 m (15 ft). The largest grows in Lake County and is 6.7 m (22 ft) tall and 38 cm (15 in.) in diameter. **LEAVES** are deciduous; alternate; pinnately compound, with 3 leaflets; and 4 cm (1.5 in.) to 14 cm (5.5 in.) long. Petioles are 2 cm (.75 in.) to 6 cm (2.5 in.) long. Leaflets have a broad elliptical shape, are dotted with glands, and measure 2 cm (.75 in.) to 7.5 cm (3 in.) long. Leaflet margins are finely scalloped to entire. Leaflet tips are pointed and bases are wedge shaped to round. Upper leaflet surfaces are shiny green and hairless, while lower surfaces are somewhat hairy. Lateral leaflets do not have stalks. **INFLORESCENCES** are flat-topped clusters. **FLOWERS** have 4 or 5 greenish white petals each, which measure 6 mm (.25 in.) long. **FRUITS** are samaras dotted with glands; each has a round, papery wing surrounding 2 seeds. **TWIGS** are brown and glandular. **BARK** is brown and smooth. **HABITAT AND RANGE:** Grows in woodlands in the foothills and interior mountains of northern and central California below 600 m (2,000 ft). It is commonly found on moist sites in canyons and flats.

REMARKS: California hop tree is planted as an ornamental because of its distinctive leaves and fruits. Its foliage has a strong odor when bruised.

PURSHIA (BITTERBRUSH)

The genus *Purshia* has 5 species found in western North America, 2 of which are native to California. *Purshia* is a collection of shrubs and small trees with hard wood and simple leaves that are alternate in arrangement and clustered on short twigs. Blades have 3 lobes with tightly rolled-under margins. Flowers are solitary on lateral twigs. The many stamens of each are attached to a cup surrounding 1 to 10 pistils. Each fruit is an achene ending in a beaklike style or a long style that is covered with shiny hairs at maturity.

1. Flowers and fruits have fewer than 3 pistils per flower, commonly 1 pistil. Leaves are commonly 3-lobed and occasionally 5- lobed . . . . **bitterbrush** (*P. tridentata*)
1. Flowers and fruits have fewer than 12, commonly 4, pistils per flower. Leaves are commonly 5- to 9-lobed and occasionally 3-lobed . . . . **cliffrose** (*P. mexicana*)

CLIFFROSE — *Purshia mexicana* var. *stansburyana*

(Pl. 28)

DESCRIPTION: A multistemmed shrub or small tree that commonly grows to be between 30 cm (1 ft) and 2.75 m (9 ft) tall. The largest lives on Spruce Mountain in Elko County, Nevada, and is 3.6 m (12 ft) tall and 1.6 m (62 in.) in diameter. **LEAVES** are alternate, mostly deciduous, simple, and clustered on short twigs. Blades are pinnately lobed and 6 mm (.25 in.) to 2.5 cm (1 in.) long, with 5 to 7 lobes (sometimes 3 or 9 lobes). The upper surface is green and dotted with glands above, with long white hairs on the lower surface. The margins are tightly rolled under. **INFLORESCENCES** are single flowers set on short lateral twigs. **FLOWERS** are cream, yellow, or white. **FRUITS** occur as 5 to 10 achenes per flower. At maturity, each style is

Figure 156
Bitterbrush,
Purshia tridentata

long and feathery. **TWIGS** are gray or brown. **BARK** is brown and shredding.

HABITAT AND RANGE: Grows in the intermountain West. In California it is found in desert mountain woodlands at elevations from 1,000 m (3,300 ft) to 2,400 m (8,000 ft).

REMARKS: Traditionally, cliffrose has been included in the genus *Cowania* because of its tailed fruits. More recently, it has been placed into the genus *Purshia,* even though *Purshia* lacks tailed fruits, largely because cliffrose hybridizes with the beaked desert bitterbrush (*P. tridentata* var. *glandulosa*), creating plants with intermediate fruits.

BITTERBRUSH (Fig. 156) ***Purshia tridentata***

DESCRIPTION: A many-branched shrub that may grow to be between 1 m (3.3 ft) and 2.5 m (8 ft) tall. **LEAVES** are alternate, mostly deciduous, simple, and clustered on short twigs.

Blades are pinnately 3-lobed and 6 mm (.25 in.) to 25 mm (1 in.) long. Upper leaf surfaces are green and have fine hairs or are hairless; lower leaf surfaces have long white hairs. The margins are rolled under. **INFLORESCENCES** are single flowers set on short lateral twigs. **FLOWERS** are cream or pale yellow. **FRUITS** occur as 1 to 3 beaked achenes per flower. **TWIGS** are gray or brown.

HABITAT AND RANGE: A common species of dry coniferous forests, woodlands, and scrubs in California's mountains and the Great Basin. It ranges from 750 m (2,500 ft) to 2,400 m (8,000 ft). The species is found throughout the intermountain West.

REMARKS: Bitterbrush is common as an understory shrub in the Jeffrey and ponderosa pine forests east of the Cascade-Sierra crest. Two varieties of bitterbrush are recognized. Antelope bitterbrush (*P. t.* var. *tridentata*) occurs in the northern part of the range and is characterized as being nonglandular and having hairy leaves. To the south this form grades into desert bitterbrush (*P. t.* var. *glandulosa*), which has glandular, mostly hairless leaves. Bitterbrush is a nitrogen fixer and provides critical forage for wildlife.

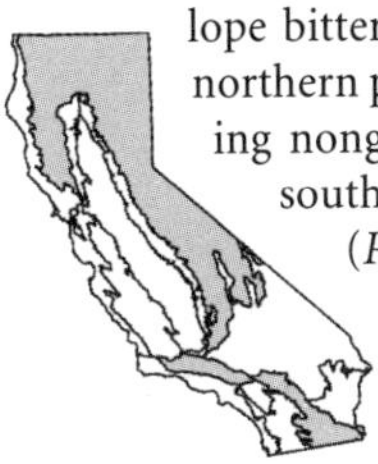

QUERCUS (OAK)

The genus *Quercus* has about 600 species; about 60 are native to the United States, and 19 of these are native to California. Mexico has over 150 native species. Most species are found in the temperate habitats of the Northern Hemisphere, while some species grow at higher elevations in the tropics. California has 9 oak species that are trees, 7 that are shrubs, and 3 that are usually shrubs but can also be treelike. In addition, several tree species have shrub varieties and forms. Numerous hybrids between California oak species have been recognized.

Leaves can be either winter-deciduous, drought-deciduous, or evergreen. They are alternate and have small, usually deciduous stipules. Margins can be lobed, entire, serrated, or toothed. Shape and size are variable. Small pollen-bearing flowers are clustered in droopy, slender catkins. The small seed-bearing flowers are either solitary or few in number and clustered in spikes. Fruits are acorns: the nut is enclosed by a scaly cup.

Acorns take 1 or 2 years to mature. Bark is often thick and furrowed but can be thin and scaly.

California oaks occur in a wide variety of habitats. They are typically found growing in woodlands and chaparrals in the warmer, drier parts of the state. Several species grow in montane coniferous forests. They can be found growing on moist prairies in northwestern California, exposed granite in the Sierra Nevada, deep alluvium in the Central Valley, serpentine soils, chaparrals, and dry slopes and ridges. Oaks develop deep, penetrating roots in well-drained soils.

Oak wood is widely used as lumber and to make tool handles, flooring, and barrels. Tannin and cork are products extracted from the bark of oaks. Acorns are eaten by a variety of wildlife species. Some Native Americans made acorns a staple in their diet.

1. Leaves are deciduous 2
1. Leaves are evergreen 6
   2. Leaves are dull gray or blue-green, with shallow lobes or wavy margins 3
   2. Leaves are dull or shiny green, with large lobes 4
3. Leaves are dull blue-green, with shallow lobes . . . . **blue oak (*Q. douglasii*)**
3. Leaves are gray-green, with wavy margins . . . . **Engelmann oak (*Q. engelmannii*)**
   4. Leaf lobes have 1 to 4 coarse, bristle-tipped teeth and pointed lobes . . . . . . **California black oak (*Q. kelloggii*)**
   4. Leaf lobes lack bristle-tipped teeth and have round lobes 5
5. Upper leaf surfaces are shiny green and waxy. Leaves have 5 to 7 lobes. Acorn cups are 4 mm (.15 in.) to 9 mm (.35 in.) deep, and nuts are 2 cm (.75 in.) to 3 cm (1.2 in.) long . . . . **Oregon white oak (*Q. garryana*)**
5. Upper leaf surfaces are dull green and have minute hairs. Leaves have 6 to 10 lobes. Acorn cups are 10 mm (.4 in.) to 30 mm (1.2 in.) deep, and nuts are 3 cm (1.2 in.) to 5 cm (2 in.) long . . . . **valley oak (*Q. lobata*)**
   6. Leaves have prominent, parallel lateral veins on their lower surfaces. Leaves look like those of chestnuts . . . . **Sadler oak (*Q. sadleriana*)**
   6. Leaves lack prominent, parallel lateral veins on their lower surfaces. Leaves do not look like those of chestnuts 7

7. Leaves are cupped 8
7. Leaves are flat or wavy 10
   8. Lower leaf surfaces have tufts of hair where lateral veins join the mid vein . . . . **coast live oak** (*Q. agrifolia*)
   8. Lower leaf surfaces lack tufts of hair where lateral veins join the mid vein 9
9. Most leaf margins are entire . . . . **island scrub oak** (*Q. parvula*)
9. Most leaf margins have spiny teeth . . . . **leather oak** (*Q. durata*)
   10. Leaves have shiny lower surfaces . . . . **interior live oak** (*Q. wislizenii*)
   10. Leaves have dull lower surfaces 11
11. Leaves have blue-green upper surfaces and mostly entire margins . . . . **Engelmann oak** (*Q. engelmannii*)
11. Leaves have other than blue-green upper surfaces and can have spiny, toothed, or entire margins 12
   12. Leaves have dark green or green upper surfaces 13
   12. Leaves have olive green, gray-green, or yellow upper surfaces 14
13. Leaf margins have teeth or spines and have round tips . . . . **scrub oak** (*Q. berberidifolia*)
13. Leaf margins may have teeth or spines, and they have acute tips . . . . **canyon live oak** (*Q. chrysolepis*)
   14. Leaves usually have entire margins, sometimes serrated, but not spiny . . . . . . **huckleberry oak** (*Q. vacciniifolia*)
   14. Most leaves have toothed or spiny margins 15
15. Leaves are regularly spiny. Plants occur in the eastern desert mountains . . . . **desert scrub oak** (*Q. turbinella*)
15. Leaves are irregularly spiny. Plants do not occur in the eastern desert mountains 16
   16. Lower leaf surfaces have a dense covering of white hair that obscures lateral veins . . . . **Muller oak** (*Q. cornelius-mulleri*)
   16. Lower leaf surfaces do not have a dense covering of hair that obscures lateral veins . . . . **Tucker oak** (*Q. john-tuckeri*)

COAST LIVE OAK (Fig. 157) *Quercus agrifolia*

DESCRIPTION: An erect, small- to medium-sized, single-stemmed tree. In chaparrals the tree can have a shrubby growth

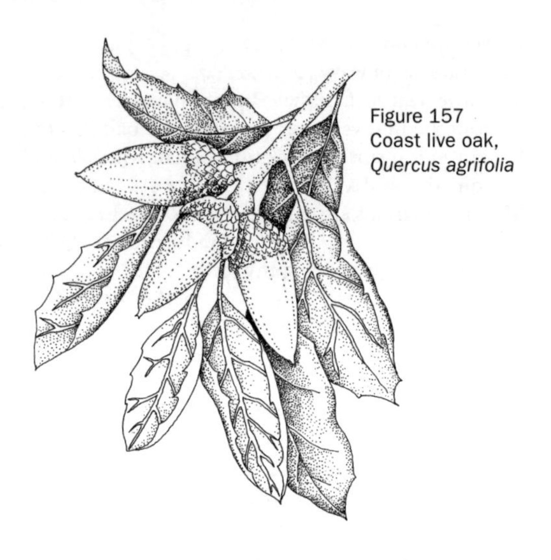

Figure 157
Coast live oak,
Quercus agrifolia

form. Mature trees are typically 9 m (30 ft) to 24 m (80 ft) tall and 30 cm (1 ft) to 1 m (3.3 ft) in diameter. The largest grows in Encino and is 21.3 m (70 ft) tall and 2.3 m (92 in.) in diameter. Trunks are typically short with large, crooked, ascending or spreading branches. Crowns are dense and broadly rounded, with foliage reaching the ground. The trees can live to be more than 250 years old. **LEAVES** are evergreen, simple, alternate, leathery, oval, strongly convex, and 2.5 cm (1 in.) to 7.5 cm (3 in.) long. Margins are spiny-toothed and leaf tips round. Upper surfaces are dark green, while lower surfaces are pale green. Tufts of hair are in axils close to the petiole on the undersides of leaves. **FRUITS** are acorns. Nuts are brown, egg shaped to conical, and 2 cm (.75 in.) to 4 cm (1.5 in.) long. Acorn cups have thin, papery scales, measure about 1 cm (.4 in.) in diameter, and enclose about 25% of the nut. **BARK** is thick, has broad ridges and furrows, and is dark gray.

HABITAT AND RANGE: Grows in woodlands and chaparrals from Mendocino County south along the Coast Ranges into Baja California. It is prevalent on the coastal side of the Coast Ranges, although populations can also be found along streams draining into the Central Valley. The species grows on deep, well-

Figure 158
Scrub oak,
Quercus berberidifolia

drained soils below 1,500 m (5,000 ft), often on lower slopes and in riparian areas.

REMARKS: Coast live oak is shade tolerant. It is resistant to fire because of its thick bark and sprouting ability. Since the mid 1990s, thousands of coast live oak trees have been killed by an unnamed bark-invading fungus in the genus *Phytophthora.* Coast live oak provides important habitat for wildlife. Its acorns feed woodpeckers, bear, deer, pigs, and small mammals. Native Americans used its acorns for food. Two varieties are recognized: *Q. a.* var. *agrifolia,* characterized by hairless to sparsely hairy leaf vein axils; and *Q. a.* var. *oxyadenia,* characterized by densely hairy leaf vein axils. The latter grows on granitic soils at elevations from 600 m (2,000 ft) to 1,500 m (5,000 ft) in the Peninsular Ranges. Coast live oak hybridizes with California black oak, interior live oak, and island scrub oak.

SCRUB OAK (Fig. 158) ***Quercus berberidifolia***

DESCRIPTION: A small- to medium-sized shrub or small tree that grows to be between 1 m (3.3 ft) and 4.5 m (15 ft) tall. Its densely packed twigs and leaves can form a nearly impenetrable barrier. **LEAVES** are evergreen, simple, alternate, leathery, oblong to elliptical, and 15 mm (.6 in.) to 30 mm (1.2 in.) long. Leaf tips are more or less round. Margins are spiny to toothed. Upper surfaces are green and flat to wavy, while lower surfaces are grayish green and slightly hairy. **FRUITS** are acorns. Nuts are brown, egg shaped, and 1 cm (.4 in.) to 3 cm (1.2 in.) long. Acorn cups are bowl shaped, measure 12 mm (.5 in.) to 20 mm (.75 in.) long, and enclose about 33% of the nut. Cup scales are warty. **BARK** is smooth and gray.

HABITAT AND RANGE: Grows in chaparrals and woodlands at elevations from 300 m (1,000 ft) to 1,500 m (5,000 ft) in the Coast Ranges, Sierra Nevada, and southern California mountains. It is also found in Baja California. The species is typically found on steep, dry slopes and shallow soils. It does not grow on serpentine soils.

REMARKS: Scrub oak is shade intolerant. It sprouts vigorously and will quickly dominate after fire. Scrub oak is planted as an ornamental on well-drained soils. It hybridizes with Engelmann oak, leather oak, Oregon white oak, Tucker oak, and valley oak. In older books, scrub oak was given the name *Q. dumosa,* a name now used for a rare southern California species.

CANYON LIVE OAK (Fig. 159) *Quercus chrysolepis*

DESCRIPTION: An erect, small- to medium-sized, single- or multistemmed tree. On harsh sites it can have a shrubby growth form. Mature trees are usually 4.5 m (15 ft) to 21 m (70 ft) tall and 75 cm (2.5 ft) to 150 cm (5 ft) in diameter. The largest grows in Idyllwild and is 23.4 m (77 ft) tall and 2.7 m (105 in.) in diameter. Trunks are single-stemmed on better sites and multistemmed on canyon walls, cliffs, and rocky sites. Trunks and branches are crooked and spread broadly. Crowns are dense and expansively rounded. Trees can be more than 300 years old. **LEAVES** are evergreen, simple, alternate, leathery, elliptical to egg shaped, and 2.5 cm (1 in.) to 10 cm (4 in.) long. Leaves that are entire and others with spiny-toothed margins may be found on the same tree. Sprouts and young leaves are more apt to have spiny margins, while older leaves are usually entire. Upper surfaces are dark green, while lower surfaces are grayish green to yellowish green. **FRUITS** are acorns. Nuts are light brown, egg

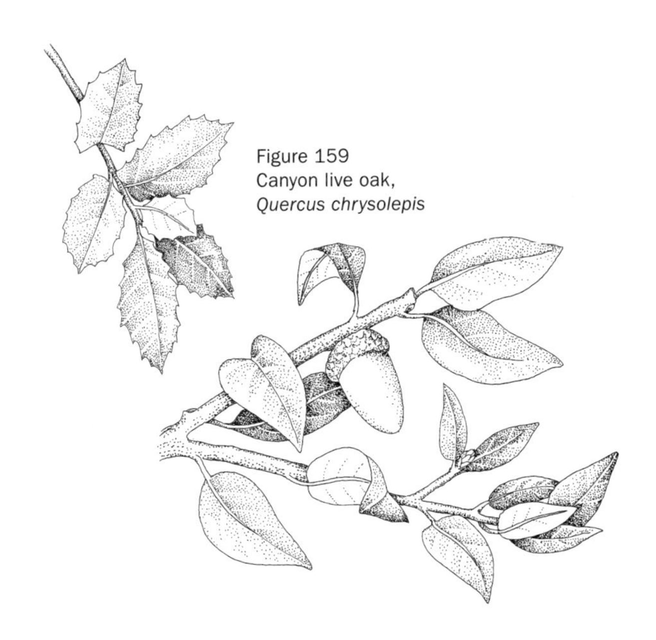

Figure 159
Canyon live oak,
Quercus chrysolepis

shaped, 2.5 cm (1 in.) to 4 cm (1.5 in.) long, and 1.2 cm (.5 in.) to 2 cm (.75 in.) wide. Acorn cups are saucer shaped, measure 1.5 cm (.6 in.) to 3 cm (1.2 in.) in diameter, and enclose about 25% of the nut. Cup scales are thick and slightly warty, with golden hairs. **BARK** is about 4 cm (1.5 in.) thick, smooth to slightly furrowed, and grayish brown.

HABITAT AND RANGE: Grows in woodlands and chaparrals from southwestern Oregon to Arizona and Baja California. In California it usually grows west of the Sierra Nevada on canyon walls, cliffs, rocky outcrops, or shallow soils, at elevations from 100 m (300 ft) to 2,700 m (9,000 ft).

REMARKS: Canyon live oak is shade and drought tolerant. Young stems are readily killed by fire, and repeated fires will convert canyon live oaks to shrubs. The species is used as firewood and to make furniture, pallets, and paneling. Wildlife species use canyon live oak for roosting, nesting, foraging, and cover. Birds and large and small mammals eat its acorns. Native Americans leached tannins from the acorns, making an edible mush. Canyon live oak hybridizes with Palmer oak, island oak, and huckleberry oak.

MULLER OAK *Quercus cornelius-mulleri*

DESCRIPTION: A small- to medium-sized shrub that grows to be between 1 m (3.3 ft) and 2.5 m (8 ft) tall. The species' densely packed twigs and leaves can form a nearly impenetrable barrier. **LEAVES** are evergreen, simple, alternate, leathery, oblong to egg shaped, and 2.5 cm (1 in.) to 4 cm (1.4 in.) long. Leaf tips are round to sharp pointed. Margins mostly have a few spines or teeth or can be entire. Upper surfaces are yellowish green to grayish green and slightly hairy, while lower surfaces are whitish and densely hairy. The midrib is yellow. **FRUITS** are acorns. Nuts are brown, have a broad conical shape, and measure 2 cm (.75 in.) to 4 cm (1.5 in.) long. Acorn cups are cup shaped, measure 12 mm (.5 in.) to 20 mm (.75 in.) in diameter, and enclose about 33% of the nut. Cup scales are flat and covered with grayish hairs. **TWIGS** have fine hairs. **BARK** is smooth and gray.

HABITAT AND RANGE: Grows in chaparrals and dry coniferous woodlands, at elevations from 1,000 m (3,200 ft) to 2,000 m (6,600 ft), generally on granitic soils. It is found between the coast and the desert from the San Bernardino Mountains, along the crest of the Peninsular Ranges, and into Baja California.

REMARKS: Muller oak is shade intolerant. It hybridizes with Engelmann oak and valley oak.

BLUE OAK (Fig. 160; Pl. 29) *Quercus douglasii*

DESCRIPTION: An erect, small- to medium-sized, single-stemmed tree. Mature trees are typically 6 m (20 ft) to 18 m (60 ft) tall and 30 cm (1 ft) to 90 cm (3 ft) in diameter. The largest grows in southern Alameda County and is 28.7 m (94 ft) tall and 1.9 m (77 in.) in diameter. Trunks are typically short with large, crooked, ascending or spreading branches. Crowns of open-grown trees are dense and broadly rounded, with foliage reaching the ground. Trees can live to be 400 years old. **LEAVES** are deciduous, alternate, simple, and 4 cm (1.5 in.) to 10 cm (4 in.)

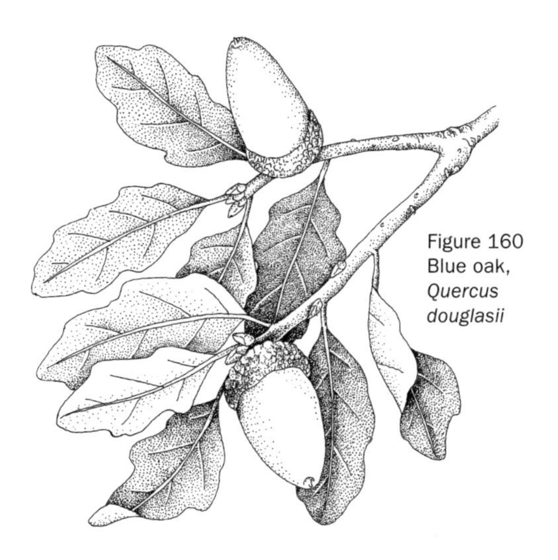

Figure 160 Blue oak, *Quercus douglasii*

long. The 5 to 7 irregular lobes are round and oblong to egg shaped. Leaves are wider above the middle. Lobe sinuses are usually shallow, and some leaves can be entire. Upper surfaces are dull bluish green, while lower surfaces are pale bluish green. **FRUITS** are acorns. Nuts are brown, egg shaped, and 2 cm (.75 in.) to 3 cm (1.25 in.) long. Acorn cups are saucer shaped, measure 12 mm (.5 in.) to 20 mm (.75 in.) in diameter, and enclose the base of the nut. Cup scales are small and warty. **BARK** is about 2.5 cm (1 in.) thick, grayish, and checkered with flaky, thin scales.

HABITAT AND RANGE: Grows in woodlands and valleys in the foothills of western California mountains and is well represented in the foothills arising from the Central Valley. It occurs from near sea level to 1,800 m (5,900 ft) in the southern part of its range. The species is endemic to California.

REMARKS: Blue oak is drought tolerant and shade intolerant. Roots are extensive and deep. The species sprouts following injury from fire or cutting. Firewood and fence posts are the primary commercial uses of blue oak. Blue oak woodlands provide critical winter range for deer and other wildlife. The species' foliage serves as browse, and many wildlife species eat its acorns. Native Americans used the acorns

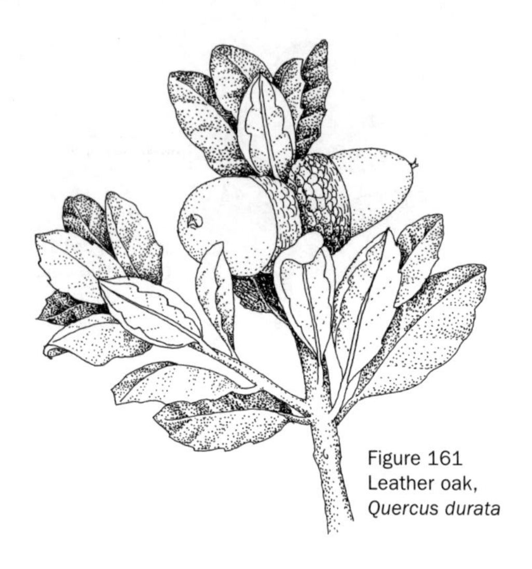

Figure 161
Leather oak,
Quercus durata

to make an edible mush. Blue oak is planted as an ornamental. It hybridizes with Oregon white oak, valley oak, and Tucker oak. The woodlands of the interior Coast Ranges and Tehachapi Mountains have semideciduous, scrubby trees (*Q.* ×*alvordiana*). These conspicuous oaks are considered hybrids (as indicated by ×) between *Q. douglasii* and *Q. john-tuckeri.*

LEATHER OAK (Fig. 161) ***Quercus durata***

DESCRIPTION: A small- to medium-sized shrub that grows to be between 1 m (3.3 ft) and 3 m (10 ft) tall. Its crown is rounded and has a dense tangle of branches. **LEAVES** are evergreen, simple, alternate, leathery, convex, oblong to elliptical, and 12 mm (.5 in.) to 3 cm (1.25 in.) long. Leaf tips are sharply pointed. Margins are spiny to entire and rolled over. Upper surfaces are dull green and have small hairs, while lower surfaces are grayish green and hairy. **FRUITS** are acorns. Nuts are brown, stoutly egg shaped, round at their tips, and 15 mm (.6 in.) to 25 mm (1 in.) long. Acorn cups are bowl shaped, measure 12 mm (.5 in.) to 18 mm (.7 in.) in diameter, and enclose about 33% of the nut. Cup scales are warty. **TWIGS** are hairy, sometimes smooth. **BARK** is smooth and gray.

HABITAT AND RANGE: Grows in coniferous woodlands and chaparrals, mostly on serpentine soils at elevations from 150 m (500 ft) to 1,500 m (5,000 ft).

REMARKS: Leather oak is apparently intolerant of shade. It recovers slowly from fire. There are 2 varieties of leather oak: *Q. d.* var. *durata,* which characteristically has strongly convex leaves and grows on serpentine soils; and *Q. d.* var. *gabrielensis,* which has slightly convex leaves and grows on granitic soils. Leather oak hybridizes with Oregon white oak and scrub oak.

ENGELMANN OAK (Fig. 162) *Quercus engelmannii*

DESCRIPTION: An erect, small- to medium-sized, single-stemmed tree. Mature trees are usually 4.5 m (15 ft) to 15 m (50 ft) tall and 30 cm (1 ft) to 60 cm (2 ft) in diameter. The largest grows in Pasadena and is 23.8 m (78 ft) tall and 104 cm (41 in.) in diameter. Trunks are typically short, with crooked, ascending or spreading branches. Crowns of open-grown trees are sparse, irregular, and broadly rounded, with foliage reaching the ground. Trees can live to be at least 350 years old. **LEAVES** are drought-deciduous to evergreen, alternate, simple, leathery, and 2.5 cm (1 in.) to 7.5 cm (3 in.) long. They are oblong to egg shaped, with the widest part above the middle. Margins are entire, wavy, or toothed. Upper surfaces are dull bluish green, while lower surfaces are pale bluish green and, when young, are hairy. Leaves typically persist until they are replaced by new foliage in the spring. During droughts, plants drop their leaves earlier. **FRUITS** are acorns. Nuts are brown, cylindrical to egg shaped, and 15 mm (.6 in.) to 25 mm (1 in.) long. Acorn cups are cup shaped to bowl shaped, measure 10 mm (.4 in.) to 15 mm (.6 in.) in diameter, and enclose less than 40% of the nut. Cup scales are warty. **BARK** is thick and furrowed. Scales are thin and grayish.

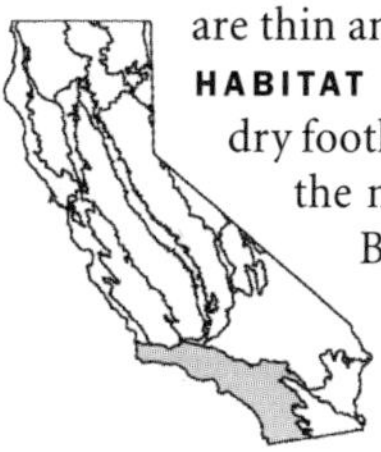

HABITAT AND RANGE: Grows in woodlands on dry foothill slopes and mesas between the coast and the mountains in southwestern California and Baja California. It is found below 1,300 m (4,200 ft).

REMARKS: Engelmann oak is resistant to fire. It is considered to be an un-

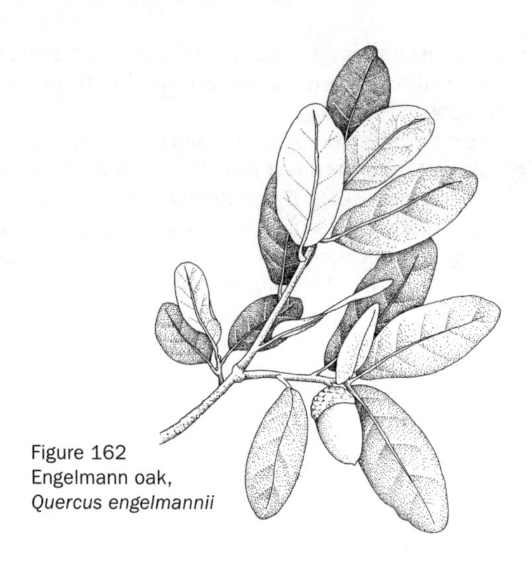
Figure 162
Engelmann oak,
Quercus engelmannii

common species, largely because of loss of habitat to residential and agricultural development. One of the best examples of a relatively undisturbed Engelmann oak landscape is at the Nature Conservancy's Santa Rosa Plateau in southwestern Riverside County. Engelmann oak hybridizes with scrub oak and Muller oak.

OREGON WHITE OAK (Fig. 163) *Quercus garryana*

DESCRIPTION: An erect, small- to medium-sized, single-stemmed tree. Mature trees are usually 7.5 m (25 ft) to 27 m (90 ft) tall and 30 cm (1 ft) to 90 cm (3 ft) in diameter. The largest grows in El Dorado County and is 37.2 m (122 ft) tall and 2.4 m (8 ft) in diameter. Open-grown trees have trunks that are short, with large, crooked, ascending or spreading branches. Crowns of open-grown trees are rounded, with foliage reaching the ground. Trees grown with more competition develop longer, straight trunks and narrow crowns. Trees can live to be 500 years old. **LEAVES** are deciduous, alternate, simple, elliptical to egg shaped, and 7.5 cm (3 in.) to 10 cm (4 in.) long. The 5 to 9 lobes are round and entire to coarsely toothed. Leaves are wider above the middle. Lobe sinuses are deep. Upper surfaces are dark green and shiny, while lower surfaces are light green. **FRUITS**

Figure 163
Oregon white oak,
Quercus garryana

are acorns. Nuts are brown, egg shaped, and 2 cm (.75 in.) to 3 cm (1.25 in.) long. Acorn cups are deep and bowl shaped, measure 12 mm (.5 in.) to 25 mm (1 in.) in diameter, and enclose less than 33% of the nut. Cup scales are slightly warty. **BARK** is thin, scaly, and whitish gray.

HABITAT AND RANGE: Grows in woodlands and coniferous forests from western British Columbia south to California. In California it is found in the central and northern Coast Ranges and in the foothills of the Cascades and Sierra Nevada. It ranges from near sea level to 1,800 m (5,900 ft).

REMARKS: Oregon white oak is moderately shade tolerant, but it can be out-competed by overtopping conifers. It sprouts following injury by fire or cutting. The species' foliage is palatable to wildlife and livestock, and its acorns provide important nutrition for many species of birds and mammals. Native Americans ground the acorns into meal. Brewer oak (*Q. g.* var. *breweri*) is a shrub form of Oregon white oak that reaches 4.5 m (15 ft) tall and is commonly found at elevations higher than those at which *Q. g.* var. *garryana* is found, from 900 m (3,000 ft) to 2,300 m (7,500 ft). Oregon white oak hybridizes with scrub oak, blue oak, leather oak, valley oak, and Sadler oak.

TUCKER OAK *Quercus john-tuckeri*

DESCRIPTION: A small- to medium-sized shrub or small tree that reaches between 2 m (6.5 ft) and 6 m (20 ft) tall. **LEAVES** are evergreen, simple, alternate, leathery, oblong to elliptical, and 12 mm (.5 in.) to 28 mm (1.1 in.) long. Leaf bases have a round to broad wedge shape. Leaf tips are round. Margins are irregularly spiny or toothed. Upper surfaces are grayish green, while lower surfaces are pale grayish green and covered with fine hairs. **FRUITS** are acorns. Nuts are brown, egg shaped to conical, and 2 cm (.75 in.) to 3 cm, (1.2 in.) long. Acorn cups are bowl shaped, measure 10 mm (.4 in.) to 15 mm (.6 in.) wide, and enclose about 25% of the cup. Cup scales are thin and flat to slightly warty. **TWIGS** have fine hairs. **BARK** is smooth and gray.

HABITAT AND RANGE: Grows in chaparrals and dry coniferous woodlands, at elevations from 900 m (3,000 ft) to 2,000 m (6,500 ft) in the interior southern Coast Ranges and Transverse Ranges.

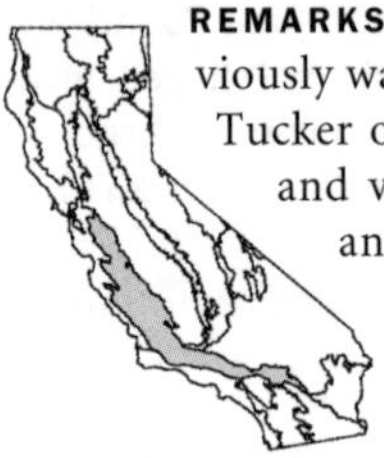

REMARKS: Tucker oak is shade intolerant. It previously was known as *Q. turbinella* ssp. *californica*. Tucker oak hybridizes with scrub oak, blue oak, and valley oak. Hybridization of Tucker oak and blue oak may explain the extensive semideciduous shrubby oaks of the Tehachapi Mountains and interior Coast Ranges.

CALIFORNIA BLACK OAK *Quercus kelloggii*

(Fig. 164; Pls. 30 and 31)

DESCRIPTION: An erect, small- to medium-sized, single-stemmed tree. On infertile sites, it is a shrub. Mature trees are often 9 m (30 ft) to 24 m (80 ft) tall and 30 cm (1 ft) to 120 cm (4 ft) in diameter. The largest grows in Siskiyou National Forest and is 37.8 m (124 ft) tall and 2.7 m (9 ft) in diameter. Open-grown trees have short, often forked trunks and large, crooked, ascending or spreading branches. Crowns are broad and rounded, with foliage reaching the ground. Trees grown with more competition develop longer, somewhat crooked trunks and narrow crowns. Trees can live to be 500 years old.

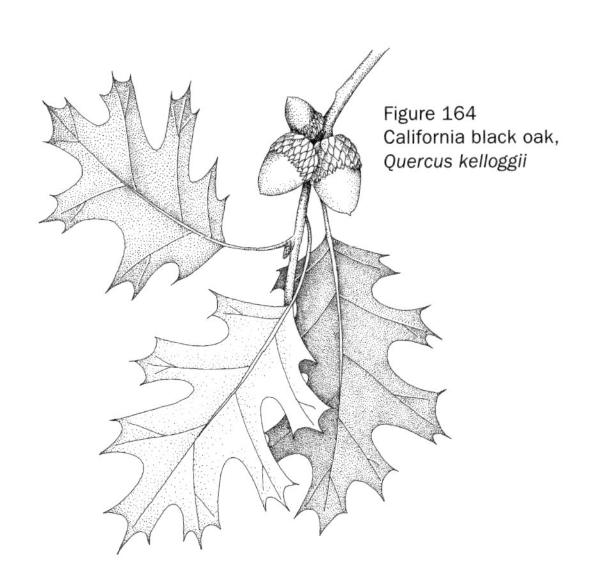

Figure 164
California black oak,
Quercus kelloggii

LEAVES are deciduous, alternate, simple, elliptical to egg shaped, and 9 cm (3.5 in.) to 20 cm (8 in.) long. Leaves are wider above the middle. The 7 to 9 lobes are pointed and bristle tipped and have deep sinuses. Upper surfaces are shiny dark green, while lower surfaces are pale green. **FRUITS** are acorns. Nuts are brown, oblong to egg shaped, and 20 mm (.75 in.) to 34 mm (1.3 in.) long. Acorn cups are deep and cup shaped, measure 18 mm (.67 in.) to 25 mm (1 in.) in diameter, and enclose about 50% of the nut. Cup scales are thin and flat. **BARK** is moderately thick, furrowed, checkered, and grayish black to black.

HABITAT AND RANGE: Grows in woodlands and coniferous forests from western Oregon south to Baja California. In California it is found in pure stands at lower montane elevations or in mixed coniferous forests, at elevations from 60 m (200 ft) to 2,400 m (8,000 ft).

REMARKS: California black oak is initially shade tolerant but becomes shade intolerant as it grows older. It sprouts following injury from fire or cutting. Trees growing in the central and north Coast Ranges are susceptible to a recently discovered lethal bark-invading fungus (an unnamed species in the genus *Phytophthora*). California black oak provides excellent forage, cover, and mast for a variety of wildlife. Native Americans made meal out of its acorns and dyes from its bark. It is the primary hardwood timber species in California. California black oak hybridizes with coast live oak and interior live oak. The latter hybrid is known as oracle oak.

VALLEY OAK (Fig. 165) *Quercus lobata*

DESCRIPTION: An erect, medium-sized, single-stemmed tree. Mature trees are typically 9 m (30 ft) to 27 m (90 ft) tall and 60 cm (2 ft) to 120 cm (4 ft) in diameter. This species is the largest California oak. The largest individual grows near Covelo and is 49.7 m (163 ft) tall and 2.8 m (110 in.) in diameter. Trunks are typically short and stout, with large, crooked, ascending or spreading branches. Crowns of open-grown trees are dense and broadly rounded, with foliage reaching the ground. Trees live to be 400 to 600 years old. **LEAVES** are deciduous, alternate, simple, and 5 cm (2 in.) to 10 cm (4 in.) long. The 6 to 10 lobes are round and egg shaped, with the widest part above the middle. Lobe sinuses are deep. Upper surfaces are shiny dark green, while lower surfaces are pale green with yellowish veins. **FRUITS** are acorns. Nuts are brown, narrow and conical, and 3 cm (1.25 in.) to 5 cm (2 in.) long. Acorn cups are deep and bowl shaped, measure 12 mm (.5 in.) to 30 mm (1.25 in.) in diameter, and enclose only the base of the nut. Cup scales are warty. **BARK** is from 5 cm (2 in.) to 15 cm (6 in.) thick on mature trees. It is light gray and deeply reticulated into alligator-like rectangular patches.

HABITAT AND RANGE: The species is endemic to California and grows in valley and foothill woodlands in the Central Valley, in the foothills of the Sierra, and west into the Coast Ranges. It is typically found on deep, alluvial soils but can occur on shallower soils as long as its roots can reach sufficient moisture. The species has deep-reaching vertical root systems that can tap into groundwater, and some trees have roots that are 24 m (80 ft) deep. It ranges from near sea level to 1,700 m (5,500 ft).

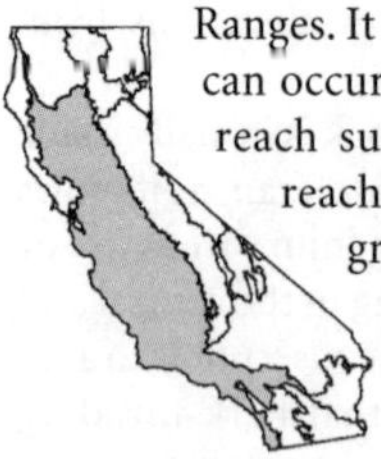

REMARKS: Valley oak is intermediate

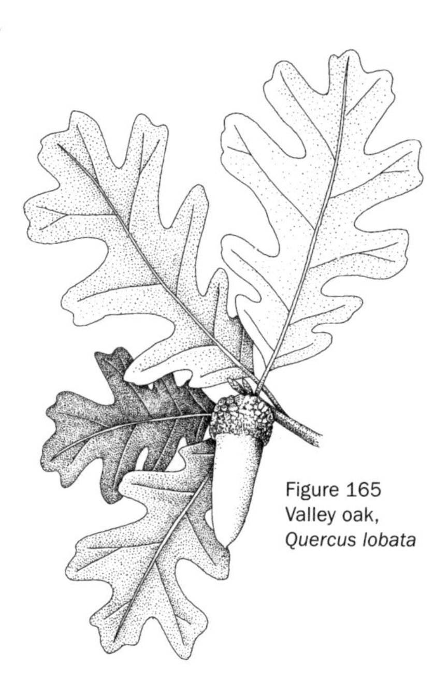

Figure 165
Valley oak,
Quercus lobata

in its shade tolerance. It sprouts following injury from fire or cutting. Valley oak is considered to be an uncommon species, largely because of loss of habitat to agriculture and urbanization. Where it does occur, it provides important habitat for wildlife. Native Americans ground its acorns into meal. Valley oak wood is used mostly as firewood and to make charcoal. The species hybridizes with scrub oak, Muller oak, blue oak, Engelmann oak, Oregon white oak, and Tucker oak.

ISLAND SCRUB OAK *Quercus parvula*

DESCRIPTION: A small- to medium-sized shrub or small tree. Shrubs grow to be between 1 m (3.3 ft) and 3 m (10 ft) tall and trees can reach 15 m (50 ft) tall. **LEAVES** are evergreen; simple; alternate; leathery; oblong, lance shaped, or egg shaped; and 3 cm (1.2 in.) to 9 cm (3.5 in.) long. Leaf tips are sharp pointed. Margins are usually entire, occasionally toothed. Upper surfaces are hairless and olive green, while lower surfaces are hairless and light olive green. **FRUITS** are acorns. Nuts are brown, barrel shaped to egg shaped, and 3 cm (1.2 in.) to 4.5 cm (1.8 in.) long. Acorn cups are bowl shaped, measure about 12 mm (.5 in.) in diameter, and enclose about 25% of the nut. Cup scales are thin and flat. **BARK** is smooth and gray.

HABITAT AND RANGE: Grows in chaparrals and woodlands on the Channel Islands and in moist coastal canyons and slopes on the mainland below 1,000 m (3,300 ft). Channel Islands populations occur in chaparrals and closed-cone pine forests. Mainland populations are found in mixed forests and woodlands.

REMARKS: Two varieties of island scrub oak have been recognized: the Santa Cruz Island oak (*Q. p.* var. *parvula*), an uncommon oak that occurs as a shrub measuring from 1 m (3.3 ft) to 3 m (10 ft) tall; and the Shreve oak (*Q. p.* var. *shrevei*), which occurs as a tree measuring from 4.5 m (15 ft) to 15 m (50 ft) tall. The Shreve oak hybridizes with coast live oak and California black oak.

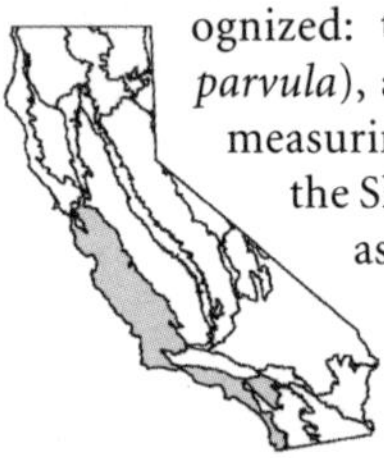

SADLER OAK (Fig. 166) *Quercus sadleriana*

DESCRIPTION: A small- to medium-sized shrub that grows to be between 1 m (3.3 ft) and 3 m (10 ft) tall. It has a rounded and spreading canopy that develops from ascending branches arising from the base of the main stem. **LEAVES** are evergreen, simple, alternate, leathery, oblong to elliptical, and 5 cm (2 in.) to 12.5 cm (5 in.) long. Leaf tips are sharp pointed. Margins have 20 to 28 teeth arising from evenly spaced, prominent, parallel veins. Upper surfaces are shiny and green, while lower surfaces are pale green and slightly hairy. **FRUITS** are acorns. Nuts are brown, egg shaped, and 15 mm (.6 in.) to 20 mm (.8 in.) long. Acorn cups are cup shaped, measure 10 mm (.4 in.) to 17 mm (.7 in.) in diameter, and enclose about 33% of the nut. Cup scales are flat, slightly warty, and hairy. **TWIGS** are hairless, slender, and flexible. **BARK** is smooth and gray.

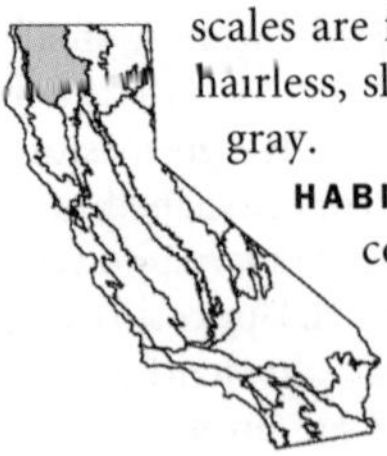

HABITAT AND RANGE: Grows in montane coniferous forests and chaparrals in the Klamath Mountains of California and Oregon, at elevations from 600 m (2,000 ft) to 2,200 m (7,200 ft). It is often an un-

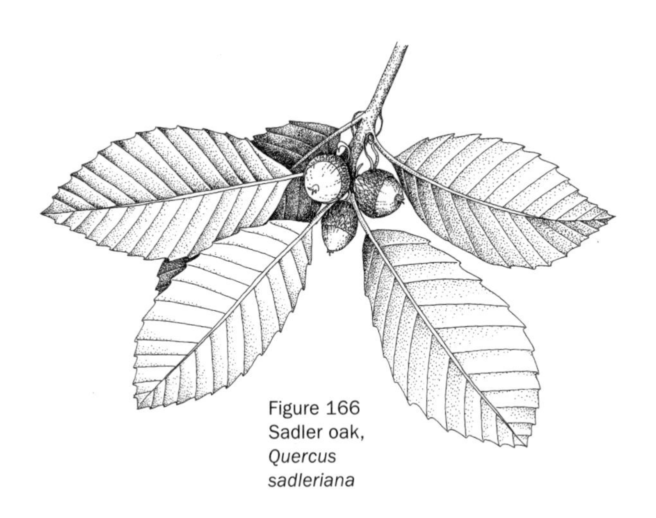

Figure 166
Sadler oak,
Quercus sadleriana

derstory species beneath conifer trees, but it can grow on sunny, dry ridges and on serpentine soils.

REMARKS: Sadler oak is shade tolerant. It hybridizes with *Q. garryana* var. *breweri*, which grows in the same montane chaparrals. Another common name for this species is deer oak.

DESERT SCRUB OAK *Quercus turbinella*

DESCRIPTION: A medium-sized shrub or small tree that forms thickets. Shrubs grow to be between 2 m (6.5 ft) and 5 m (16 ft) tall and trees can reach 7 m (23 ft) tall. **LEAVES** are evergreen, simple, alternate, leathery, oblong to elliptical, and 15 mm (.6 in.) to 30 mm (1.2 in.) long. Leaf tips are sharp pointed to broad. Margins are spiny to coarsely toothed. Upper surfaces are grayish green or yellowish green, while lower surfaces are yellowish green and have yellowish hairs. **FRUITS** are acorns. Nuts are yellowish brown, cylindrical to egg shaped, and 12 mm (.5 in.) to 25 mm (1 in.) long. Acorn cups are bowl shaped, measure 8 mm (.35 in.) to 12 mm (.5 in.) in diameter, and enclose about 25% of the nut. Cup scales are thin, flat, and slightly warty. **TWIGS** are hairy. **BARK** is gray.

Figure 167
Huckleberry oak,
Quercus vacciniifolia

HABITAT AND RANGE: Grows in interior woodlands along the desert fringe in southeastern California. Its range extends eastward to western Texas and southward into Baja California.
REMARKS: Desert scrub oak is shade intolerant.

HUCKLEBERRY OAK *Quercus vacciniifolia*

(Fig. 167)

DESCRIPTION: A prostrate or low, spreading shrub. Shrubs are dense and 30 cm (1 ft) to 120 cm (4 ft) high. **LEAVES** are evergreen, simple, alternate, leathery, oblong to elliptical, and 12 mm (.5 in.) to 30 mm (1.25 in.) long. Leaf tips are round to pointed. Margins are entire and occasionally have a few spiny teeth. Upper surfaces are green and hairless, while lower surfaces are pale green and either are hairless or have sparse hairs. **FRUITS** are acorns. Nuts are brown, round to egg shaped, and about 12 mm (.5 in.) long. Acorn cups are cup shaped, measure about 12 mm (.5 in.) in diameter, and enclose about 25% of the nut. Cup scales are flat and slightly warty, with fine white hairs. **TWIGS** are willowy and flexible and may have hairs. **BARK** is smooth and gray.
HABITAT AND RANGE: Grows in montane and subalpine forests, woodlands, and chaparrals. It ranges from Oregon south

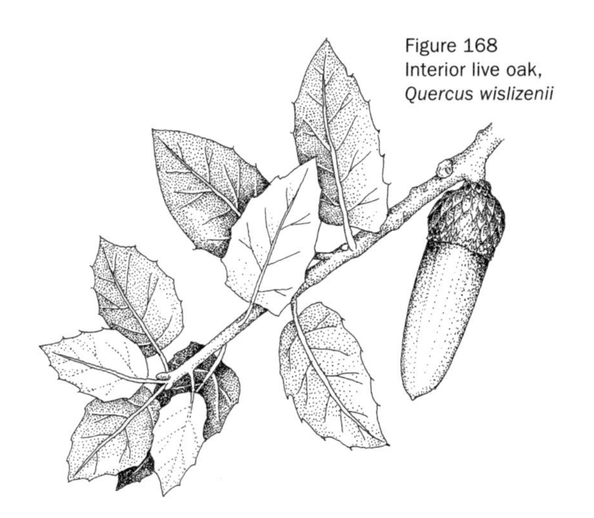
Figure 168
Interior live oak,
Quercus wislizenii

to the Sierra Nevada and northwestern California. The species can be found at elevations from 900 m (3,000 ft) to 3,000 m (10,000 ft), higher than other California oaks. It typically grows on cliffs, ridges, rocky slopes, or shallow soils. It can grow on serpentine soils.

REMARKS: Huckleberry oak is shade intolerant. It is self-perpetuating on rocky sites and areas prone to repeated fire. Huckleberry oak will sprout following injury from fire or cutting. Deer browse the leaves, and acorns are consumed by a variety of wildlife species. Huckleberry oak hybridizes with canyon live oak.

INTERIOR LIVE OAK (Fig. 168) *Quercus wislizenii*

DESCRIPTION: An erect, small- to medium-sized, single-stemmed tree. It also has a shrub form. Mature trees are usually 9 m (30 ft) to 23 m (75 ft) tall and 30 cm (1 ft) to 90 cm (3 ft) in diameter. The largest grows near Stockton and is 27.4 m (90 ft) tall and 2.2 m (85 in.) in diameter. Trunks are typically short, with large, crooked, ascending or spreading branches. Crowns of open-grown trees are dense and broadly rounded, with foliage reaching the ground. Forest-grown trees have irregular, narrow crowns. Trees can live to be more than 200 years old. **LEAVES**

are evergreen, simple, alternate, leathery, oblong to elliptical, mostly flat, and 25 mm (1 in.) to 40 mm (1.5 in.) long. Margins are spiny toothed to entire and leaf tips acute to pointed. Upper surfaces are shiny dark green, while lower surfaces are shiny, light to yellowish green. **FRUITS** are acorns. Nuts are brown, slender, conical to egg shaped, and 2 cm (.75 in.) to 4 cm (1.5 in.) long. Acorn cups have thin, papery scales, are 12 mm (.5 in.) to 20 mm (.75 in.) in diameter, and enclose about 33% of the nut. **BARK** is thick, ridged, furrowed, and dark gray.

HABITAT AND RANGE: Grows in woodlands and chaparrals from Siskiyou County south along the Sierra Nevada foothills to Baja California, at elevations from near sea level to 2,000 m (6,500 ft). It is prevalent in the Coast Ranges and southern California mountains.

REMARKS: Interior live oak is shade tolerant. It is resistant to fire because of its thick bark and sprouting ability. The species provides important wildlife forage and habitat, and its shade gives welcome summertime relief to livestock. Native Americans used the acorns to make various foods. Interior live oak is cut for firewood. In areas subject to recurring fire, the species forms shrubby thickets. Interior live oak hybridizes with California black oak, huckleberry oak, and coast live oak.

RHAMNUS (BUCKTHORN)

Rhamnus is a large genus with about 100 species that grow throughout the temperate and tropical areas of the Northern Hemisphere. Nine species are native to California.

Shrubs and trees in this genus have simple, evergreen or deciduous leaves. Flowers are small and inconspicuous but fragrant. Fruits are red or black drupes.

1. Terminal buds are covered with scales. Leaf margins can be spiny 2
1. Terminal buds are not covered with scales. Leaf margins are not spiny 3
    2. Leaves are flat . . . . **spiny redberry (*R. crocea*)**
    2. Leaves are concave . . . . **hollyleaf redberry (*R. ilicifolia*)**
3. Leaves are thick and evergreen 4
3. Leaves are thin and deciduous 5

4. Leaves lack hair on lower surfaces .... **California coffeeberry** (*R. californica*)
4. Leaves have dense hair on lower surfaces .... **hoary coffeeberry** (***R. tomentella***)

5. Leaves are narrowly elliptical and from 12 mm (.5 in.) to 38 mm (1.5 in.) long .... **Sierra coffeeberry** (***R. rubra***)
5. Leaves are broadly oblong and from 2 cm (.75 in.) to 6 cm (2.5 in.) long .... **cascara** (***R. purshiana***)

CALIFORNIA COFFEEBERRY *Rhamnus californica*

DESCRIPTION: A shrub with flexible stems and evergreen leaves. Plants typically grow to be 2 m (6.5 ft) tall. The largest lives in Sunol Regional Park in Alameda County and is 9 m (30 ft) tall and 19 cm (7.5 in.) in diameter. **LEAVES** are alternate, evergreen, simple, and hairless. Leaf blades are elliptical and from 2.5 cm (1 in.) to 7.5 cm (3 in.) long, with margins that are toothed or entire. **INFLORESCENCES** consist of clusters of 1 to 60 flowers set in the axils of leaves. **FLOWERS** are small, with 5 petals and abundant nectar. **FRUITS** are black drupes about 12 mm (.5 in.) in diameter, with 2 or 3 stones. **TWIGS** are flexible and light gray to brown. The terminal bud is not covered with scales.
HABITAT AND RANGE: Grows extensively in forests, woodlands, and chaparrals from Oregon south through northwestern California to southern California. It can be found at elevations below 2,300 m (7,500 ft).

REMARKS: The leaf color, shape, and size of California coffeeberry's 2 subspecies vary, but the shrubs are differentiated according to habitat. *R. c.* ssp. *californica* has dark green upper blade surfaces and green or yellow lower surfaces; it grows on nonserpentine soils. In *R. c.* ssp. *occidentalis,* both blade surfaces are similarly green; it grows on serpentine soils.

SPINY REDBERRY *Rhamnus crocea*

DESCRIPTION: A shrub with stiff stems and round, spiny leaves. Plants usually grow to be 1 m (3.3 ft) tall. The largest lives in Greenfield, California, and is 6.7 m (22 ft) tall and 50 cm (19.5 in.) in diameter. **LEAVES** are alternate, evergreen, simple, and hairless. Leaf blades are oval and from 6 mm (.25 in.) to 12 mm (.5 in.) long, with margins that may have spines. Depending on habitat, the leaves may be dull or glossy. **INFLORESCENCES** consist of small clusters of 1 to 6 flowers found in leaf axils. **FLOWERS** are small and green, lack petals, and have abundant nectar. **FRUITS** are bright red drupes about 6 mm (.25 in.) in diameter, with 2 stones. **TWIGS** are rigid, gray, and often thorn tipped. Terminal buds are covered with scales.

HABITAT AND RANGE: Grows extensively in woodlands, coastal scrubs, and chaparrals of coastal California below 1,000 m (3,300 ft). It ranges into Baja California.

REMARKS: A closely related species is hollyleaf redberry (*R. ilicifolia*), which has larger, hollylike leaves and thornless branches. Both species are found more frequently in chaparrals but are commonly encountered in woodlands as well.

HOLLYLEAF REDBERRY *Rhamnus ilicifolia*

DESCRIPTION: A shrub with stiff stems and round leaves with spiny or toothed margins. Plants can grow to be 4.5 m (15 ft) tall. **LEAVES** are alternate, evergreen, and simple and may have hairs. Leaf blades are oval and measure from 13 mm (.5 in.) to 30 mm (1.25 in.) long. Margins may be spiny. Depending on habitat the leaves may be dull or glossy. **INFLORESCENCES** consist of small clusters of 1 to 6 flowers found in the axils of leaves. **FLOWERS** are small and green, lack petals, and have abundant nectar. **FRUITS** are bright red drupes with 2 stones. **TWIGS** are rigid and gray. Terminal buds are covered with scales.

HABITAT AND RANGE: Grows extensively in woodlands, coastal scrubs, and chaparrals of California below 1,800 m (6,000 ft), including the Mojave Desert. It ranges into Baja California.

REMARKS: A closely related species is spiny redberry (*R. crocea*), which has shorter leaves and thorny branches. Both species are found more frequently in chaparrals, but both are commonly encountered in woodlands as well.

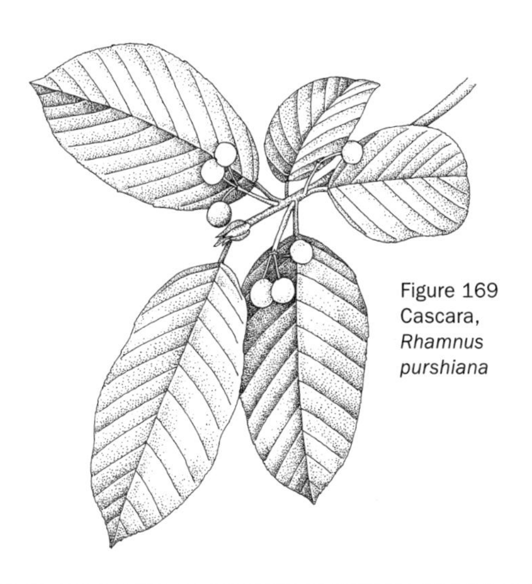

Figure 169
Cascara,
Rhamnus purshiana

CASCARA (Fig. 169; Pl. 32) ***Rhamnus purshiana***

DESCRIPTION: A large shrub or small tree with deciduous leaves tufted at the ends of branches. Plants often grow to be 6 m (20 ft) tall. The largest lives in Sedro Woolley, Washington, and is 11.3 m (37 ft) tall and 62 cm (24.5 in.) in diameter. **LEAVES** are alternate, simple, and hairless or have fine hairs. Leaf blades have a broad elliptical shape and are from 5 cm (2 in.) to 20 cm (8 in.) long. They have fine-toothed or entire margins. The undersides of leaves have conspicuous parallel veins. **INFLORESCENCES** consist of clusters of up to 25 flowers found in the leaf axils. **FLOWERS** are small and have 5 petals and abundant nectar. **FRUITS** are black drupes about 10 mm (.4 in.) in diameter, with 3 stones. **TWIGS** are brown or red. The terminal bud is not covered with scales.

HABITAT AND RANGE: Grows extensively in coniferous forests, coastal scrubs, and chaparrals along the coast and in the mountains of northern California, at elevations up to 2,000 m (6,500 ft). It ranges north to Canada.

REMARKS: In coastal areas, cascara is often tree sized, with oblong leaves. In the mountains, the species usually has wedge-

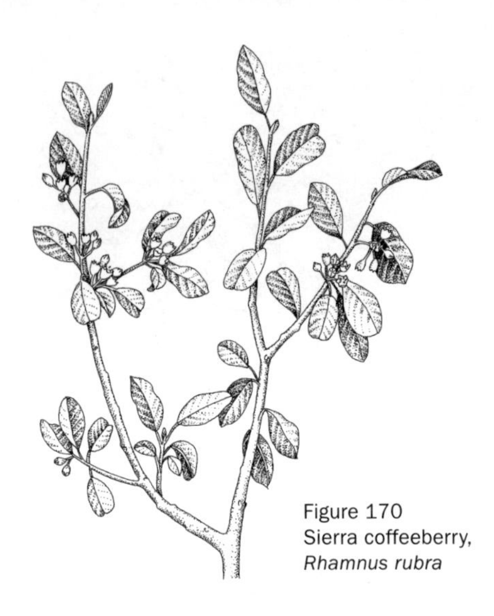

Figure 170
Sierra coffeeberry,
Rhamnus rubra

shaped leaves (*R. p.* var. *annonifolia*). The bark of either form yields cathartic drugs. Bark and fruit are toxic in excess, especially to children.

SIERRA COFFEEBERRY (Fig. 170) ***Rhamnus rubra***

DESCRIPTION: A many-branched shrub with deciduous leaves. Plants grow to be 2 m (6.5 ft) tall. **LEAVES** are alternate, simple, and hairless or have fine hairs. Leaf blades are narrow and elliptical and measure from 12 mm (.5 in.) to 4 cm (1.5 in.) long; they have fine-toothed or entire margins. The leaves may be set on short shoots. **INFLORESCENCES** consist of clusters of 1 to 15 flowers found in the leaf axils. **FLOWERS** are small and have 5 petals and abundant nectar. **FRUITS** are black drupes 12 mm (.5 in.) in diameter, with 2 stones. **TWIGS** are slender and bright gray or red. The terminal bud is not covered with scales.

HABITAT AND RANGE: Grows extensively in montane chaparrals, forests, and woodlands of the mountains of northern California, at elevations between 900 m (3,000 ft) and 2,100 m (7,000 ft). It ranges east into Nevada.

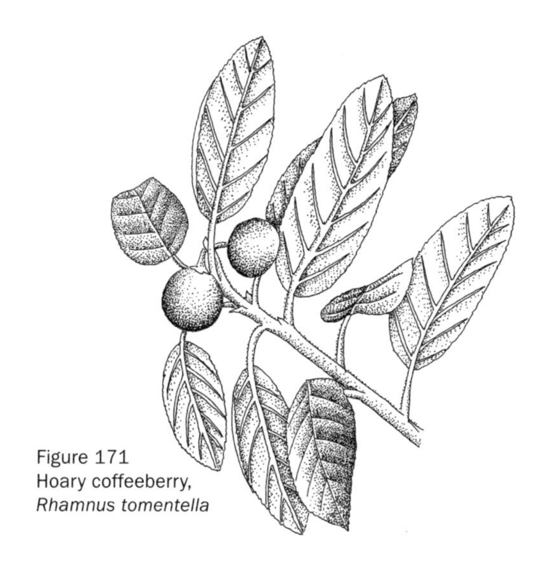

Figure 171
Hoary coffeeberry,
Rhamnus tomentella

REMARKS: Sierra coffeeberry's leaf color, shape, and size vary among habitats. Five varieties have been proposed based on single characteristic differences (though this is not a good taxonomic practice).

HOARY COFFEEBERRY *Rhamnus tomentella*

(Fig. 171)

DESCRIPTION: A shrub with flexible stems and evergreen leaves. Plants grow to be 2 m (6.5 ft) tall. **LEAVES** are alternate, evergreen, simple, and thick and have a dense covering of fine hair on the lower surfaces. The upper surfaces may be green and may either be hairless or have a dense covering of fine white hair. Leaf blades are elliptical and measure from 2.5 cm (1 in.) to 7.5 cm (3 in.) long; margins are scalloped, toothed, or entire. **INFLORESCENCES** consist of clusters of 1 to 60 flowers found in the leaf axils. **FLOWERS** are small and have 5 petals and abundant nectar. **FRUITS** are black drupes about 12 mm (.5 in.) in diameter, with 2 stones. **TWIGS** are flexible and bright gray to red, with velvety hairs. The terminal bud is not covered with scales.

HABITAT AND RANGE: Occurs extensively in chaparrals,

forests, and woodlands throughout most of California, at elevations below 2,300 m (7,500 ft). It ranges south into Baja California and east to New Mexico.

REMARKS: Hoary coffeeberry's leaf color, shape, and size vary among habitats, but all forms have leaves that are white (hoary) from a dense covering of fine hair on the lower blade surfaces. Many plants have leaves with entire margins. Among them, some have green upper leaf surfaces (*R. t.* ssp. *tomentella*), and some have white upper leaf surfaces (*R. t.* ssp. *crassifolia*). Many plants have leaves with toothed margins. Some have scalloped margins (*R. t.* ssp. *cuspidata*), and others have serrated margins (*R. t.* ssp. *ursina*).

RHODODENDRON (RHODODENDRON)

The genus *Rhododendron* has about 800 tree and shrub species; about 21 are native to North America, and of these, 2 shrubs are native to California. In the Northern Hemisphere, *Rhododendron* species are restricted to cool, temperate forests, where they may be canopy trees.

Leaves are evergreen or deciduous, alternate, and simple. Margins can be entire or toothed. Inflorescences and individual flowers are large and showy. Flowers are funnel to bell shaped. Fruits are capsules. Buds are large, with overlapping scales. Flower buds are located at the ends of twigs and are considerably larger than vegetative buds.

Rhododendrons are widely used as ornamentals, and many cultivars have been developed. Some rhododendron species have toxic foliage.

1. Leaves are thick and evergreen. Margins lack fine hairs . . . . **Pacific rhododendron** (*R. macrophyllum*)
1. Leaves are thin and deciduous. Margins have fine hairs. . . . . **western azalea** (*R. occidentale*)

PACIFIC RHODODENDRON *Rhododendron macrophyllum*

(Fig. 172; Pl. 33)

DESCRIPTION: An erect, single- or multistemmed, straggly shrub that typically ranges in height between 1 m (3.3 ft) and

Figure 172
Pacific rhododendron,
Rhododendron macrophyllum

4 m (13 ft). The largest grows in Mendocino County and is 10 m (33 ft) tall and 16.5 cm (6.5 in.) in diameter. **LEAVES** are evergreen, alternate, simple, elliptical, and 7.5 cm (3 in.) to 15 cm (6 in.) long. Margins are entire. Leaf tips have broad points and bases are wedge shaped. Upper surfaces are dark green and hairless and have a sunken mid vein, while lower surfaces are pale green and occasionally have rusty hairs. **INFLORESCENCES** form large, showy clusters of flowers. **FLOWERS** are pink to rose colored and about 4 cm (1.5 in.) long. **FRUITS** are capsules about 2 cm (.75 in.) long. **BUDS** are large and pointed with overlapping scales. **BRANCHES** are stout.

HABITAT AND RANGE: Grows in moist coniferous forests from British Columbia south to Monterey County. In California it is found in coastal forests below 1,000 m (3,300 ft).

REMARKS: Pacific rhododendron sprouts following injury, and its foliage may be poisonous to sheep. The species is the state flower of the state of Washington. It is widely planted as an ornamental because of its showy inflorescences and resistance to deer browsing.

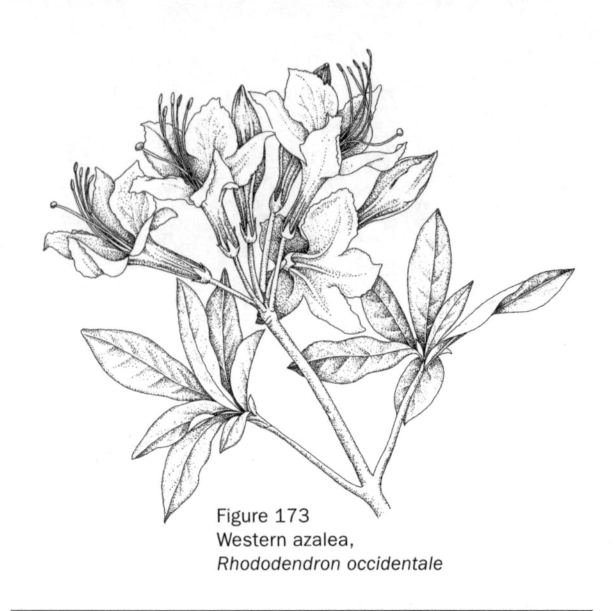

Figure 173
Western azalea,
Rhododendron occidentale

WESTERN AZALEA ***Rhododendron occidentale***
(Fig. 173)

DESCRIPTION: An erect, multistemmed, loosely branched shrub that ranges in height between 1 m (3.3 ft) and 3 m (10 ft). **LEAVES** are deciduous, alternate, simple, elliptical, and 4 cm (1.5 in.) to 10 cm (4 in.) long. Margins have small hairs. Leaf tips have broad points and bases are wedge shaped. Upper surfaces are green and hairless, while lower surfaces are light green and either hairless or covered with fine hairs. **INFLORESCENCES** form large, showy clusters of flowers. **FLOWERS** are white, tinged with pink and orange, and about 4 cm (1.5 in.) long. **FRUITS** are capsules about 12 mm (.5 in.) long. **BUDS** are egg shaped with red, overlapping scales. **BRANCHES** are slender.

HABITAT AND RANGE: Grows in moist coniferous forests from Oregon to southern California, at elevations from sea level to 2,100 m (7,000 ft). It can often be found near streams and in wet meadows and springs.

REMARKS: Western azalea vigorously sprouts following fire or cutting. It is renowned for its fragrant blossoms in the late spring and early summer. Horticulturists have developed numerous cultivars of western azalea. Its foliage is toxic to livestock.

RHUS (SUMAC)

The genus *Rhus* has about 150 species of trees and shrubs that grow in temperate and subtropical areas worldwide. Of these, 3 shrub species are native to California. *Rhus* is the Greek word for sumac.

1. Leaves are compound, with 3 leaflets . . . . **skunkbush** (*R. trilobata*)
1. Leaves are simple 2
   2. Leaves are flat and fruits are white . . . . **lemonadeberry** (*R. integrifolia*)
   2. Leaves are folded along the mid vein and fruits are red . . . . **sugarbush** (*R. ovata*)

LEMONADEBERRY *Rhus integrifolia*

DESCRIPTION: A medium-sized shrub or small tree that can grow to be 1 m (3.2 ft) to 3 m (10 ft) tall. **LEAVES** are alternate, simple, evergreen, aromatic, and leathery. Blades are elliptical to egg shaped, flat, usually round at both ends, and from 25 mm (1 in.) to 63 mm (2.5 in.) long. Margins are entire to toothed. Upper surfaces are smooth and shiny, while lower surfaces have prominent veins. Petioles are 3 mm (.1 in.) to 10 mm (.4 in.) long. **INFLORESCENCES** form terminal clusters of small, hairy, whitish to pinkish flowers. **FRUITS** are hairy, reddish drupes about 10 mm (.4 in.) long. Fruits mature in late summer. **TWIGS** are stout and reddish.

HABITAT AND RANGE: Grows in chaparrals and coastal scrubs from Santa Barbara County to northern Baja California. It can be found on coastal slopes, valleys, mesas, and foothills in southern California, at elevations below 750 m (2,500 ft).

REMARKS: Lemonadeberry gets its name from the pinkish, lemon-flavored drink that Native Americans made from its "berries." Along the southern California

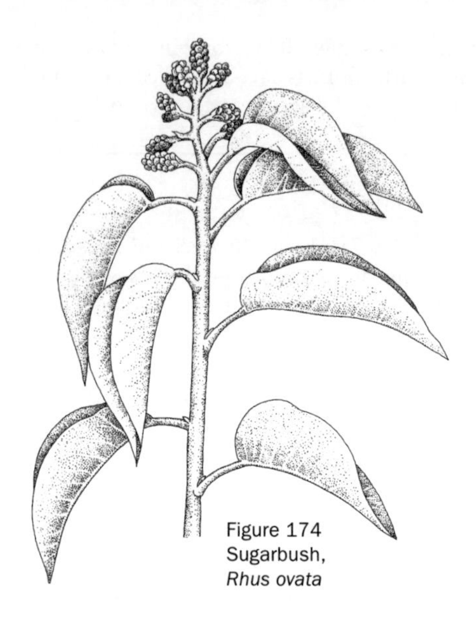

Figure 174
Sugarbush,
Rhus ovata

coast, the species forms conspicuous thickets. Lemonadeberry is an ornamental species in southern California, and it hybridizes with sugarbush.

SUGARBUSH (Fig. 174) ***Rhus ovata***

DESCRIPTION: A medium-sized shrub or small tree that can grow to be 1 m (3.2 ft) to 3 m (10 ft) tall. **LEAVES** are alternate, simple, evergreen, nonaromatic, and leathery. Blades are elliptical to egg shaped, folded sharply along the midrib, and from 3.8 cm (1.5 in.) to 7.6 cm (3 in.) long. Leaf tips are pointed. Margins are entire to toothed or, rarely, lobed. Both surfaces are smooth and shiny. Petioles are 10 mm (.4 in.) to 25 mm (1 in.) long. **INFLORESCENCES** form terminal clusters of small, hairy, whitish to pinkish flowers. **FRUITS** are hairy, glandular, sticky, reddish drupes about 6 mm (.25 in.) long. **TWIGS** are stout and reddish.

HABITAT AND RANGE: Grows in foothill and lower montane chaparrals from Santa Barbara County to northern Baja California, at elevations below 1,200 m (4,000 ft).

Figure 175
Skunkbush,
Rhus trilobata

It is usually absent from the coast and can extend to the desert slopes of the Peninsular Ranges and into central Arizona.

REMARKS: Sugarbush is a desirable ornamental species in southern California and is sometimes planted to help prevent erosion on dry soils. It gets its name from the sugary coating on its fruits. Sugarbush hybridizes with lemonadeberry.

SKUNKBUSH (Fig. 175; Pl. 34) *Rhus trilobata*

DESCRIPTION: A diffusely branched, spreading shrub that forms thickets and may grow to be 2.5 m (8 ft) tall. **LEAVES** are alternate, deciduous, and thin and typically have 3 leaflets. Leaflets taper from the base to a broad tip. The terminal leaflets are larger than the lateral ones and measure 2 cm (.75 in.) to 5 cm (2 in.) long; they have fine hairs. **INFLORESCENCES** are terminal spikes rising above the leaves. The flowers bloom before leaves develop. **FLOWERS** are pale yellow. **FRUITS** are orange or red drupes covered with sticky hairs. **TWIGS** are smooth and brown.

HABITAT AND RANGE: Grows in chaparrals, scrubs, and woodlands below

1,000 m (3,300 ft) in much of California. It ranges throughout the central and western United States and northern Mexico. **REMARKS:** There is a great deal of variation in leaf and stem hairiness in skunkbush. Seven varieties have been recognized throughout its range. The leaves resemble those of poison-oak, but they do not cause contact dermatitis. Indeed, the fruit can be made into a drink and its stems are used in basketry.

RIBES (CURRANT AND GOOSEBERRY)

The genus *Ribes* has about 120 species of shrubs; 30 of these are native to California. The genus is found in the Northern Hemisphere and in temperate habitats in South America.

Its species have simple, palmately lobed leaves that are clustered on short, lateral branchlets, which are alternate in arrangement. Flowers have floral tubes from which arise 4 or 5 sepals and petals. Flowers are arranged in open, branched inflorescences. Fruits are berries and form from inferior ovaries. Branches may have nodal spines and/or internodal prickles, bristles, or hairs.

Ribes is often encountered as individual plants or small populations in the state's forests, woodlands, and chaparrals. At one time currants, which have spineless fruits, and gooseberries, which have spiny fruits, were treated as 2 genera: *Ribes* and *Grossularia,* respectively. Some 12 taxa (species or varieties) are sufficiently rare to be listed by various agencies. Some species are cultivated for their flowers and fruits. Wildlife species eat the berries throughout the summer. *Ribes* is an alternate host for white pine blister rust (*Cronartium ribicola*), which infests 5-needled pines. Various eradication efforts in California have been unsuccessful in reducing infection of the pines.

This book treats only about a third of the *Ribes* species that grow in California.

1. Stems lack nodal spines 2
1. Stems have nodal spines 5
   2. Flowers are saucer shaped. Leaf blades are the size of bigleaf maple leaves, 5 cm (2 in.) to 10 cm (4 in.) wide . . . . **stink currant (*R. bracteosum*)**

2. Flowers are tubular in shape. Leaf blades are less than 5 cm (2 in.) wide 3

3. Flowers are yellow . . . . **golden currant** (*R. aureum*)

3. Flowers are white, pink, or red 4

4. Leaf blades are waxy on upper surfaces. Berries are red . . . . **wax currant** (*R. cereum*)

4. Leaf blades are dull on the upper surfaces. Berries are bluish black . . . . **red flowering currant** (*R. sanguineum*)

5. Flowers have 4 erect sepals . . . . **fuchsia-flowered gooseberry** (*R. speciosum*)

5. Flowers have 4 spreading or down-turned sepals 6

6. Flowers are saucer shaped. Inflorescences have more than 5 flowers 7

6. Flowers are cuplike or tubular in shape. Inflorescences have fewer than 5 flowers 8

7. Leaf blades are hairy and berries are orange . . . . **mountain gooseberry** (*R. montigenum*)

7. Leaf blades are not hairy and berries are black . . . . **swamp currant** (*R. lacustre*)

8. Flowers have yellow petals and sepals . . . . **oak gooseberry** (*R. quercetorum*)

8. Flowers have white petals and green, red, or purple sepals 9

9. Berries are black and lack hairs and spines . . . . **straggly gooseberry** (*R. divaricatum*)

9. Berries are red or purple and have many hairs or spines 10

10. Berries are covered only with glandular hairs . . . . **gummy gooseberry** (*R. lobbii*)

10. Berries are covered with stiff spines, some of which are gland tipped 11

11. Stems have prickles between nodes, as well as nodal spines . . . . **canyon gooseberry** (*R. menziesii*)

11. Stems lack prickles between nodes but have nodal spines 12

12. Flower tubes are as long as they are wide . . . . **hillside gooseberry** (*R. californicum*)

12. Flower tubes are longer than they are wide . . . . **Sierra gooseberry** (*R. roezlii*)

GOLDEN CURRANT *Ribes aureum*

DESCRIPTION: An erect shrub that grows to be 2.5 m (8 ft) tall. **LEAVES** are firm and 2 cm (.75 in.) to 5 cm (2 in.) wide, with 3 or 5 lobes. Leaf surfaces are light green, glossy, and hairless. **INFLORESCENCES** are clusters of a few to 15 flowers. **FLOWERS** are yellow, turning orange or red with age. **FRUITS** are spherical; about 6 mm (.25 in.) in diameter; yellow, orange, red, or black; and hairless. **TWIGS** are hairless. The ascending **BRANCHES** are unarmed.

HABITAT AND RANGE: Grows in moist habitats in the inland Coast Ranges southward to Riverside County and in the Great Basin and adjacent areas with similar conditions. It occurs from 30 m (100 ft) to 3,000 m (9,500 ft). Outside of California it ranges throughout the Great Basin and Rocky Mountains.

REMARKS: There are 2 geographic forms of golden currant. The flowers of plants in the southern Coast Ranges (*R. a.* var. *gracillimum*) lack the spicy odor of the flowers found on eastern California plants (*R. a.* var. *aureum*). The yellow flowers of the coastal form turn orange with age, and those of the inland form turn red. Both forms are popular in gardens for their flower and fruit color.

STINK CURRANT *Ribes bracteosum*

DESCRIPTION: An erect shrub that grows to be 3.5 m (12 ft) tall. **LEAVES** are thin, large, and 6.5 cm (2.5 in.) to 20 cm (8 in.) wide; they resemble maple leaves with saw-toothed margins. Both blade surfaces are alike and are sparsely hairy. **INFLORESCENCES** are long clusters of 20 to 50 flowers subtended by large, leaflike lower bracts. **FLOWERS** are saucer shaped and green. **FRUITS** are spherical, 6 mm (.25 in.) in diameter, and black, with whitish waxy surfaces. **TWIGS** are sparsely hairy. The ascending **BRANCHES** are unarmed.

HABITAT AND RANGE: Grows in wet habitats in coniferous forests from Alaska to northwestern California. It occurs from sea level to 1,400 m (4,500 ft).

REMARKS: Stink currant is easily recognized by its large, maplelike leaves and disagreeable odor.

HILLSIDE GOOSEBERRY *Ribes californicum*

DESCRIPTION: A compact shrub that grows to be 1 m (3.3 ft) tall. **LEAVES** are thin, 12 mm (.5 in.) to 4 cm (1.5 in.) wide, and 3- or 5-lobed, with round teeth. Leaf surfaces are hairless. **INFLORESCENCES** are small clusters of 1 to 3 flowers. **FLOWERS** are tubular and bicolored. The reflexed sepals are green, grading to purple. The petals are white. **FRUITS** are spherical, about 1 cm (.4 in.) in diameter, red, and covered with long bristles and short glandular hairs. **TWIGS** are generally hairless. **BRANCHES** have 3 nodal spines.

HABITAT AND RANGE: Grows in chaparrals and woodlands below 1,000 m (3,500 ft) in California's coastal ranges south of Mendocino County and in southern California's mountains.

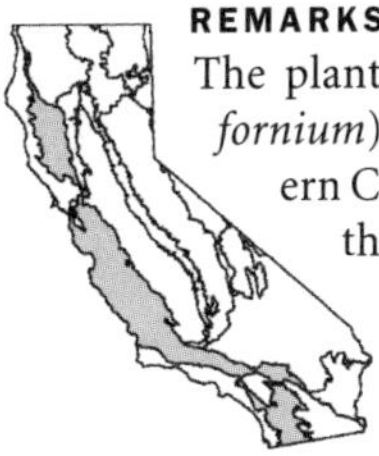

REMARKS: Hillside gooseberry comes in 2 forms. The plants of northern California (*R. c.* var. *californium*) have hairless leaves, while those of southern California (*R. c.* var. *hesperium*) are hairy. In the northern part of the species' range, intermediate specimens suggest hybridization of hillside gooseberry with canyon gooseberry.

WAX CURRANT *Ribes cereum*

DESCRIPTION: An erect shrub that grows to be 2.5 m (8 ft) tall. **LEAVES** are potently fragrant, firm, 12 mm (.5 in.) to 25 mm (1 in.) wide, with shallow lobes. Upper leaf surfaces are glossy green, waxy, and most commonly sticky. **INFLORESCENCES** have few to 7 flowers. **FLOWERS** are tubular and white, cream, or pink. **FRUITS** are 12 mm (.5 in.) in diameter and bright red; they lack bristles. **TWIGS** have fine hairs when young. The stiff **BRANCHES** are unarmed, gray, and hairless.

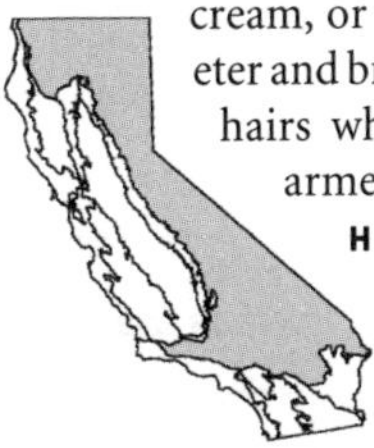

HABITAT AND RANGE: Grows in dry, open habitats in California's inland mountains, at elevations from 1,500 m (4,800 ft) to 4,000 m (13,000 ft). Outside of California it ranges throughout the Great Basin and the Rocky Mountains.

REMARKS: Wax currant has 2 forms. The commonly encountered one, *R. c.* var. *cereum,* which has sticky leaves, can be differentiated from the nonsticky plant, *R. c.* var. *inebrians.* In California, *R. c.* var. *inebrians* is restricted to the highest elevations of the southern Sierra Nevada and the White and Inyo Mountains. Its common name comes from the white coating on its leaves.

STRAGGLY GOOSEBERRY *Ribes divaricatum*

DESCRIPTION: A diffuse shrub that grows to be 2.5 m (8 ft) tall. **LEAVES** are thin, 2 cm (.75 in.) to 5 cm (2 in.) wide, 3- or 5-lobed, and coarsely toothed. Leaf surfaces are light green and either hairy or hairless. **INFLORESCENCES** are small clusters of 1 to 4 flowers. **FLOWERS** are tubular and bicolored. The reflexed sepals are purple or green and petals are white. **FRUITS** are 6 mm (.25 in.) to 10 mm (.4 in.) in diameter, dark purple, and hairless. **TWIGS** may have bristles. The straggly or droopy **BRANCHES** have 1 to 3 nodal spines.

HABITAT AND RANGE: Grows among other shrubs, especially along forest edges and in thickets. It ranges from British Columbia south to California's Coast Ranges, Klamath Mountains, and Cascades, at elevations below 650 m (2,200 ft).

REMARKS: Straggly gooseberry is most often found as individual plants trailing through the branches of other shrubs. Its coastal range, straggly habit, and bicolored flowers make it easy to identify. Three varieties are recognized. *R. d.* var. *divaricatum* grows north of California, from Oregon to British Columbia. *R. d.* var. *pubiflorum,* an uncommon variety, is widely dispersed in northern California. *R. d.* var. *parishii* is rare and grows in southern California. The similar-looking white-stemmed gooseberry (*R. inerme*) is found in the Cascades, Sierra Nevada, and Warner Mountains.

SWAMP CURRANT *Ribes lacustre*

DESCRIPTION: A prostrate or straggly shrub that may grow to be 1 m (3.3 ft) tall. **LEAVES** are thin, deciduous, and 2.5 cm

(1 in.) to 5 cm (2 in.) wide, with shallow or deep lobes. Upper leaf surfaces are glossy dark green and have a few hairs or are hairless. **INFLORESCENCES** are open, droopy clusters of 5 to 15 flowers. **FLOWERS** are saucer shaped and purple to greenish white. **FRUITS** are 6 mm (.25 in.) in diameter and black or purple, with scattered glandular hairs. **TWIGS** may have prickles between the nodes. **BRANCHES** are armed with 5 to 9 spines at the nodes.

HABITAT AND RANGE: Grows in swampy areas and along streams in forests and meadows from Alaska south to the Klamath Mountains. It is found at elevations from 1,200 m (3,900 ft) to 1,800 m (6,000 ft).

REMARKS: Swamp currant does not easily fit the concept of either a currant or a gooseberry, as it has nodal spines and the fruits have only a scattering of hairs. Its habitat and shiny leaves make it an easy species to identify, though. Birds and mammals eat swamp currant berries, and its foliage is moderately palatable to ungulates. Native Americans ate the currants. Swamp currant is an alternate host for white pine blister rust.

GUMMY GOOSEBERRY (Pl. 35) *Ribes lobbii*

DESCRIPTION: An erect shrub that may grow to be 2 m (6.5 ft) tall. **LEAVES** are thin and 2 cm (.75 in.) to 5 cm (2 in.) wide, with 3- or 5-toothed lobes. Leaf surfaces are light green and are either hairy or hairless. **INFLORESCENCES** are small clusters of 1 to 3 flowers. **FLOWERS** are tubular and bicolored. The reflexed sepals are red and the petals are white. **FRUITS** are 12 mm (.5 in.) to 25 mm (1 in.) in diameter, dark red, and covered with glandular hairs. **TWIGS** may have a scattering of hairs. **BRANCHES** have 1 to 3 nodal spines.

HABITAT AND RANGE: Grows in forests from British Columbia south to California's northwestern mountains, at elevations between 1,400 m (4,500 ft) to 2,000 m (6,500 ft).

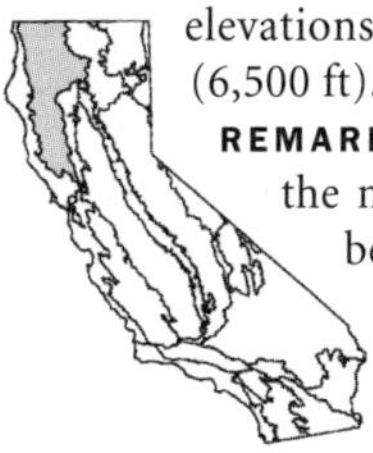

REMARKS: Gummy gooseberry is distinctive as the montane gooseberry with the big, gummy berries. Hybrids between gummy gooseberry and Sierra gooseberry are found in logged lands. Sierra gooseberry, however, has bristly, nonsticky berries.

CANYON GOOSEBERRY *Ribes menziesii*

DESCRIPTION: An erect shrub that may grow to be 2.5 m (8 ft) tall. **LEAVES** are firm, 12 mm (.5 in.) to 4 cm (1.5 in.) wide, and 3- or 5-lobed, with round teeth. Leaf surfaces are covered with glandular hairs. **INFLORESCENCES** are small clusters of 1 to 3 flowers. **FLOWERS** are tubular and bicolored. The reflexed sepals are purple and the petals are white. **FRUITS** are 12 mm (.5 in.) in diameter, purple, and covered with bristles. **TWIGS,** especially the young growth, are densely covered with bristles, some of which are gland tipped. **BRANCHES** have 3 nodal spines.

HABITAT AND RANGE: Grows in forests and woodlands of California's Coast Ranges from Del Norte to San Luis Obispo County, at elevations below 300 m (1,000 ft). Outside of California it is found in southwestern Oregon.

REMARKS: There is a great amount of variation in leaf and fruit characteristics among populations of this species. Some populations have aromatic leaves. Intermediate plants in the central Coast Ranges suggest hybridization between canyon gooseberry and hillside gooseberry.

MOUNTAIN GOOSEBERRY *Ribes montigenum*

(Fig. 176)

DESCRIPTION: An open, straggly shrub that grows to be 60 cm (2 ft) tall. **LEAVES** are deciduous and 12 mm (.5 in.) to 25 mm (1 in.) wide, with shallow or deep lobes. Upper leaf surfaces are green, with many or a few sticky hairs. **INFLORESCENCES** are short clusters of 3 to 7 flowers. **FLOWERS** are saucer shaped and reddish brown to greenish white. **FRUITS** are 6 mm (.25 in.) in diameter and red, with sticky bristles. **TWIGS** may have bristly hairs between the nodes. **BRANCHES** are armed with 1 to 5 spines at the nodes.

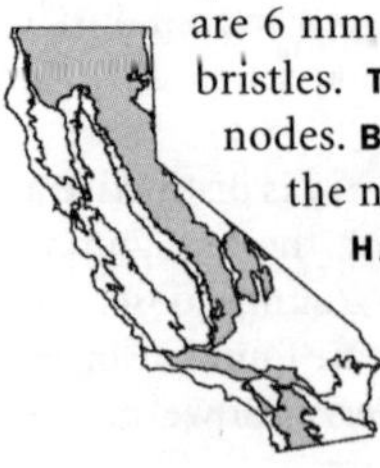

HABITAT AND RANGE: Grows in dry subalpine and alpine habitats of California's inland mountains. It is found at elevations from 2,000 m (6,800 ft) to 4,200 m

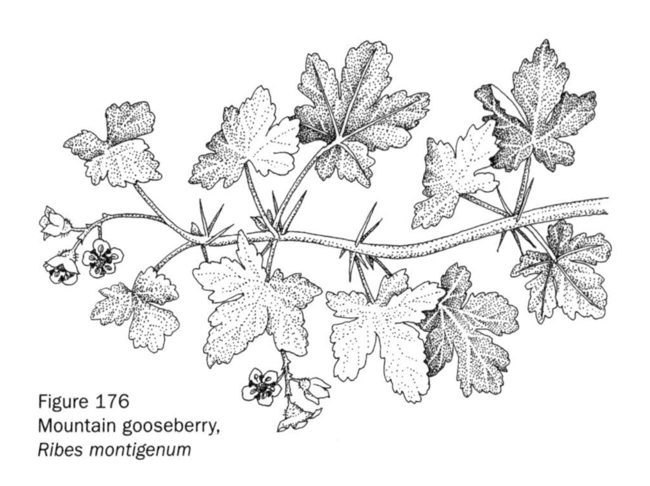

Figure 176
Mountain gooseberry,
Ribes montigenum

(14,000 ft). Outside of California it ranges throughout the mountains of the West.

REMARKS: Mountain gooseberry is most easily found in the lodgepole pine forests of the Sierra Nevada and among alpine talus. Its dull, sticky flowers and its spines, and the fact that it grows at high elevations, make it an easy species to identify.

OAK GOOSEBERRY *Ribes quercetorum*

DESCRIPTION: A rounded shrub that may grow to be 1 m (3.3 ft) tall. **LEAVES** are thin and 1 cm (.4 in.) to 2 cm (.75 in.) wide, with 3 or 5 lobes. Leaf surfaces are light green, with gland-tipped hairs. **INFLORESCENCES** are tight clusters of 2 or 3 flowers. **FLOWERS** are tubular and yellow or white. **FRUITS** are 6 mm (.25 in.) in diameter, black, and hairless. **TWIGS** have short, fine hairs. The arching **BRANCHES** usually have 1 curved nodal spine.

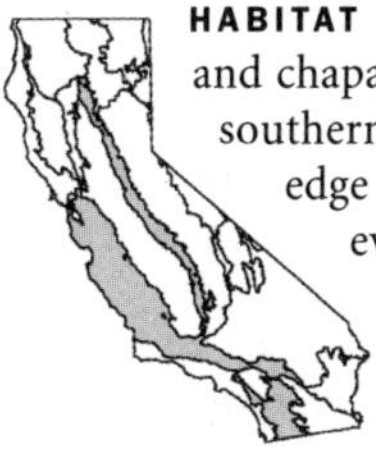

HABITAT AND RANGE: Grows in oak woodlands and chaparrals surrounding the San Joaquin Valley, southern California mountains, and the western edge of the Mojave and Colorado Deserts, at elevations below 1,200 m (4,000 ft). It also ranges into Baja California.

REMARKS: Oak gooseberry develops conspicuous rounded thickets under the

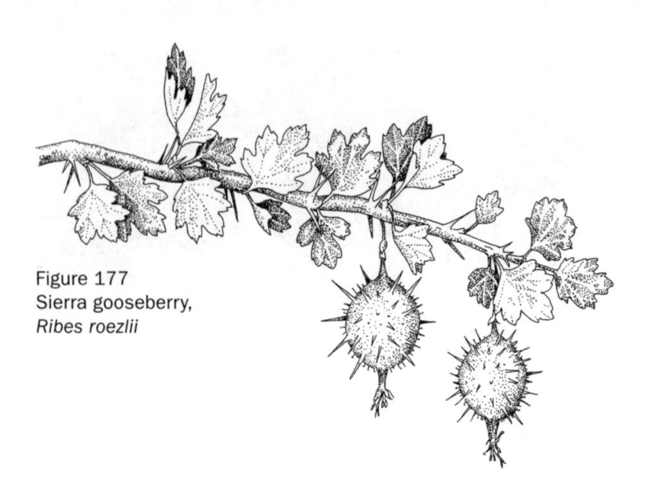

Figure 177
Sierra gooseberry,
Ribes roezlii

oaks of the foothill woodlands. The other common yellow-flowered *Ribes*, golden currant, lacks nodal spines and has darker leaves and smaller flowers.

SIERRA GOOSEBERRY (Fig. 177) ***Ribes roezlii***

DESCRIPTION: An erect shrub that may grow to be 1 m (3.3 ft) tall. **LEAVES** are thin and 12 mm (.5 in.) to 25 mm (1 in.) wide, with 3- or 5-toothed lobes. Leaf surfaces are light green and may have hairs. **INFLORESCENCES** are small clusters of 1 to 3 flowers. **FLOWERS** are tubular and bicolored. The reflexed sepals are dull red and the petals white. **FRUITS** are 12 mm (.5 in.) to 20 mm (.75 in.) in diameter, dark red, and covered with bristles. **TWIGS** lack hairs. **BRANCHES** have 1 to 3 nodal spines.

HABITAT AND RANGE: Grows in the coniferous forests in most of California's mountains. It occurs at elevations between 1,000 m (3,500 ft) and 2,500 m (8,500 ft). It also occurs in Oregon.

REMARKS: Sierra gooseberry is distinctive as the montane gooseberry with the big, bristly berries. In the northwestern mountains, plants may have leaf blades with white, hairy lower surfaces (*R. r.* var. *amictum*) or hairless upper surfaces (*R. r.* var. *cruentum*). Inland

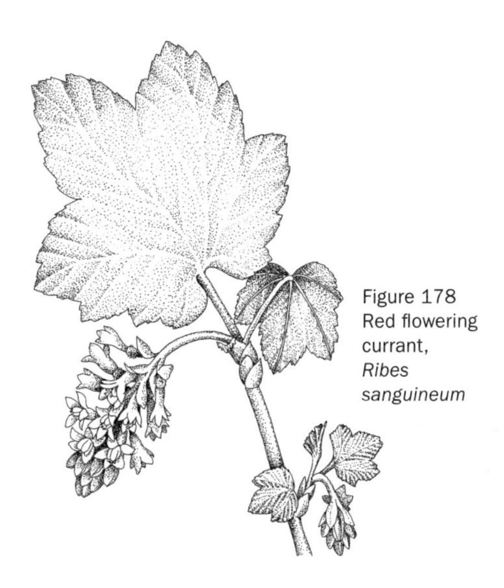

Figure 178
Red flowering currant,
Ribes sanguineum

plants have leaf blades with hairy upper surfaces (*R. r.* var. *roezlii*). Hybrids between the Sierra and gummy gooseberries are found in logged lands. They have bristles as well as glandular hairs. Birds and mammals eat Sierra gooseberry fruits. Native Americans ate the gooseberries. The species is an alternate host for white pine blister rust.

RED FLOWERING CURRANT *Ribes sanguineum*

(Fig. 178; Pl. 36)

DESCRIPTION: An open, erect shrub that may grow to be 3.5 m (12 ft) tall. **LEAVES** are firm and 2.5 cm (1 in.) to 5 cm (2 in.) wide, with shallow lobes. Upper leaf surfaces are dark green, with scattered hairs. **INFLORESCENCES** are open clusters of 10 to 15 flowers. **FLOWERS** are tubular and crimson, pink, or white. **FRUITS** are 6 mm (.25 in.) to 12 mm (.5 in.) in diameter and bluish black, with a white, waxy covering. **TWIGS** have fine hairs when they are young. **BRANCHES** are unarmed, with brown or gray bark.
HABITAT AND RANGE: Grows in and around wooded habitats of northern California's Coast Ranges. It is found from sea

level to 2,200 m (7,200 ft). Outside of California it ranges north to British Columbia.

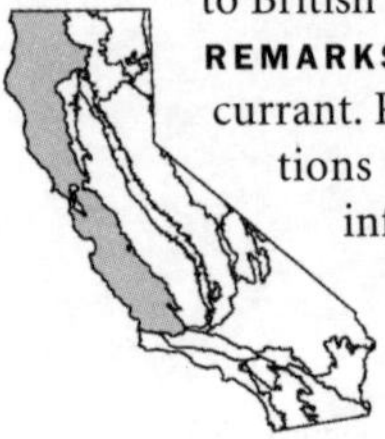

REMARKS: There are 2 forms of red flowering currant. Plants of the montane and subalpine elevations (*R. s.* var. *sanguineum*) have upturned inflorescences with crimson flowers. Plants along the coast at low elevations (*R. s.* var. *glutinosum*) have droopy inflorescences of pink to white flowers.

FUCHSIA-FLOWERED GOOSEBERRY *Ribes speciosum*

DESCRIPTION: A bristly, compact shrub that grows to be 3 m (10 ft) tall. **LEAVES** are thick, leathery, semideciduous, 12 mm (.5 in.) to 40 mm (1.5 in.) wide, 3- or 5-lobed, and toothed. Leaf surfaces are hairless and upper ones are shiny dark green. **INFLORESCENCES** are small clusters of 1 to 4 flowers. **FLOWERS** are tubular. The erect sepals and petals are deep crimson. **FRUITS** are spherical, about 1 cm (.4 in.) in diameter, and covered with long glandular bristles that dry with age. **TWIGS** are densely covered with bristles. **BRANCHES** have 3 nodal spines.

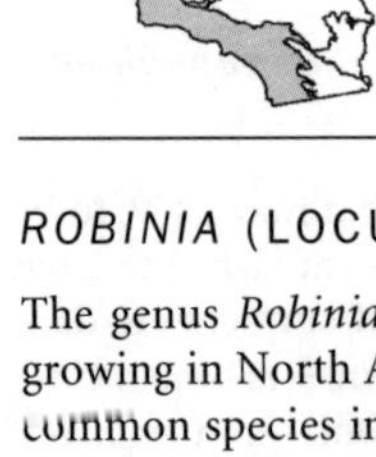

HABITAT AND RANGE: Grows in chaparrals and woodlands below 500 m (1,600 ft) in California's Coast Ranges south of Santa Clara County into southern California.

REMARKS: Fuchsia-flowered gooseberry has stunning flowers and is easily cultivated for use in dry gardens.

ROBINIA (LOCUST)

The genus *Robinia* has about 10 species of trees and shrubs growing in North America; 1 is native to California. The more common species in California is one that is naturalized, a native to the eastern United States.

BLACK LOCUST (Fig. 179) *Robinia pseudoacacia*

DESCRIPTION: A single-stemmed, medium-sized tree. Mature trees are typically 12 m (40 ft) to 18 m (60 ft) tall and 30

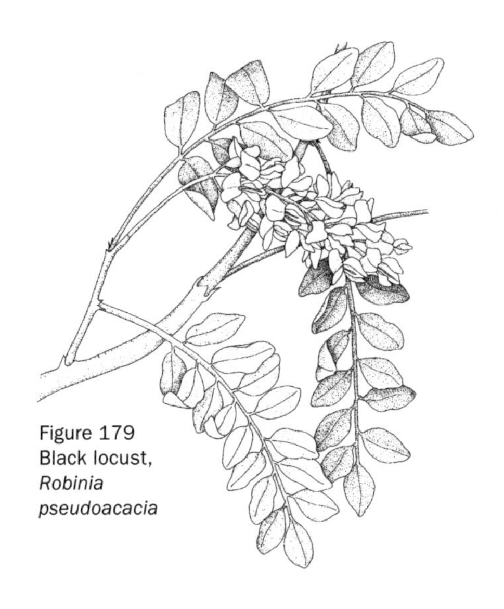

Figure 179
Black locust,
Robinia pseudoacacia

cm (1 ft) to 90 cm (3 ft) in diameter. The largest grows in Dansville, New York, and is 29.3 m (96 ft) tall and 2.2 m (89 in.) in diameter. Crowns are open and irregular. Trunks can be straight or crooked. **LEAVES** are deciduous, alternate, pinnately compound (usually 7 to 19 leaflets), and 15 cm (6 in.) to 35 cm (14 in.) long. Leaflets are elliptical to egg shaped and about 5 cm (2 in.) long. Leaflet margins are entire. Leaflet tips have small notches with tiny bristles, and bases are round. Leaflet surfaces are green and hairless. **FLOWERS** are showy and white and hang in droopy clusters that measure 10 cm (4 in.) to 15 cm (6 in.) long. Individual flowers are similar to pea flowers. **FRUITS** are legumes that measure 5 cm (2 in.) to 10 cm (4 in.) long. **TWIGS** are stout and reddish brown, with paired spines at the bases of leaves. These spines become evergreen and enlarged on branches and smaller stems. **BARK** is dark reddish brown, with interlacing ridges separated by deep furrows.

HABITAT AND RANGE: The species' native range is in the eastern United States. It has been naturalized in California and is widespread along streams, lower slopes, and roads, at elevations from 60 m (200 ft) to 1,800 m (6,000 ft).

REMARKS: Black locust is shade intolerant and is a pioneering species in disturbed habitats. It sprouts vigorously following injury from fire or cutting. Nitrogen-fixing bacteria form root nodules in black locust, adding considerable nitrogen to ecosystems via decaying plant parts. Black locust is used as firewood and for making fence posts, tool handles, and novelty items. All plant parts, especially seeds and inner bark, contain a toxin that can be fatal to humans and livestock. In California, black locust provides shade and hiding cover for wildlife. The native species, *R. neomexicana,* grows in the mountains of the eastern Mojave Desert.

ROSA (ROSE)

The genus *Rosa* has about 100 species growing throughout the temperate parts of the Northern Hemisphere. Of these, 9 are native to California.

Rosa is a collection of shrubs and vines with pinnately compound leaves that are alternate in arrangement. Stipules are attached to the petiole. Stems are covered with a few to many prickles (called rose thorns by many) between the nodes. Flowers are showy and pink, and they appear in the spring. Each flower has 5 petals and numerous stamens attached to the top of a saucer surrounding the many pistils. Fruits are achenes surrounded by a saucer that becomes leathery and red with age. This kind of fruit is called a hip and is characteristic of the genus.

California's roses, with their simple pink flowers, are easy to recognize at the genus level but are harder to discern at the species level. The 3 species in the key are the more commonly encountered species growing in the state. The 2 escaped exotics, *R. canina* and *R. eglanteria,* unlike the natives, have sepals with toothed lobes. The native *R. bridgesii, R. pinetorum,* and *R. spithamea* share the few-stemmed habit of wood rose. *R. nutkana* and *R. pisocarpa* share the thicket-forming habit with California rose. *R. minutifolia* is a rare rose that grows in chaparrals. Roses hybridize, making identification even more difficult. Hips are a source of vitamin C.

1. Plants have a few stems, and sepals are lost as fruits mature .... **wood rose (*R. gymnocarpa*)**
1. Plants form thickets, and sepals are maintained as fruits mature 2

2. Prickles are slender and straight . . . . **interior rose** (*R. woodsii*)
2. Prickles are thick and curved . . . . **California rose** (*R. californica*)

CALIFORNIA ROSE *Rosa californica*

DESCRIPTION: A thicket-forming shrub that may grow to be 3 m (10 ft) tall. It has straight or curved, stout prickles. **LEAVES** are deciduous and compound, with 5 to 7 leaflets. Leaflet blades are elliptical and 12 mm (.5 in.) to 40 mm (1.5 in.) long, with fine hairs. Margins have 1 or 2 sets of teeth. **INFLORESCENCES** consist of many-branched clusters of a few to 30 flowers. **FLOWERS** are bright rose to light pink. **FRUITS** are round, reddish hips with long, tapering sepals.

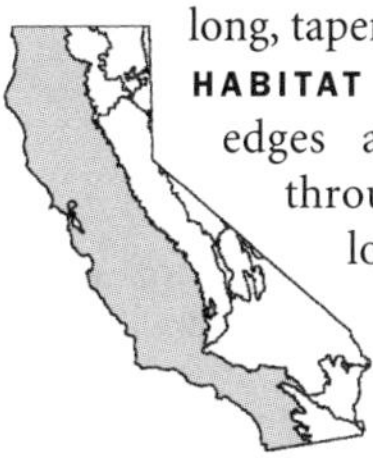

HABITAT AND RANGE: Grows along woodland edges associated with streams, from Oregon throughout much of California, at elevations below 1,800 m (6,000 ft).

REMARKS: California rose is one of the most common, abundant, and variable of the native roses.

WOOD ROSE (Fig. 180) *Rosa gymnocarpa*

DESCRIPTION: A slender shrub that may grow to be 1 m (3.3 ft) tall. **LEAVES** are deciduous and compound, with 5 to 7 leaflets. Leaflet blades are round, measure 5 cm (2 in.) to 7.5 cm (3 in.) long, and have 2 sets of teeth on the margins. **INFLORESCENCES** consist of 1 to 4 flowers. **FLOWERS** are pink. **FRUITS** are red, berrylike hips that lose their sepals with age. **STEMS** have straight, slender prickles.

HABITAT AND RANGE: Grows under forest canopies of northern California's mountains and in the Palomar Mountains in southern California, at elevations from 30 m (100 ft) to 2,000 m (6,500 ft). It ranges throughout the western mountains.

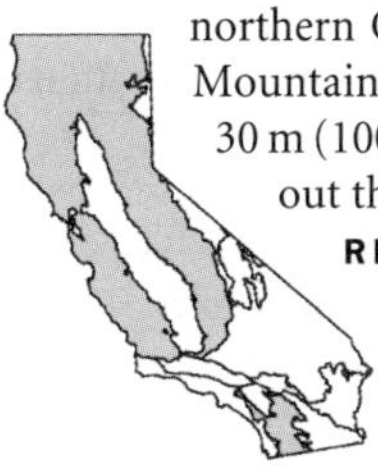

REMARKS: There is much variation in hairiness and leaf size in wood rose throughout its range. *R. bridgesii, R. pinetorum, R. spithamea* have the same growth

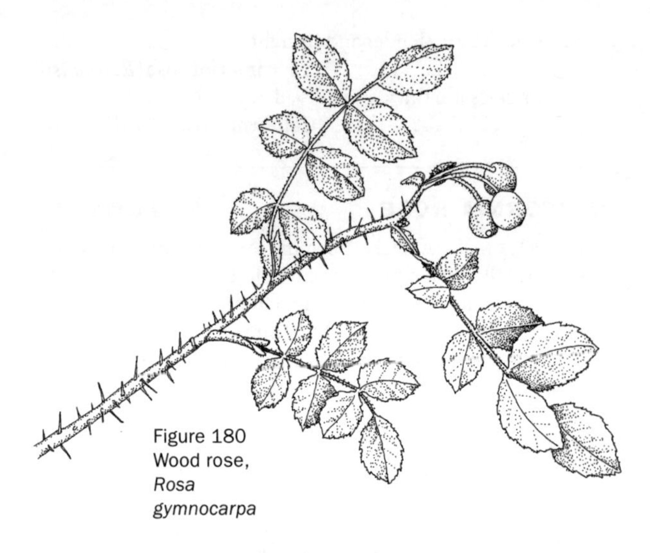

Figure 180 Wood rose, *Rosa gymnocarpa*

habit, but their fruits and prickles differ from those of wood rose. Remember that this species is the rose with the naked (*gym*) fruit (*carpel*).

INTERIOR ROSE (Fig. 181) ***Rosa woodsii***

DESCRIPTION: A thicket-forming shrub that may grow to be between 60 cm (2 ft) and 3 m (10 ft) tall. **LEAVES** are deciduous and compound, with 5 to 7 leaflets. Leaflet blades are round, measure 12 mm (.5 in.) to 40 mm (1.5 in.) long, and have 1 set of teeth on their margins. **INFLORESCENCES** consist of clusters of 1 to 5 flowers. **FLOWERS** are rose to light pink. **FRUITS** are round, red hips with long, tapering sepals. **STEMS** may have slender, straight prickles.

HABITAT AND RANGE: Found in the eastern Cascades, the eastern Sierra Nevada, the desert mountains, and some southern California mountains, at elevations from 900 m (3,000 ft) to 3,000 m (10,000 ft).

REMARKS: In California, interior rose is represented by *R. w.* var. *ultramontana,* 1 of 4 varieties of this wide-ranging species. It is an easy rose to identify: it is the only one with an interior range.

Figure 181
Interior rose,
Rosa woodsii

RUBUS (BLACKBERRY)

The genus *Rubus* has 700 species worldwide; 11 are native to California. *Rubus* is a collection of brambles, shrubs, and vines with simple or compound leaves that are alternate in arrangement. Stems of most species are covered with a few to many, thin to stout prickles between the nodes. Flowers are commonly white or pink, and each flower has 5 petals and many stamens attached to the top of a cup surrounding many pistils. Fruits are sweet, fleshy drupelets attached to an expanded receptacle. If the drupelets are easily detached from the receptacle at maturity, the plants are called raspberries. If the receptacle is removed with the drupelets at maturity, the plants are called blackberries.

The bramble life-form requires some explanation. In many *Rubus* species, plants have 2 stem types called "canes." The first-year canes are vegetative, each extending as vines or long, arching shoots. The second-year canes produce flowers and fruits and then die. The result is a mound-building thicket called "brambles." Not all *Rubus* species have this life-form; some are erect shrubs and others are strawberry-like, with ground-hugging stems and small leaves.

1. Leaves are simple and maplelike, with 5 lobes .... **thimbleberry** (*R. parviflorus*)
1. Leaves are compound or, if simple, have 3 lobes 2
   2. Plants are erect shrubs with red flowers and salmon-colored fruits .... **salmonberry** (*R. spectabilis*)
   2. Plants are brambles or vines with white or pink flowers and black or red fruits 3
3. Plants create a prostrate ground cover .... **dwarf bramble** (*R. lasiococcus*)
3. Plants are erect brambles or climbing vines 4
   4. The lower surfaces of leaf blades are green .... **California blackberry** (*R. ursinus*)
   4. The lower surfaces of leaf blades are white 5
5. Stems are thin and round .... **western raspberry** (*R. leucodermis*)
5. Stems are stout and ribbed .... **Himalayan blackberry** (*R. discolor*)

HIMALAYAN BLACKBERRY *Rubus discolor*

(Fig. 182)

DESCRIPTION: A mound of arching canes that may grow to be 6 m (20 ft) long. **LEAVES** are semideciduous; palmately compound, with 3 to 5 leaflets; and 2.5 cm (1 in.) to 7.5 cm (3 in.) long. Lower surfaces have a dense covering of short white hair. **INFLORESCENCES** consist of large, terminal clusters of 10 or more flowers. **FLOWERS** are white or pink. **FRUITS** are blackberry-like, shiny black or purple at maturity, and edible. **STEMS** are stout, ribbed, and hairless, with a thin covering of wax, especially on the recent vegetative growth. Stems have stout, straight, or curved prickles.

HABITAT AND RANGE: Grows in northern California below 1,500 m (5,000 ft). This widely naturalized exotic commonly occurs in fencerows and along roadsides. Increasingly, it is invading open forests, woodlands, and riparian corridors.

REMARKS: Himalayan blackberry is not the only escaped exotic blackberry, just the most pernicious one. Other exotics include *R. laciniatus,* which has deeply di-

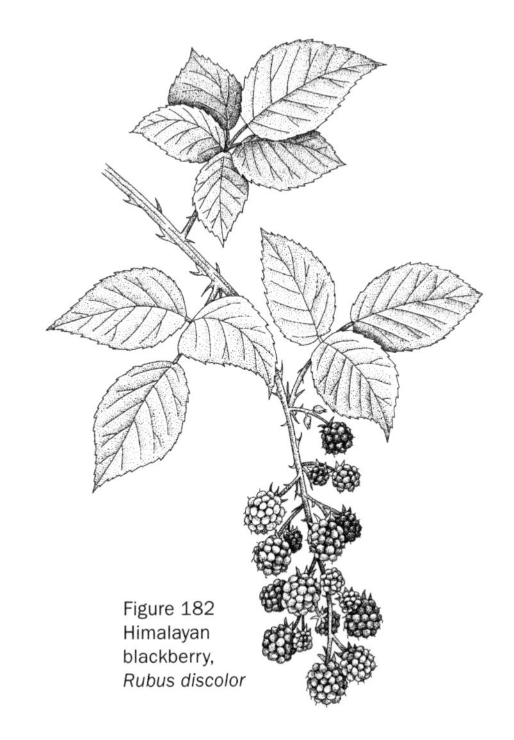

Figure 182
Himalayan
blackberry,
Rubus discolor

vided leaflets; *R. pensilvanicus,* which is found extensively on disturbed areas; and *R. ulmifolius,* which is common in southern California. People and animals eat Himalayan blackberry fruits. The brambles provide wildlife habitat, but unfortunately, natural habitats are being lost to these invasive plants.

DWARF BRAMBLE *Rubus lasiococcus*

DESCRIPTION: A perennial, prostrate trailing plant whose leaves are 10 cm (4 in.) tall. **LEAVES** are evergreen, compound, and 2 cm (.75 in.) to 5 cm (2 in.) long; they have 3 to 5 leaflets. Leaflets have wedge-shaped bases. Margins are sharply toothed. **INFLORESCENCES** consist of clusters of 1 or 2 flowers. **FLOWERS** are white. **FRUITS** are blackberry-like, red at maturity, and edible. **STEMS** are green, unarmed, and ground-hugging and root at the nodes.

HABITAT AND RANGE: Grows in open montane coniferous

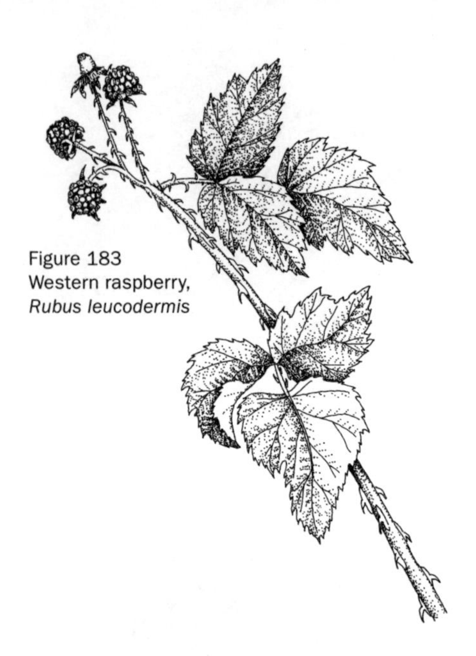

Figure 183
Western raspberry,
Rubus leucodermis

forests in the northwestern mountains of California, at elevations from 450 m (1,500 ft) to 2,000 m (6,500 ft). It is found north into British Columbia.

REMARKS: Dwarf bramble's growth habit superficially resembles that of the strawberry, but its fruits and dark leaves differ. Snow dwarf bramble (*R. nivalis*) is another prostrate bramble, but with prickly stems. It grows from the northern Rocky Mountains to Oregon and has only recently been reported in the western Klamath Mountains in California; it is considered uncommon in California. Dwarf bramble has ornamental value as a ground cover.

WESTERN RASPBERRY *Rubus leucodermis*

(Fig. 183)

DESCRIPTION: A mounded shrub with arching canes that may grow to be 2 m (6.5 ft) long. **LEAVES** are deciduous; com-

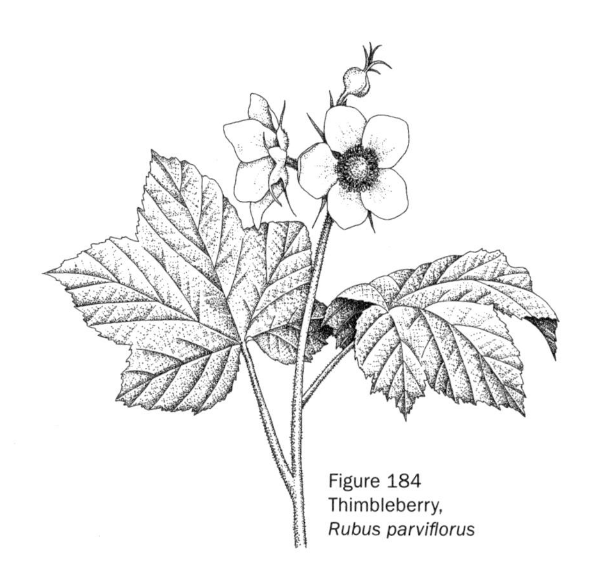

Figure 184
Thimbleberry,
Rubus parviflorus

pound, with 5 to 7 leaflets; and 2.5 cm (1 in.) to 7.5 cm (3 in.) long. Lower leaflet surfaces have a dense covering of short white hair. **INFLORESCENCES** consist of clusters of 3 to 10 flowers. **FLOWERS** are white. **FRUITS** are raspberry-like, black or dark purple at maturity, and edible. **STEMS** are white and waxy, especially on the recent vegetative growth. Stems have straight to curved prickles.

HABITAT AND RANGE: Grows throughout most of California, except in the deserts, in forests at elevations below 2,000 m (6,500 ft).

REMARKS: Fruits of western raspberry are called blackcaps and are a summertime treat.

THIMBLEBERRY (Fig. 184) ***Rubus parviflorus***

DESCRIPTION: An erect shrub whose canes may grow to be 2 m (6 ft) long. **LEAVES** are deciduous, simple, palmately lobed, 5 cm (2 in.) to 15 cm (6 in.) long, and 5- to 7-lobed, with heart-shaped bases. Margins are sharply toothed. Upper and lower surfaces are covered with soft hairs. **INFLORESCENCES** consist of clusters of 4 to 7 flowers. **FLOWERS** are white or pink. **FRUITS** are raspberry-like, red at maturity, edible, and thimblelike. **STEMS** have gray-brown bark that peels off in thin strips and lacks prickles.

HABITAT AND RANGE: Grows throughout California, except in the deserts, from sea level to 2,700 m (9,000 ft). It occurs in and along the edges of forests. Over much of its range, it is associated with moist habitats under tree canopies.

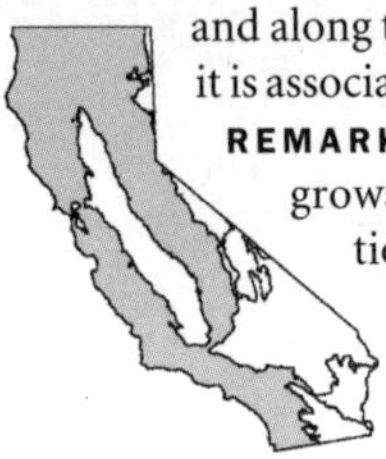

REMARKS: At lower elevations thimbleberry grows to be a tall shrub, but at montane elevations tall stems rarely survive the winter and the plants are distinctively shorter. Thimbleberry can also dominate the bluffs of California's treeless northern coastline.

SALMONBERRY *Rubus spectabilis*

DESCRIPTION: An erect shrub with canes that may grow to be 3.5 m (12 ft) long. **LEAVES** are deciduous, 4 cm (1.5 in.) to 11 cm (4.5 in.) long, and compound, with 3 leaflets. Margins are sharply doubly toothed. Lower surfaces may be covered with soft hairs. **INFLORESCENCES** consist of clusters of 1 to 4 flowers. **FLOWERS** are pinkish red. **FRUITS** are blackberry-like; yellow, salmon, or red at maturity; and edible. **STEMS** have reddish brown bark that peels off in strips. Prickles are loosely held and fall off with the peeling bark.

HABITAT AND RANGE: Grows from Alaska to California's northern coastal coniferous forests below 1,400 m (4,500 ft). In California it is often an indicator of moist soils, such as in riparian areas.

REMARKS: Fruit color and palatability vary among plants and habitats.

CALIFORNIA BLACKBERRY *Rubus ursinus*

(Fig. 185)

DESCRIPTION: A plant with straggling, viny canes that may grow to be 6 m (20 ft) long. The plant has thin, straight prickles. **LEAVES** are semideciduous; compound, with 3 leaflets; and 2.5 cm (1 in.) to 7.5 cm (3 in.) long. Lower leaflet

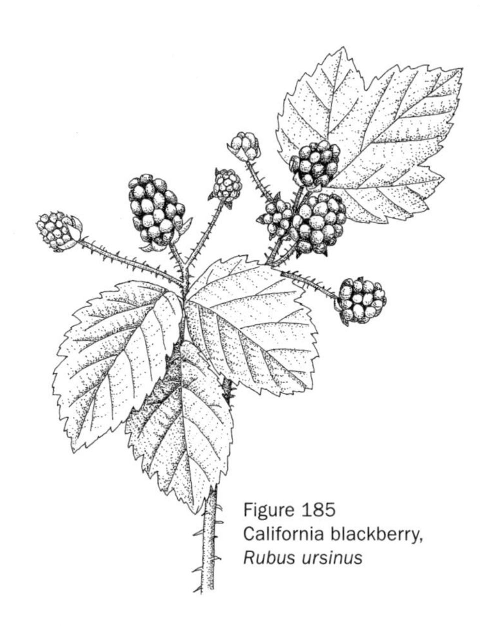

Figure 185
California blackberry,
Rubus ursinus

surfaces are hairy or hairless and light green. **INFLORESCENCES** are small, with 2 to 15 flowers. **FLOWERS** are white and may be complete or incomplete. **FRUITS** are blackberry-like, shiny black or purple at maturity, and edible. **STEMS** are flexible, round, and hairless and may have a thin covering of wax, especially on recent vegetative growth. Stems have slender, straight prickles.

HABITAT AND RANGE: Grows in California, except in the deserts, at elevations below 1,500 m (5,000 ft). It is found in forests and woodlands as well as open habitats.

REMARKS: California blackberry is a variable species, especially in its flowers. Populations vary in vegetative characteristics as well, especially in whether the leaves are hairy, and whether the hairs end in glands. In many books, the names *R. macropetalus* and *R. vitifolius* are used to describe extremes in the species. California blackberry is the ancestor of the cultivated loganberry and boysenberry. Native populations are worth seeking in July for their appetizing fruits.

SALIX (WILLOW)

The genus *Salix* grows throughout the world and has more than 400 species of trees and shrubs. Thirty species are native to California.

Leaves are simple, deciduous, and alternate. Most are thin and have stipules that fall with age. Individual flowers are not striking. Flowers bearing only seeds are found on separate plants from those bearing only pollen. Flowers are arranged in catkins that can be showy. Fruits are capsules that produce many seeds. Each capsule bears a tuft of long hair that assists in wind dispersal of seed. Lateral buds generally have 1 caplike bud scale that is pressed against the stem.

Willows are often encountered in wet habitats, where they may dominate. Leaf size and hairiness, which often change as plants mature, are influenced by environmental conditions such as shade. Hybridization between species is common. In many species, plants bloom before they leaf out. In many keys, both floral and vegetative characteristics are used, so plants must be tagged when flowering and collected again when leaves are mature. The key in this book uses only vegetative characteristics of the more commonly encountered species.

Willows are short-lived, fast-growing trees and shrubs that readily sprout new branches when heavily pruned, a process called "coppicing." These branches, called wands, have long been used for making boats (known as coracles), baskets, and furniture and as fodder. Salicylic acid, the active ingredient in aspirin, is synthesized from a glucoside in willow's bitter-tasting sap. Willows play an important role in restoration, since stems readily root and are fast growing. Willow wood is used for making boxes, cricket bats, and pulp. In parts of northern Europe, willow wood chips serve as fuel, replacing fossil fuels.

1. Margins of bud scales are free and they overlap 2
1. Margins of bud scales are fused 3
    2. Leaf blades are waxy and white on lower surfaces . . . . **red willow (*S. laevigata*)**
    2. Leaf blades are not waxy and white on lower surfaces . . . . **black willow (*S. gooddingii*)**
3. Glands are present at the bases of leaf blades 4
3. Glands are absent from the bases of leaf blades 6

4. Plants are trees with long, droopy branches . . . . **weeping willow (*S. babylonica*)**
4. Plants are trees or shrubs with erect branches 5
5. Leaf blades have acute tips . . . . **Sierra willow (*S. eastwoodiae*)**
5. Leaf blades have long, tapering tips . . . . **shining willow (*S. lucida*)**
6. Leaf blades are long and linear . . . . **narrowleaf willow (*S. exigua*)**
6. Leaf blades are not long and linear 7
7. Leaf blade margins are rolled under 8
7. Leaf blade margins are flat 9
8. Leaf blades are widest above the middle . . . . **Scouler willow (*S. scouleriana*)**
8. Leaf blades are widest at the middle or lower . . . . **Sitka willow (*S. sitchensis*)**
9. Lower surfaces of leaf blades are not white or waxy . . . . **dusky willow (*S. melanopsis*)**
9. Lower surfaces of leaf blades are white and waxy 10
10. Upper surfaces of leaf blades are glossy 11
10. Upper surfaces of leaf blades are not glossy 12
11. Twigs are white and waxy . . . . **Hooker willow (*S. hookeriana*)**
11. Twigs are not white or waxy . . . . **arroyo willow (*S. lasiolepis*)**
12. Twigs are yellow or gray . . . . . . **yellow willow (*S. lutea*)**
12. Twigs are brown 13
13. Both blade surfaces are covered with gray hairs . . . . **Sierra willow (*S. eastwoodiae*)**
13. Upper blade surfaces are green, and the lower blade surfaces have a dense covering of white wax and hair . . . . **Jepson willow (*S. jepsonii*)**

WEEPING WILLOW *Salix babylonica*

DESCRIPTION: A tree that usually grows to be 20 m (65 ft) tall. The largest lives in Detroit, Michigan, and is 35.6 m (117 ft) tall and 3.8 m (98 in.) in diameter. **LEAVES** are linear to elliptical, measure 6 cm (2.5 in.) to 12.5 cm (5 in.) long, and have long, tapering tips and entire margins. Both blade surfaces are green and may have hairs. A pair of glands occurs where the blade meets the petiole. Stipules are lacking. **INFLORESCENCES**

appear on leafy stalks as the leaves emerge. Fruiting catkins are 6 mm (.25 in.) to 5 cm (2 in.) long. **TWIGS** are gray to yellowish brown, with hair that they lose as they age. Bud-scale margins are fused. **BRANCHES** are long and droop to the ground.
HABITAT AND RANGE: A commonly planted ornamental tree that has escaped cultivation. It occurs in disturbed locales throughout the state at elevations below 900 m (3,000 ft).
REMARKS: Weeping willow plants are part of a clone from China that has been propagated for 300 years. Many ornamental forms exist, especially hybrids with the European white willow (*S. alba*). Golden weeping willow (*S. a.* var. *vitellina*) is also commonly planted in California. These species and their many forms naturalize and hybridize with native willows, adding to identification problems.

SIERRA WILLOW *Salix eastwoodiae*

DESCRIPTION: A shrub that may grow to be 2 m (6.5 ft) tall. **LEAVES** are elliptical and 2.5 cm (1 in.) to 10 cm (4 in.) long, with fine-toothed or gland-tipped margins. Both blade surfaces are covered with gray, tightly pressed hairs. Older blades become hairless. A pair of glands generally occurs where the blade meets the petiole. Stipules are small and round. **INFLORESCENCES** appear on leafy stalks along with the leaves. Fruiting catkins are 6 mm (.25 in.) to 25 mm (1 in.) long. **FRUITS** are capsules covered with fine gray hairs. **TWIGS** are dark brown and waxy and have fine hairs. Older twigs become hairless. Bud-scale margins are fused.
HABITAT AND RANGE: Grows in subalpine and alpine riparian zones and meadows in northern California's mountains, at elevations from 1,600 m (5,200 ft) to 3,000 m (10,000 ft). It ranges east into the northern Rocky Mountains.
REMARKS: In the Sierra Nevada, Sierra willow may grow with *S. orestera*, which also has gray leaves.

NARROWLEAF WILLOW (Fig. 186) *Salix exigua*

DESCRIPTION: A multistemmed shrub that typically grows

Figure 186
Narrowleaf willow,
Salix exigua

to be 6 m (20 ft) tall. The largest lives in Criglersville, Virginia, and is 11 m (36 ft) tall and 56 cm (22 in.) in diameter. **LEAVES** are linear and 5 cm (2 in.) to 12.5 cm (5 in.) long, with entire to coarsely toothed margins. Both blade surfaces are gray to green and covered thinly or thickly with long, silky hairs. Stipules are lacking. **INFLORESCENCES** appear on long, leafy stems during or after leaf emergence. Fruiting catkins are 2 cm (.75 in.) to 5 cm (2 in.) long. **FRUITS** are hairy capsules. **TWIGS** are slender and may have fine white hairs. Bud-scale margins are fused.

HABITAT AND RANGE: A common wetland plant that occurs throughout North America. It grows best in moist, sandy soils. In California it is found from sea level to 2,400 m (8,000 ft).

REMARKS: Narrowleaf willow has a variety of forms over its broad range. Some books have considered some forms to be species, subspecies, or varieties. Think of this willow as the silvery willow of sandy habitats.

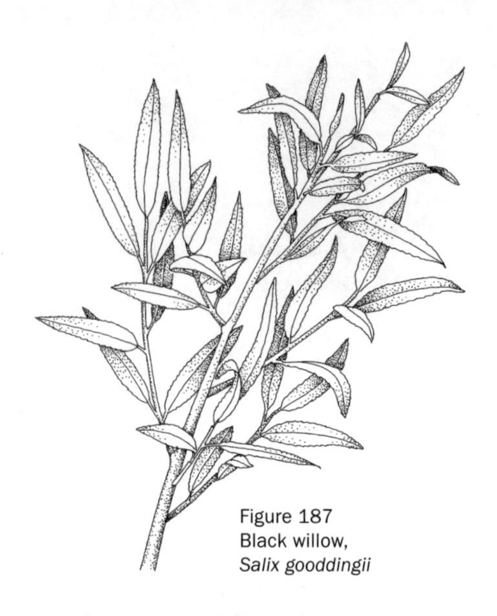
Figure 187
Black willow,
Salix gooddingii

BLACK WILLOW (Fig. 187) ***Salix gooddingii***

DESCRIPTION: This species is a tree that typically grows to be 30 m (100 ft) tall. The largest lives in Luna County, New Mexico, and is 13.7 m (45 ft) tall and 2.8 m (112 in.) in diameter. **LEAVES** are elliptical and measure 6 cm (2.5 in.) to 12.5 cm (5 in.) long; they have long, tapering tips and smooth margins. Both blade surfaces are pale green and hairless. Stipules are lacking. **INFLORESCENCES** appear on leafy stalks along with the leaves. Fruiting catkins droop and are 2.5 cm (1 in.) to 6 cm (2.5 in.) long. **FRUITS** are hairy or hairless capsules. **TWIGS** are yellow, hairy when young, and hairless later. Bud-scale margins overlap. **BRANCHES** are dark brown and hairless. **BARK** is dark and rough.

HABITAT AND RANGE: Grows throughout California's wetlands, at elevations below 600 m (2,000 ft). Outside of California it ranges throughout the western states and into Mexico.

REMARKS: Black willow is one of the California willows that have unfused bud-scale margins. The other species, red willow, is differentiated by its glossy leaves. *S. gooddingii* is very closely related to another black willow (*S. nigra*) of the eastern United States. They differ, though, in twig color and leaf shape.

HOOKER WILLOW *Salix hookeriana*

DESCRIPTION: A shrub or small tree that usually grows to be 7.5 m (25 ft) tall. The largest lives in Warrenton, Oregon, and is 10 m (32 ft) tall and 40 cm (16 in.) in diameter. **LEAVES** are elliptical, measure 4 cm (1.25 in.) to 7.5 cm (5 in.) long, and have fine-toothed or entire margins. The upper blade surfaces are glossy green, with a covering of fine hair. The lower blade surfaces have a dense covering of white wax and hair. Stipules have lobes that surround the stem. **INFLORESCENCES** appear before the leaves and are set on short leafy stems. Fruiting catkins are 4 cm (1.5 in.) to 7 cm (2.75 in.) long. **FRUITS** are hairy or hairless capsules. **TWIGS** are brown and covered with white wax. Bud-scale margins are fused.

HABITAT AND RANGE: A common willow of the coastal forest wetlands below 150 m (500 ft).

REMARKS: In California, Sitka willow has a range similar to that of Hooker willow, but its blades have rolled-under margins and are not glossy. Along the northern California coast, Hooker willow widely hybridizes with arroyo willow.

JEPSON WILLOW *Salix jepsonii*

DESCRIPTION: A shrub that may grow to be 2 m (6.5 ft) tall. **LEAVES** are lance shaped and 4 cm (1.5 in.) to 7.5 cm (3 in.) long, with entire margins. The upper blade surfaces are green, with scattered hairs. The lower blade surfaces have a dense covering of white wax and hair. Stipules are small, if present. **INFLORESCENCES** appear with the leaves and are set on short leafy stems. Fruiting catkins are 4 cm (1.5 in.) to 7 cm (2.75 in.) long. **FRUITS** are hairy capsules. **TWIGS** are brown, with silky hair that is lost at maturity. Bud-scale margins are fused.

HABITAT AND RANGE: Found in the

mountains of northern California, at elevations from 1,000 m (3,500 ft) to 3,500 m (11,500 ft). It grows along montane and subalpine creeks and meadows.

REMARKS: Jepson willow appears to be a Sierra Nevada relative of Sitka willow, but with smaller and narrower leaves. The two species' ranges meet in the Klamath Mountains.

RED WILLOW *Salix laevigata*

DESCRIPTION: A tree that may grow to be 13.5 m (45 ft) tall. **LEAVES** have a broad elliptical shape, are 6 cm (2.5 in.) to 15 cm (6 in.) long, and have finely scalloped margins. The upper blade surfaces are pale green and glossy. The lower blade surfaces are white and waxy. Stipules are small with glandular teeth. **INFLORESCENCES** appear on leafy stalks as or after the leaves emerge. Fruiting catkins droop and are 4 cm (1.5 in.) to 10 cm (4 in.) long. **FRUITS** are capsules covered with fine gray hairs. **TWIGS** are red to yellowish brown and initially hairy, but they lose the hair with age. Bud-scale margins are free and overlapping. **BRANCHES** are dark brown and hairless.

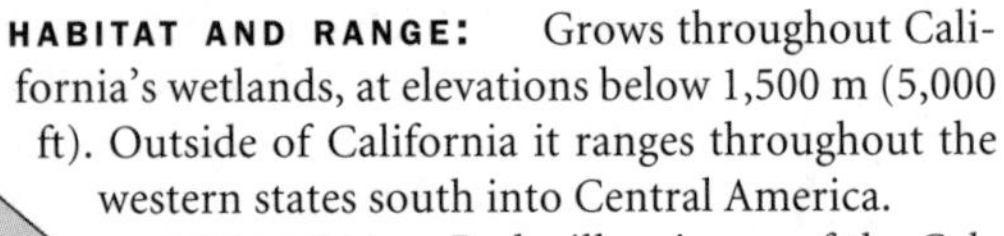

HABITAT AND RANGE: Grows throughout California's wetlands, at elevations below 1,500 m (5,000 ft). Outside of California it ranges throughout the western states south into Central America.

REMARKS: Red willow is one of the California willows in which the bud-scale margins are not fused. The other is black willow, differentiated by its glossy leaves.

ARROYO WILLOW *Salix lasiolepis*

DESCRIPTION: A shrub or small tree that often grows to be 12 m (40 ft) tall. The largest lives in Wallowa County, Oregon, and is 8.2 m (27 ft) tall and 34 cm (13.5 in.) in diameter. **LEAVES** are lance shaped and 5 cm (2 in.) to 12.5 cm (5 in.) long, with fine-toothed or smooth margins. The upper blade surfaces are dark green and shiny and covered with fine hairs that are lost with age. The lower blade surfaces have a waxy covering. Stipules are small or lacking. **INFLORESCENCES** appear on short leafy stems before the leaves emerge. Fruiting catkins are 3 cm (1.25 in.) to 7 cm (2.75 in.) long. **FRUITS** are hairless

Figure 188
Shining willow,
Salix lucida

capsules. **TWIGS** are yellow or brown and covered with fine hairs that are lost with age. Bud-scale margins are fused.

HABITAT AND RANGE: An abundant species in wetlands below 2,100 m (7,000 ft) in the western United States.

REMARKS: Over its broad range, Arroyo willow varies sufficiently that various forms have been given names; few of these names are now used. Arroyo willow widely hybridizes with Hooker willow along the north coast.

SHINING WILLOW (Fig. 188) ***Salix lucida***

DESCRIPTION: A multistemmed tree or shrub that typically grows to be 18 m (60 ft) tall. The largest lives in Traverse City, Michigan, and is 22.5 m (74 ft) tall and 104 cm (41 in.) in diameter. **LEAVES** are lance shaped, measure 2.5 cm (1 in.) to 10 cm (4 in.) long, and have fine-toothed, gland-tipped margins. The upper blade surfaces are shiny green. The lower blade surfaces are pale green. Glands can be found where the blade meets the petiole. The small, round stipules have glandular teeth. **INFLORESCENCES** appear on short leaf stems along with the leaves. Fruiting catkins are 4 cm (1.5 in.) to 15 cm (6 in.) long. **FRUITS** are light brown, hairless capsules. **TWIGS** are very glossy; green, yellow, or brown; and hairless. Bud-scale margins are fused.

HABITAT AND RANGE: A wetland willow, it grows throughout the United States and southern Canada, at elevations below 3,200 m (10,500 ft).

REMARKS: Over its broad range, shining willow varies sufficiently that various subspecies and varieties have been recognized. Many times, these forms have been given species rank, so comparing books is difficult. Shining willow (*S. l.* var. *lucida*) grows in the eastern United States. California has 2 subspecies: Pacific willow (*S. l.* var. *lasiandra*), a common low-elevation tree whose lower leaf surfaces are waxy white; and Bryant willow (*S. l.* var. *caudata*), a montane shrub whose leaf blade surfaces are not waxy white.

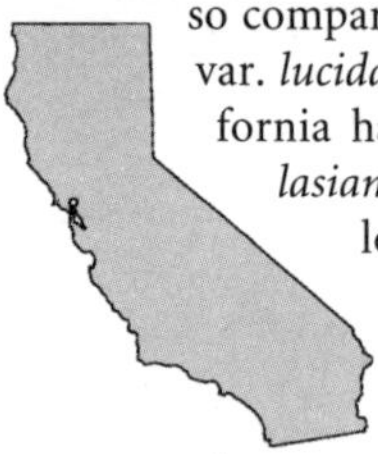

YELLOW WILLOW *Salix lutea*

DESCRIPTION: A shrub that often grows to be 4.5 m (15 ft) tall. The largest lives in Lemhi County, Idaho, and is 7.3 m (24 ft) tall and 20 cm (8 in.) in diameter. **LEAVES** are lance shaped and 3 cm (1.25 in.) to 12.5 cm (5 in.) long, with fine-toothed or entire margins. The upper blade surfaces are shiny dark green and hairless. The lower blade surfaces have a dense covering of white hair. Stipules are round, with toothed or entire margins. **INFLORESCENCES** appear on short leafy stems just before or as the leaves emerge. Fruiting catkins are 3 cm (1.25 in.) to 7 cm (2.75 in.) long. **FRUITS** are hairless capsules. **TWIGS** are yellow or brown and hairless. Bud-scale margins are fused.

HABITAT AND RANGE: Found in the mountains of the western United States, east through southern Canada and the northern United States to New England. In California it grows along montane creeks at elevations from 1,500 m (5,000 ft) to 2,900 m (9,500 ft).

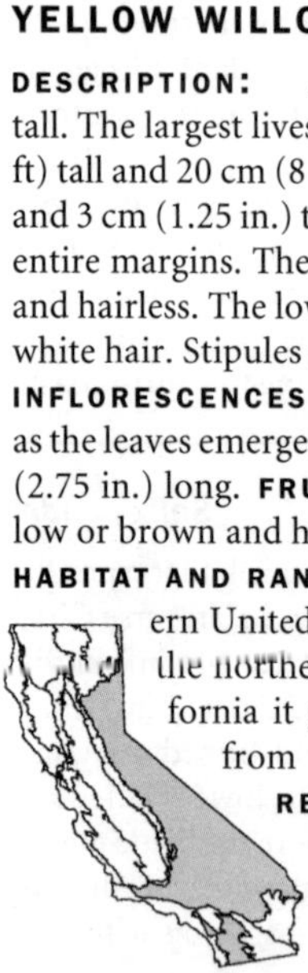

REMARKS: Over yellow willow's broad range, plants vary sufficiently that various forms have been given names, but unlike the case with most other willows few of these names are now used. Think of yellow willow as the montane, shiny-leaved willow.

DUSKY WILLOW *Salix melanopsis*

DESCRIPTION: A shrub that forms thickets and may grow to be 4.5 m (15 ft) tall. **LEAVES** are narrow and lance shaped and 4 cm (1.5 in.) to 10 cm (4 in.) long, with a few teeth on the margins. The upper surfaces are dark green, with scattered silky hairs that are lost with age, while the lower surfaces are hairless. Stipules are toothed. **INFLORESCENCES** appear on short leafy stems before or with the leaves. Flowering catkins are 3 cm (1.25 in.) to 7 cm (2.75 in.) long. **FRUITS** are hairy capsules that become hairless later. **TWIGS** are brown and covered with silky hairs that are lost with age. Bud-scale margins are fused.

HABITAT AND RANGE: A montane species of the western mountains of the United States and Canada. It occurs in sandy or rocky wetlands, at elevations from 800 m (2,600 ft) to 1,800 m (6,000 ft).

REMARKS: Dusky willow is a montane plant that can be thought of as a less hairy relative of narrowleaf willow.

SCOULER WILLOW *Salix scouleriana*

DESCRIPTION: A multistemmed shrub or small tree that typically grows to be 9 m (30 ft) tall. The largest lives in Lincoln City, Oregon, and is 12 m (40 ft) tall and 1.3 m (53 in.) in diameter. **LEAVES** are widest above the middle, tapering to tips and bases. They measure 2.5 cm (1 in.) to 10 cm (4 in.) long and are entire or fine-toothed, with rolled-under margins. The upper blade surfaces are dark green and hairless. The lower blade surfaces may have silvery hairs. Stipules are small. **INFLORESCENCES** appear on short leafy stems before the leaves emerge. Fruiting catkins are 12 mm (.5 in.) to 25 mm (1 in.) long. **FRUITS** are capsules covered with white hairs. **TWIGS** are white, yellow, or brownish and covered with hairs that are lost with age. Bud-scale margins are fused.

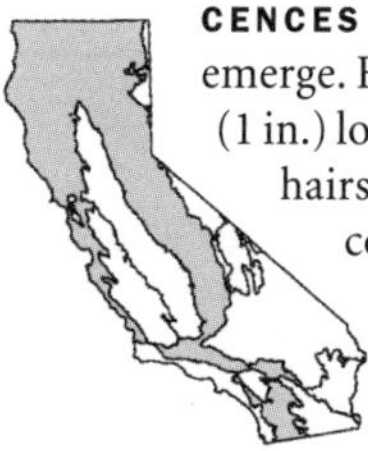

HABITAT AND RANGE: A common upland willow that grows in forests, meadows, and wetlands in the mountains of the western United States and Canada, including those of California, at elevations below 3,000 m (10,000 ft).

REMARKS: Over its broad range, Scouler willow varies sufficiently so that various forms have been given names, but unlike the case with many other willows few of these names are now used. Think of this willow as the montane willow of upland habitats.

SITKA WILLOW *Salix sitchensis*

DESCRIPTION: A multistemmed shrub or small tree that often grows to be 6 m (20 ft) tall. The largest lives in Seattle and is 11 m (36 ft) tall and 14 cm (5.5 in.) in diameter. **LEAVES** are lance shaped and 4.5 cm (1.75 in.) to 12.5 cm (5 in.) long. Margins are entire or fine-toothed and tightly rolled under. The upper blade surfaces are dull green, with a few or many silky hairs that are lost with age. The lower blade surfaces are pale green, with a gleam of silky hair. Stipules are small, if present. **INFLORESCENCES** appear on short leafy stems before or with the leaves. Fruiting catkins are 3 cm (1.25 in.) to 7 cm (2.75 in.) long. **FRUITS** are capsules covered with silky hairs. **TWIGS** are yellowish or reddish brown and covered with scattered hairs that are lost with age. Bud-scale margins are fused.

HABITAT AND RANGE: A common willow of Pacific Coast wetlands below 300 m (1,000 ft).

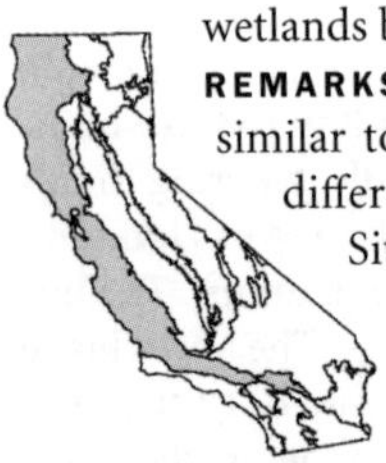

REMARKS: In California, Sitka willow has a range similar to that of Hooker willow. The two can be differentiated by the rolled-under margins of Sitka willow compared to Hooker willow's flat margins, and the glossy leaves of Hooker willow as opposed to the dull green, hairy leaves of Sitka willow.

SALVIA (SAGE)

The genus *Salvia* has about 900 species of strongly aromatic shrubs and herbs. Seventeen species grow in California, 8 of which are associated with forests and woodlands. Many species are found in tropical and subtropical America.

Leaves are simple, drought-deciduous, and generally lance shaped to oval. Leaves are opposite in arrangement. Flowers, which are various colors, are tubular, 2-lipped corollas with 2 exserted stamens. Flowers are associated with leafy bracts and are arranged in showy, elongated inflorescences. Fruits are nutlets from superior pistils. Stems are square in cross section.

Salvia species are the "true sages," as opposed to *Artemisia* species, which are called "sagebrushes." Both are strongly aromatic. Sages have important culinary and medicinal uses and are important bee plants. In California, Native Americans used the seeds as food. Many species are cultivated as ornamentals. In southern California considerable hybridization makes identification a challenge.

1. Flowers grow in long clusters up to 1 m (3.3 ft) long. Leaves are white . . . . **white sage (*S. apiana*)**
1. Flowers grow in compact whorls, forming interrupted spikes. Leaves are gray or grayish green 2
   2. Bracts below flower whorls are paperlike 3
   2. Bracts below flower whorls are not paperlike 6
3. Leaf blades are wrinkled on the upper surfaces 4
3. Leaf blades are smooth on the upper surfaces 5
   4. Flower bracts are purple or purplish green . . . . **sand sage (*S. eremostachya*)**
   4. Flower bracts are white . . . . **Mojave sage (*S. mohavensis*)**
5. Bracts are purplish and 12 mm (.5 in.) to 20 mm (.75 in.) long . . . . **mountain desert sage (*S. pachyphylla*)**
5. Bracts are rose or purplish and less than 12 mm (.5 in.) long . . . . **desert sage (*S. dorrii*)**
   6. Leaf blades have broad round bases . . . . **purple sage (*S. leucophylla*)**
   6. Leaf blade bases gradually taper to petioles 7
7. Plants form prostrate mats . . . . **creeping sage (*S. sonomensis*)**
7. Plants are erect shrubs . . . . **black sage (*S. mellifera*)**

WHITE SAGE *Salvia apiana*

DESCRIPTION: A rounded shrub that may grow to be 2.5 m (8 ft) tall. **LEAVES** are crowded at the bases of a few straight, almost herbaceous stems. Leaf blades are lance shaped and 4 cm

(1.5 in.) to 8 cm (3 in.) long. Both upper and lower leaf blade surfaces are covered with minute, tightly pressed white hairs. **INFLORESCENCES** are less than 1 m (3.3 ft) long and spikelike, with small clusters of flowers. **FLOWERS** are white and may be spotted with lavender. **FRUITS** are light brown and shiny. **BRANCHES** are white, with tightly pressed hairs.

HABITAT AND RANGE: Grows in coastal scrublands, chaparrals, and coniferous woodlands of southern California, Baja California, and the western edges of the Colorado and Mojave Deserts. It occurs below 1,500 m (5,000 ft).

REMARKS: Animals browse white sage in the winter. White sage hybridizes with 7 different species of *Salvia*, so expect regional variation in its appearance.

DESERT SAGE (Fig. 189) ***Salvia dorrii***

DESCRIPTION: A low, mounded shrub that may grow to be 75 cm (2.5 ft) tall. **LEAVES** are clustered on short shoots, 4 mm (.15 in.) to 30 mm (1.2 in.) long, and linear to spoon shaped, with smooth margins. Both upper and lower leaf surfaces are covered with fine hairs. **INFLORESCENCES** have alternating flower clusters on bare stalks. Clusters are surrounded by bracts that are purplish green to rose, round, and 6 mm (.25 in.) long. **FLOWERS** are dark blue. **FRUITS** are tan. **BRANCHES** are white and densely scurfy.

HABITAT AND RANGE: Grows in the interior scrublands and woodlands of the Great Basin, the Mojave Desert, and the east sides of the Cascades and Sierra Nevada, at elevations between 750 m (2,500 ft) and 2,700 m (8,800 ft). It ranges throughout the intermountain West.

REMARKS: Desert sage is similar to mountain desert sage, as both have purplish bracts surrounding their flowers. Desert sage has bracts that are 6 mm (.25 in.) long and leaves less than 30 mm (1.2 in.) long, whereas mountain desert sage's bracts are about 12 mm (.5 in.) long and its leaves are more than 25 mm (1 in.) long. Individual desert sage plants are

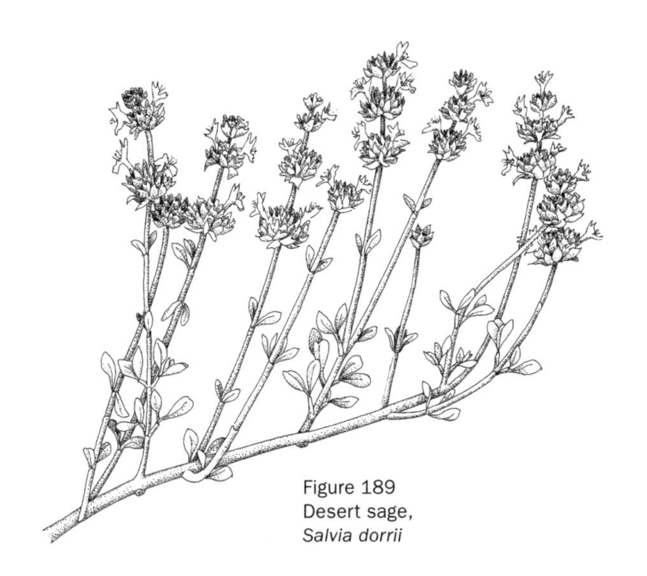

Figure 189
Desert sage,
Salvia dorrii

highly variable across the species' range. Three somewhat geographically separated varieties can be differentiated with difficulty.

SAND SAGE *Salvia eremostachya*

DESCRIPTION: A finely branched shrub that may grow to be 90 cm (3 ft) tall. **LEAVES** are of 2 kinds. Leaves on flowering shoots are lance shaped and about 3 cm (1.2 in.) long. Leaves on the short side branches are linear and about 2 cm (.75 in.) long, with tightly rolled-under margins. Both upper and lower leaf surfaces are puckered and rough to the touch. **INFLORESCENCES** have alternating clusters of 2 or 3 flowers set on bare stalks. Clusters are surrounded by papery, purplish green to rose bracts. **FLOWERS** are blue, rose, or white. **FRUITS** are yellowish brown. **BRANCHES** are covered with glandular hairs.

HABITAT AND RANGE: Grows in desert scrublands and coniferous woodlands on the western edge of the Colorado Desert and in the Peninsular Ranges, at elevations from 450 m (1,500 ft) to 1,400 m (4,500 ft). It ranges into Baja California.

REMARKS: Sand sage is considered uncommon, but where found it is locally common.

PURPLE SAGE — *Salvia leucophylla*

DESCRIPTION: An erect or prostrate shrub that may grow to be 2 m (6.5 ft) tall. **LEAVES** are lance shaped and 2 cm (.75 in.) to 8 cm (3 in.) long, with smooth margins that may be rolled under. Both upper and lower leaf surfaces are finely wrinkled and covered with tightly pressed white hairs. **INFLORESCENCES** have alternating clusters of 2 or 3 flowers set on bare stalks. Clusters are surrounded by mealy, white bracts. **FLOWERS** are light purple. **FRUITS** are brown or dark gray. **BRANCHES** are covered with grayish white hairs.

HABITAT AND RANGE: This species is characteristic of coastal scrublands, woodlands, and forests of the mountains and coast south of San Luis Obispo, at elevations below 600 m (2,000 ft). It ranges into Baja California.

REMARKS: Purple sage hybridizes with the other coastal sages: black and white sage.

BLACK SAGE (Fig. 190; Pl. 37) — *Salvia mellifera*

DESCRIPTION: An erect, aromatic shrub that grows to be 1.5 m (5 ft) tall. **LEAVES** are oblong and 2.5 cm (1 in.) to 7 cm (2.75 in.) long, with scalloped margins. Surfaces are finely puckered and hairless above, but hairy below. **INFLORESCENCES** have alternating flower clusters set on bare stalks. Flower clusters have round, down-turned bracts. **FLOWERS** are pale blue, lilac, or white. **FRUITS** are dark brown. **BRANCHES** are covered with sharp, tightly pressed glandular hairs.

HABITAT AND RANGE: Grows in coastal sage scrubs, chaparrals, and woodlands of central and southern California, at elevations below 365 m (1,200 ft). It is also found in northern Baja California.

Figure 190
Black sage,
Salvia mellifera

REMARKS: Black sage is one of the best sources of honey in the state. The species often hybridizes with purple or white sage.

MOJAVE SAGE *Salvia mohavensis*

DESCRIPTION: A low, rounded shrub that may grow to be 60 cm (2 ft) tall. **LEAVES** are lance shaped and about 2 cm (.75 in.) long, with small, round teeth on the margins. Both leaf surfaces are bright green and puckered. **INFLORESCENCES** have flowers clustered at the head and surrounded by papery, white bracts. **FLOWERS** are pale blue or lavender. **FRUITS** are light brown. **BRANCHES** are covered with minute hairs.

HABITAT AND RANGE: Grows in the scrublands and coniferous woodlands of the Mojave Desert and east to Nevada, Arizona, and Mexico. It occurs at elevations between 275 m (900 ft) and 1,500 m (5,000 ft).

MOUNTAIN DESERT SAGE *Salvia pachyphylla*

DESCRIPTION: A shrub with ascending or prostrate branches that may grow to be 60 cm (2 ft) tall. Branches in contact with the ground may root at the nodes. **LEAVES** are spoon shaped and 2 cm (.75 in.) to 5 cm (2 in.) long, with wavy, entire margins. Leaves are covered with tightly pressed white hairs. **INFLORESCENCES** have alternating clusters of 2 or 3 flowers set on bare stalks. Clusters are surrounded by rounded, purplish green to rose bracts. **FLOWERS** are dark violet-blue to rose. **FRUITS** are tan. **BRANCHES** are white and scurfy.

HABITAT AND RANGE: Grows in coniferous woodlands and forests of the montane southern Sierra Nevada and southern California mountains, at elevations between 1,500 m (5,000 ft) and 3,000 m (10,000 ft). It ranges into Nevada, Arizona, and Baja California.

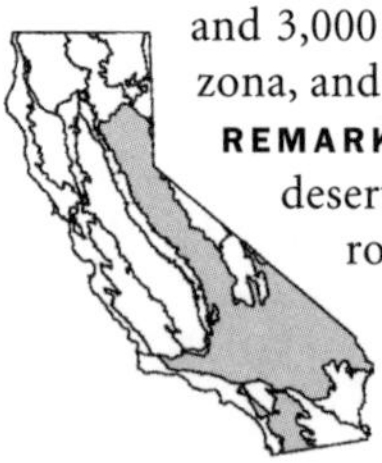

REMARKS: Mountain desert sage is similar to desert sage, as both have purplish bracts surrounding the flowers. Mountain desert sage flowers are surrounded by bracts that are about 12 mm (.5 in.) long, whereas desert sage's bracts are about 6 mm (.25 in.) long.

CREEPING SAGE *Salvia sonomensis*

DESCRIPTION: A prostrate shrub that forms extensive mats and may grow to be 30 cm (1 ft) tall. **LEAVES** are elliptical and 3 cm (1.2 in.) to 6 cm (2.4 in.) long, with smooth margins. Both leaf surfaces are finely wrinkled and covered with fine white hairs. **INFLORESCENCES** have alternating flower clusters on bare stalks. Flower clusters lack bracts. **FLOWERS** are blue, lilac, or purple. **FRUITS** are brown. **BRANCHES** are covered with dead leaf bases.

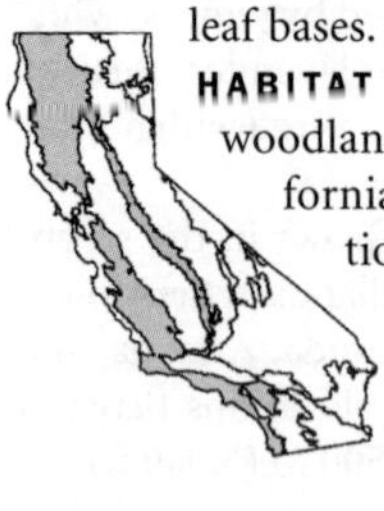

HABITAT AND RANGE: Grows in chaparrals, woodlands, and forests throughout much of California's mountains at low and montane elevations, below 2,000 m (6,500 ft).

REMARKS: Creeping sage is locally common in many widely isolated locations throughout its extensive range.

SAMBUCUS (ELDERBERRY)

Sambucus is a small genus with about 20 species that grow throughout the temperate and subtropical mountains of both the Northern and Southern Hemispheres.

Sambucus is a collection of shrubs with pinnately compound deciduous leaves. Inflorescences are arrayed in open sprays of white flowers. Individual flowers are small and not individually showy. Fruits are red, blue, or black drupes developed from inferior ovaries. Twigs have spongy piths.

Animals and people seek out elderberry fruits, but fruits of some species are considered toxic when eaten in quantity.

1. Inflorescences are flat topped, and fruits are blue with a white waxy covering ...... **blue elderberry (*S. mexicana*)**
1. Inflorescences are dome shaped, and fruits are red or purple without a white waxy covering .... **red elderberry (*S. racemosa*)**

BLUE ELDERBERRY ***Sambucus mexicana***

DESCRIPTION: A shrub that typically grows to be 7.5 m (25 ft) tall. The largest lives in Tubac, Arizona, and is 10 m (33 ft) tall and 106 cm (42 in.) in diameter. **LEAVES** are opposite, deciduous, and pinnately compound, with 3 to 9 elliptical leaflets. Each leaflet is from 2.5 cm (1 in.) to 15 cm (6 in.) long and may have hairs. Leaf axes are often bowed downward. **INFLORESCENCES** are from 5 cm (2 in.) to 20 cm (8 in.) wide and consist of flat clusters with many small flowers. **FLOWERS** are white or cream colored with 5 small petals. **FRUITS** are round, black or white, and covered with a waxy white coating that makes the black berries appear blue. **TWIGS** have spongy piths.

HABITAT AND RANGE: Grows along streams and in openings in forests throughout much of California, at elevations from sea level to 3,000 m (10,000 ft). It occurs through western Canada, the United States, and Mexico.

REMARKS: In many books, blue-fruited elderberries found in the mountains and having over 5 leaflets are assigned to *S. cerulea* (also spelled *caerulea*). Low-

Figure 191
Red elderberry,
Sambucus racemosa

elevation plants often have 5 leaflets and are called *S. mexicana*. This characteristic and others are variable within and among populations, so both forms are included here. The fruits are important food for wildlife. This is the elderberry species used to make jellies, pies, and wine.

RED ELDERBERRY (Fig. 191) ***Sambucus racemosa***

DESCRIPTION: A shrub that grows to be 7.5 m (25 ft) tall. The largest lives in Coupeville, Washington, and is 11 m (36 ft) tall and 38 cm (15 in.) in diameter. **LEAVES** are opposite, deciduous, pinnately compound, and thin. They have 5 to 7 elliptical leaflets. Each leaflet is from 4 cm (1.5 in.) to 12.5 cm (5 in.) long and may have hairs. **INFLORESCENCES** are from 4 cm (1.5 in.) to 6 cm (2.5 in.) long and consist of small domelike clusters with many small flowers. **FLOWERS** are white or cream colored, with 5 small petals. **FRUITS** are round, red or black, and lack a waxy white covering. **TWIGS** have spongy piths.
HABITAT AND RANGE: Grows along streams and in openings in forests throughout much of California, at elevations

from sea level to 3,600 m (12,000 ft). It grows from Alaska south through the western mountains. It also grows in the eastern United States, Europe, and Asia.

REMARKS: Red elderberry fruits are used by wildlife but are unpalatable, even toxic, to humans. There is a great deal of variation in this species. Coastal plants are treelike, have leaflets that are hairy on the lower surfaces, and have red berries (*S. r.* var. *arborescens*). Mountain plants have hairless leaflets with red berries (*S. r.* var. *microbotrys*) or black berries (*S. r.* var. *melanocarpa*). This species is not the elderberry used in jellies, pies, and wine.

SHEPHERDIA (BUFFALOBERRY)

The genus *Shepherdia* has 3 shrub species growing in North America; 1 of these is native to California.

SILVER BUFFALOBERRY *Shepherdia argentea*

(Fig. 192)

DESCRIPTION: An erect, multistemmed shrub or small tree that can grow in thickets. Plants typically range in height from 1 m (3.3 ft) to 6 m (20 ft) tall. The largest grows in Malheur County, Oregon, and is 6.7 m (22 ft) tall and 63 cm (25 in.) in diameter. Crowns are rounded, with spreading, thorny branches. Plants can live to be around 30 years old. **LEAVES** are deciduous, opposite, thick, simple, oblong to elliptical, and 2 cm (.75 in.) to 4.5 cm (1.75 in.) long. Margins are entire. Leaf tips are round and bases are wedge shaped. Upper surfaces are light green and covered with silvery hairs, while lower surfaces are silvery. **FRUITS** are drupelike, roughly spherical, red, and about 6 mm (.25 in.) in diameter. **TWIGS** are thorny and have silvery hairs that eventually turn grayish brown. **BARK** is thin, gray, exfoliating, and shaggy.

HABITAT AND RANGE: Grows in riparian zones and on lower slopes in conifer woodlands, grasslands, and shrublands. It is more prevalent in the northern Great Plains and the Rocky Mountains than in

Figure 192
Silver buffaloberry,
Shepherdia argentea

California. In California the species can be found scattered in Siskiyou, Ventura, Mono, Alpine, Santa Barbara, and San Bernardino Counties, at elevations from 900 m (3,000 ft) to 2,000 m (6,500 ft).

REMARKS: Silver buffaloberry leaves are eaten by deer, and its fruits by small mammals and birds. Native Americans and early settlers made a sauce out of the fruit and ate it with buffalo meat. Today the sour fruits are used in jams and pies. Silver buffaloberry is planted as an ornamental.

SORBUS (MOUNTAIN-ASH)

The genus *Sorbus* has about 80 species growing throughout the Northern Hemisphere. California has 3 native species and 1 naturalized species.

WESTERN MOUNTAIN-ASH *Sorbus scopulina* (Fig. 193)

DESCRIPTION: A shrub or small tree that typically grows to be between 1 m (3.3 ft) and 6 m (20 ft) tall. The largest lives in

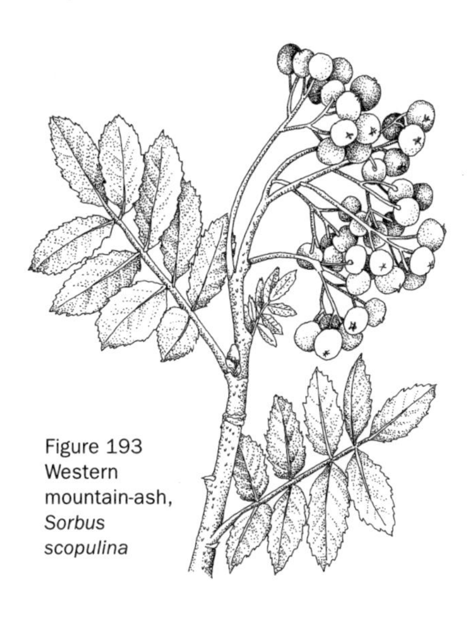

Figure 193 Western mountain-ash, *Sorbus scopulina*

Spokane, Washington, and is 4.3 m (14 ft) tall and 18 cm (7 in.) in diameter. **LEAVES** are deciduous, alternate, and compound, with 7 to 13 leaflets. Blades are lance shaped or oblong and 2.5 cm (1 in.) to 6 cm (2.5 in.) long, with fine teeth on the margins. **INFLORESCENCES** consist of many flowers in flat clusters. **FLOWERS** are white with 5 petals. **FRUITS** are orange or bright red pomes about 6 mm (.25 in.) in diameter. **TWIGS** have buds that are sticky and hairy.

HABITAT AND RANGE: Occurs in moist areas and coniferous forests and woodlands in northern California mountains, at elevations from 1,200 m (4,000 ft) to 2,700 m (9,000 ft). Outside of California it is found in the mountains of the western United States.

REMARKS: Mountain-ashes are easily recognized at the generic level by their distinctive leaves and fruits, but recognizing species is a challenge, as the vegetative characteristics used to differentiate species vary regionally. Western mountain-ash has 2 varieties in California. The one with lance-shaped leaflets (*S. s.* var. *scopulina*) is more common east of the Cascade-Sierra crest. The one with oblong leaflets (*S. s.* var. *cascadensis*) is more common west of the Cascade-Sierra crest. There are 3 additional mountain-ashes in the state. California mountain-ash (*S. californica*) has fewer leaflets than western mountain-ash. Sitka mountain-ash (*S. sitchensis*) is local in the Klamath Mountains, and the European *S. aucuparia* is an escaped ornamental.

SPARTIUM (SPANISH BROOM)

The genus *Spartium* has 60 species that grow in the Mediterranean region of Europe. In California, 1 of these species has escaped from cultivation.

SPANISH BROOM *Spartium junceum*

DESCRIPTION: A shrub with dark green, round stems that may grow to be 3 m (10 ft) tall. **LEAVES** are alternate, deciduous, ephemeral, and less than 2.5 cm (1 in.) long. Blades are simple and linear. **INFLORESCENCES** are terminal clusters of several flowers. **FLOWERS** are bright yellow, somewhat fragrant, and about 2.5 cm (1 in.) long. **FRUITS** are brown legumes that measure 5 cm (2 in.) to 10 cm (4 in.) long. **TWIGS** are dark green and rushlike.

HABITAT AND RANGE: Locally common in disturbed areas at elevations below 600 m (2,000 ft). It is native to the Mediterranean region of Europe.

REMARKS: Spanish broom is another invasive broom. See also descriptions of French broom, gorse, and Scotch broom. Stems are used for fiber, and the flowers for creating a yellow dye.

SPIRAEA (SPIRAEA)

The genus *Spiraea* has about 70 species growing throughout the temperate parts of the Northern Hemisphere. Two of these are native to California.

Spiraea is a collection of shrubs with simple leaves that are alternate in arrangement. Flowers are small, pink, and showy in mass. Each flower has 5 petals and numerous stamens attached to the top of a cup surrounding the many pistils. Fruits are 5 follicles that each break along one side.

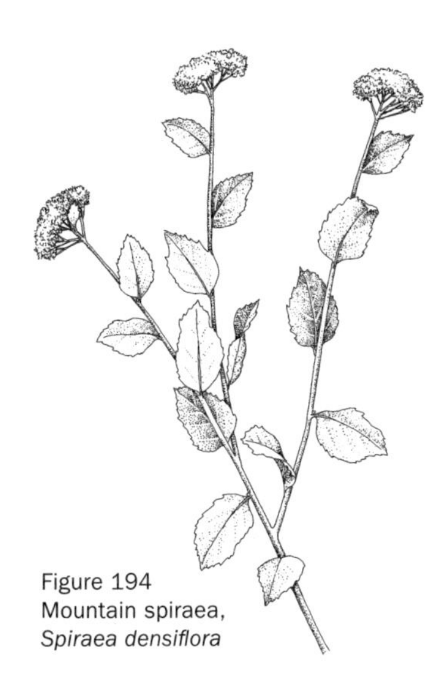

Figure 194
Mountain spiraea,
Spiraea densiflora

Many ornamentals have been developed from this genus.

1. Inflorescences are flat topped . . . . **mountain spiraea** (*S. densiflora*)
1. Inflorescences are elongated . . . . **Douglas spiraea** (*S. douglasii*)

MOUNTAIN SPIRAEA (Fig. 194) *Spiraea densiflora*

DESCRIPTION: A shrub that may grow to be between 20 cm (8 in.) and 1 m (3.3 ft) tall. **LEAVES** are alternate, deciduous, and thin. Blades have fine hairs or are hairless. They are egg shaped and 12 mm (.5 in.) to 40 mm (1.5 in.) long, with teeth on the side and top margins but not at the base. The degree of hairiness varies greatly. **INFLORESCENCES** are terminal, flat topped, and very showy when flowers are in bloom. **FLOWERS** are rose pink. **FRUITS** are 5 follicles. **TWIGS** are brown and may have hairs.

HABITAT AND RANGE: Grows in mountains away from the coast in northern California, at elevations from 600 m (2,000 ft) to 3,000 m (10,000 ft). It lives along coniferous forest edges and in moist meadows.

REMARKS: Mountain spiraea plants whose branches are covered with fine hairs are called *S. d.* var. *splendens.* These plants have flowers a deeper pink than those of the typical form, *S. d.* var. *densiflora.*

DOUGLAS SPIRAEA (Pl. 38) *Spiraea douglasii*

DESCRIPTION: A shrub that may grow to be between 1 m (3.3 ft) and 2 m (6.5 ft) tall. **LEAVES** are alternate, deciduous, and thin. Blades are oblong, 4 cm (1.5 in.) to 7.5 cm (3 in.) long, green, and hairless above but have white hairs below. Blade margins have teeth on the side and top, but not at the base. **INFLORESCENCES** are terminal, elongated, pyramidal, and showy when flowers are in bloom. **FLOWERS** are rose pink. **FRUITS** are 5 follicles. **TWIGS** are brown and have fine hairs.

HABITAT AND RANGE: Grows along coniferous forest edges and in moist meadows in the mountains of northern California, at elevations below 600 m (2,000 ft). Mountain spiraea, in contrast, grows above 600 m (2,000 ft). Douglas spiraea ranges eastward into the northern Rocky Mountains.

REMARKS: This low-elevation spiraea with lovely, long-lasting flowers is easily grown as an ornamental.

STAPHYLEA (BLADDERNUT)

The genus *Staphylea* has about 10 species of shrubs and small trees that grow in the temperate regions of the Northern Hemisphere. One is native to California.

SIERRA BLADDERNUT (Fig. 105) *Staphylea bolanderi*

DESCRIPTION: An erect shrub or small tree. It typically grows to be between 2 m (6.5 ft) and 6 m (20 ft) tall. The largest lives in Fresno County and is 8.5 m (28 ft) tall and 11 cm (4.5 in.) in diameter. **LEAVES** are compound, with 3 leaflets; opposite; and deciduous. Leaflets are round to egg shaped, 2.5 cm

Figure 195
Sierra bladdernut,
Staphylea bolanderi

(1 in.) to 6 cm (2.5 in.) long, and hairless. Leaflet margins are finely serrated. Lateral leaflets have minute stalks, while the terminal leaflet's stalks are from 6 mm (.25 in.) to 25 mm (1 in.) long. **INFLORESCENCES** are droopy clusters of flowers. **FLOWERS** are white with 5 petals and 5 sepals. The 5 stamens and style of each are exserted well beyond the petals. **FRUITS** are cream to white, bladdery, egg-shaped, inflated capsules that measure 2.5 cm (1 in.) to 5 cm (2 in.) long. The capsule tips have 3 prominent horns. **TWIGS** are hairless.

HABITAT AND RANGE: Grows in chaparrals and woodlands on steep canyon walls at elevations from 300 m (1,000 ft) to 1,400 m (4,500 ft). It is endemic to California, though uncommon, and grows in the Sierra Nevada, Cascades, and eastern Klamath Mountains.

REMARKS: Sierra bladdernut is planted as an ornamental because of its attractive flowers and foliage and its unusual fruits. It is a species you need to seek out because of its rareness and somewhat inaccessible habitat.

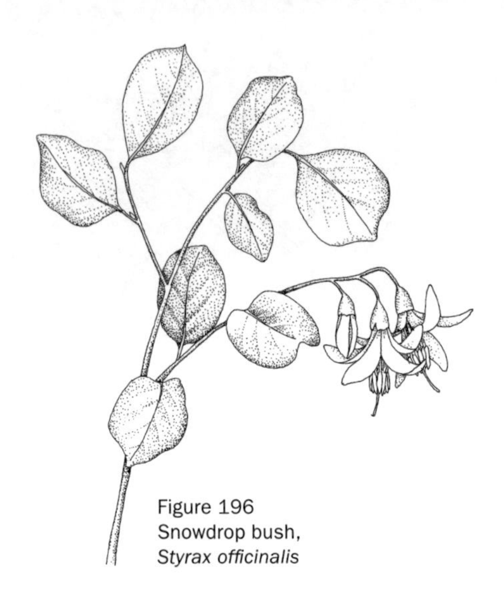

Figure 196
Snowdrop bush,
Styrax officinalis

STYRAX (STORAX)

The genus *Styrax* has about 100 species of shrubs and small trees that grow in the Americas, Mediterranean Europe, and Asia Minor. One is native to California.

SNOWDROP BUSH (Fig. 196) ***Styrax officinalis***

DESCRIPTION: A shrub that may grow to be 2.5 m (8 ft) tall. **LEAVES** are alternate, deciduous, and simple. Blades are round, from 2 cm (.75 in.) to 9 cm (3.5 in.) long, and green. They may have simple or starlike hairs on their lower surfaces or may be hairless. **INFLORESCENCES** consist of simple clusters of droopy flowers. **FLOWERS** are large, white, and up to 25 mm (1 in.) long. They have many stamens and 4 to 10 petals fused at their base. Flowers have the scent of orange flowers. **FRUITS** are round capsules. **TWIGS** are gray.

HABITAT AND RANGE: Grows sporadically in woodlands and chaparrals of the inland hills surrounding the Central

Valley and in southern California mountains; it is considered uncommon. It is found at elevations below 1,500 m (5,000 ft). This species also grows in Mediterranean Europe.
REMARKS: The resin of Mediterranean plants is the balsam used in medicine.

SYMPHORICARPOS (SNOWBERRY)

Symphoricarpos is a small genus with about 10 species that grows throughout temperate North America and China.

The genus *Symphoricarpos* is a collection of shrubs with simple, opposite, deciduous leaves. Leaves on sterile shoots may be entire or lobed. Leaves on fertile shoots are more commonly entire. Flowers are small, with floral tubes that have 4 or 5 lobes. Fruits are white or red berries that develop from inferior ovaries.

1. Plants have trailing and spreading stems and are less than 60 cm (24 in.) tall 2
1. Plants have erect stems and are more than 60 cm (24 in.) tall 3
    2. Floral tubes are bell shaped and about 4 mm (.15 in.) long . . . . **creeping snowberry** (*S. mollis*)
    2. Floral tubes are elongated and from 6 mm (.25 in.) to 9 mm (.4 in.) long . . . . **mountain snowberry** (*S. rotundifolius* var. *parishii*)
3. Floral tubes are elongated, 6 mm (.25 in.) to 10 mm (.4 in.) long, and hairy below the attachment of the stamens. Plants are usually found above 1,500 m (5,000 ft) . . . . **mountain snowberry** (*S. rotundifolius* var. *rotundifolius*)
3. Floral tubes are bell shaped, 4 mm (.15 in.) to 6 mm (.25 in.) long, and hairy near the attachment of the stamens. Plants are usually found below 600 m (2,000 ft) . . . . **common snowberry** (*S. albus*)

COMMON SNOWBERRY *Symphoricarpos albus*

DESCRIPTION: A thicket-forming shrub that may grow to be 2 m (6.5 ft) tall. **LEAVES** are opposite, deciduous, simple, and thin. Leaf blades are oval, measure from 2 cm (.75 in.) to 5 cm (2 in.) long, and may be hairy. **INFLORESCENCES** consist of small clusters of flowers in the axils of leaves or at the tops of stems. **FLOWERS** have white or pink floral tubes, are from

4 mm (.15 in.) to 6 mm (.25 in.) long, and have lobes about 50% of the length of the tubes. The lobes and upper floral tubes are densely hairy. **FRUITS** are round white berries. **TWIGS** are slender and brown when young, but gray with age.

HABITAT AND RANGE: In California the species grows extensively along streams and in forests and woodlands below 1,200 m (4,000 ft) and usually below 600 m (2,000 ft). It ranges north to Alaska and east to New England.

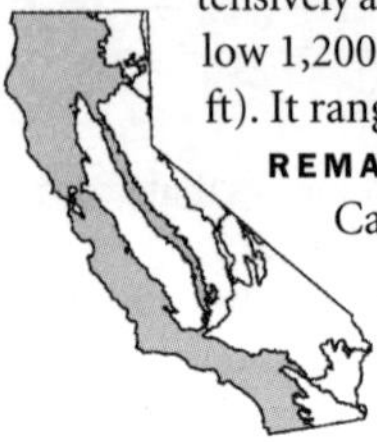

REMARKS: The common snowberry found in California, *S. a.* var. *laevigatus,* is less hairy than varieties found elsewhere. It is an important browse plant for wildlife and cattle, although the berries may be toxic to humans.

CREEPING SNOWBERRY *Symphoricarpos mollis*

(Fig. 197)

DESCRIPTION: A small shrub with trailing or creeping stems that root at the nodes. Plants grow to be 60 cm (2 ft) tall. **LEAVES** are opposite, deciduous, simple, and thin. Leaf blades are oval, measure from 12 mm (.5 in.) to 20 mm (.75 in.) long, and may have hair. **INFLORESCENCES** consist of small clusters of flowers in the axils of leaves. **FLOWERS** have pink floral tubes less than 6 mm (.25 in.) long, with short lobes about 50% of the length of the tubes. **FRUITS** are round white berries. **TWIGS** are slender and brown when young.

HABITAT AND RANGE: In California the species grows extensively in forests and woodlands below 3,000 m (10,000 ft). It ranges throughout the mountains of the West.

REMARKS: Three creeping snowberry forms, called separate species in some books, can be distinguished with difficulty in California. Plants from the Klamath Mountains have leaves with either no hair or just a few hairs on both blade surfaces (*S. m.* var. *hesperius*). Plants of the coastal hills and Sierra Nevada have leaves with dark green upper surfaces and pale green, densely hairy lower surfaces (*S. m.* var. *acutus*). Elsewhere, plants have leaves with pale green upper surfaces and densely hairy lower surfaces (*S. m.* var. *mollis*). Plants with typical varietal characteristics are easy to identify. Many plants, unfortunately,

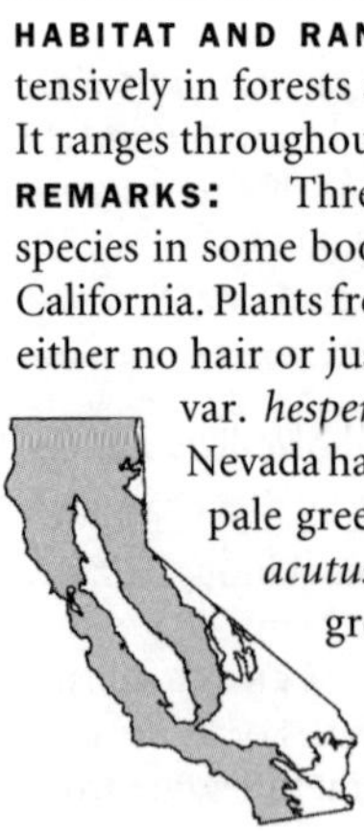

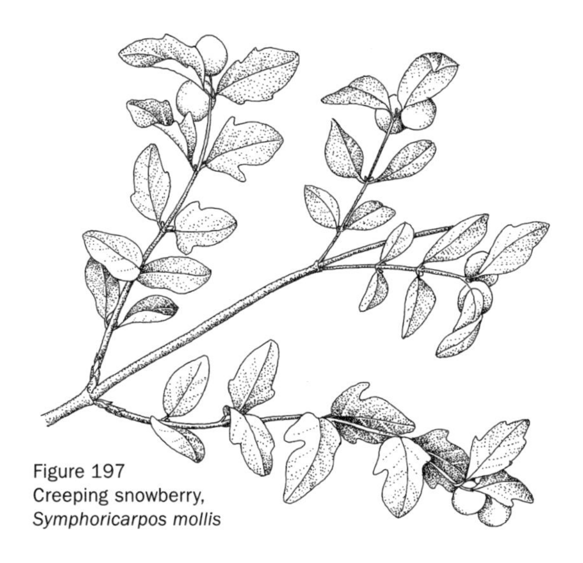

Figure 197
Creeping snowberry,
Symphoricarpos mollis

have intermediate characteristics, making identification difficult at the variety level.

MOUNTAIN SNOWBERRY *Symphoricarpos rotundifolius*

DESCRIPTION: A straggly shrub with rhizomes. Plants grow to be 1.2 m (4 ft) tall. **LEAVES** are opposite, deciduous, simple, and thin, with fine, soft hairs. Leaf blades are oval to elliptical and from 12 mm (.5 in.) to 25 mm (1 in.) long. **INFLORESCENCES** consist of 1 or 2 flowers in the axils of the leaves. **FLOWERS** have white, yellowish, or pink floral tubes that measure 6 mm (.25 in.) to 10 mm (.4 in.) long. Lobes are 25% of the length of the tubes. **FRUITS** are round white berries. **TWIGS** are slender and brown when young.

HABITAT AND RANGE: Grows extensively in the forests and woodlands of inland California, at elevations from 1,500 m (5,000 ft) to 3,000 m (10,000 ft). It is especially associated with rocky areas and forest openings. It ranges through the mountains of the West.

REMARKS: Two forms are distinctive

over much of mountain snowberry's range but are hard to distinguish from one another immediately east of the Sierra Nevada. In the southern part of the range, *S. r.* var. *parishii* has trailing stems, while in the northern part of the range *S. r.* var. *rotundifolius* has erect stems. Fragrant snowberry (*S. longiflorus*) grows in mountain snowberry's range and has similar habits, but it has longer floral tubes that are often more than 25 mm (1 in.) in length.

TAMARIX (TAMARISK)

The genus *Tamarix* has about 75 species of shrubs and trees native to the Mediterranean region and Asia. None are native to North America, but at least 5 species are naturalized in California.

SMALLFLOWER TAMARISK *Tamarix parviflora* (Fig. 198)

DESCRIPTION: A shrub or small, multistemmed tree that can form thickets. Plants typically range in height from 1.5 m (5 ft) to 5.5 m (18 ft). The largest grows in Hampton, Virginia, and is 4.3 m (14 ft) tall and 48 cm (18 in.) in diameter. Crowns are rounded and have many ascending, slender branches. **LEAVES** are deciduous, alternate, simple, awl-like, yellowish green, and from 1.5 mm (.06 in.) to 3 mm (.12 in.) long; they lack petioles. Leaf tips are sharp pointed and elongated. **INFLORESCENCES** are clustered and unbranched and measure from 12 mm (.5 in.) to 40 mm (1.5 in.) long. **FLOWERS** have 4 sepals and 4 petals that are pink and about 3 mm (.12 in.) across. **FRUITS** are capsules with hairy, tufted seeds. **TWIGS** are slender and feathery. **BARK** is rough.

HABITAT AND RANGE: In California the species grows in riparian woodlands, sand dunes, and disturbed areas from Humboldt County to the southeastern deserts, at elevations below 750 m (2,500 ft). It typically grows in the drier parts of the state, but it can be found elsewhere on exposed sand and gravel bars.

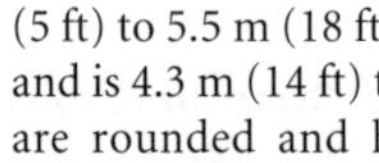

REMARKS: Smallflower tamarisk was introduced from southeastern Europe as an ornamental and for use as a windbreak. It has since become naturalized, and it in-

Figure 198
Smallflower tamarisk,
Tamarix parviflora

vades disturbed areas such as roadsides and newly exposed soils, as well as riparian zones. Botanists recognize 5 species (*T. aphylla, T. chinensis, T. gallica, T. parviflora,* and *T. ramosissima*) in California. The first 3 are considered uncommon and the last two common. *T. ramosissima* is restricted to the deserts and foothills of the southern California mountains, while *T. parviflora* ranges into central and northern California. This genus is considered an invasive pest. In riparian zones, as in other habitats, it displaces native plants and transpires prodigious quantities of water, affecting stream flows in desert regions.

TETRADYMIA (HORSEBRUSH)

The genus *Tetradymia* has 10 species growing throughout western North America. Nine are native to California.

GRAY HORSEBRUSH *Tetradymia canescens*

DESCRIPTION: A rigid shrub that may grow to be 30 cm (1 ft) tall. **LEAVES** are clustered, linear, less than 4 cm (1.5 in.) long, and covered with fine, silvery hairs. **INFLORESCENCES**

are dense clusters of heads, each head with 4 cream-colored flowers surrounded by 4 stout bracts. **FLOWERS** are 6 mm (.25 in.) to 15 mm (.6 in.) long. **FRUITS** are achenes. **STEMS** have a dense covering of matted, white hair.

HABITAT AND RANGE: Grows on the eastern slopes of the mountains of southern California, in the Mojave Desert, and throughout the intermountain West, at elevations from 1,500 m (5,000 ft) to 3,000 m (10,000 ft). The species is found in sagebrush scrubs and coniferous woodlands.

REMARKS: Gray horsebrush is one of several horsebrushes that grow in the interior West. Its distinctive set of 4 bracts surrounding the flowers makes the genus an easy one to remember. All parts of the plant are poisonous to livestock, especially sheep.

TOXICODENDRON (POISON-OAK)

The genus *Toxicodendron* has 6 species of shrubs, trees, and vines native to America and Asia. One is native to California.

POISON-OAK (Fig. 199) ***Toxicodendron diversilobum***

DESCRIPTION: An erect, multistemmed shrub, vine, or small tree that can grow in thickets. Shrubs range in height from 60 cm (2 ft) to 2 m (6.5 ft), trees can grow to be 4.5 m (15 ft) tall, and vines can climb more than 30 m (100 ft) up the trunks of large trees. **LEAVES** are deciduous; alternate; pinnately compound, with 3 leaflets (sometimes 5); and 7.5 cm (3 in.) to 15 cm (6 in.) long. Leaflets are egg shaped, shiny, green, hairless, and 2.5 cm (1 in.) to 10 cm (4 in.) long. Leaflet margins are irregularly lobed to entire. Some leaflets resemble white oak leaves. The terminal leaflet has a long stalk, while lateral leaflets have very short stalks. Leaves are red in the early spring, become green, and then turn red again in the fall. **INFLORESCENCES** are loosely clustered in a branched arrangement. **FLOWERS** are yellowish green and about 3 mm (.12 in.) across. **FRUITS** are spherical, grayish white to grayish brown drupes about 6 mm (.25 in.) in diameter. **TWIGS** are gray to orange-brown with black mottling. They can be hairless or hairy. **BUDS** are naked.

HABITAT AND RANGE: Grows in forests, woodlands, chap-

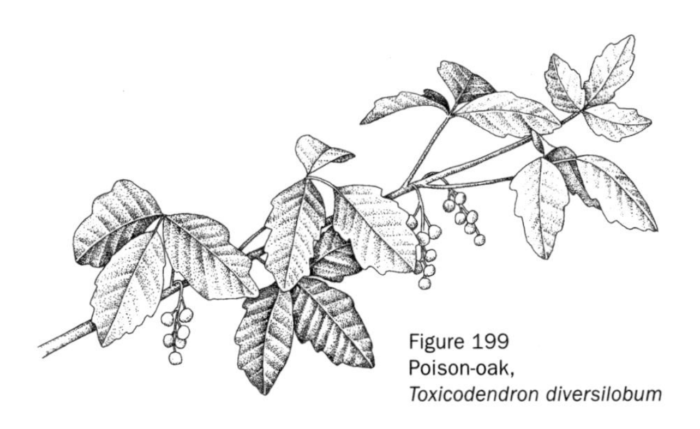

Figure 199
Poison-oak,
Toxicodendron diversilobum

arrals, and riparian zones, from British Columbia to Baja California. It is widespread in California, occurring in a great variety of habitats, such as cliffs, dry slopes, and moist forests, at elevations below 1,700 m (5,500 ft).

REMARKS: Poison-oak causes mild to severe rashes in most people. People who have not previously been affected by poison-oak may lose their resistance at a later time. Be careful of both its leaves and stems, since even leafless stems can cause dermatitis. Poison-oak sprouts vigorously following injury from fire or cutting. Eradication can be achieved through shading, herbicides, or grazing animals such as goats. Deer and other ungulates eat poison-oak leaves, and birds disperse its seeds.

ULEX (GORSE)

The genus *Ulex* has about 20 species native to western Europe and North Africa. One species is a noxious weed and has become naturalized in California.

GORSE (Fig. 200) ***Ulex europaea***

DESCRIPTION: A densely branched, spiny shrub that can form thickets. Plants range in height from 60 cm (2 ft) to 3 m (10 ft). **LEAVES** are alternate, simple, awl-like to spiny (juve-

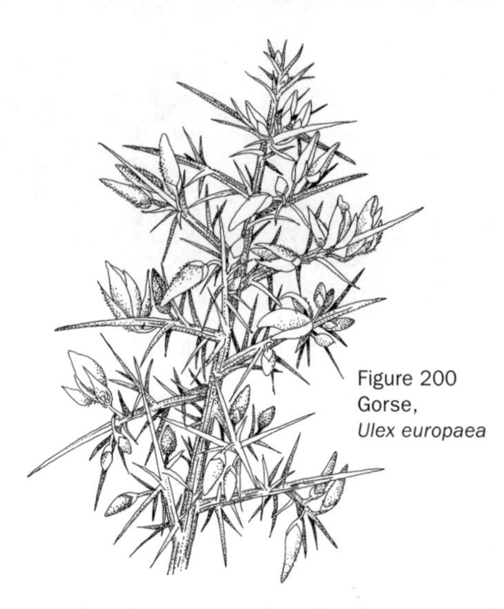

Figure 200
Gorse,
Ulex europaea

nile leaves are linear), and 3 mm (.12 in.) to 12 mm (.5 in.) long. **FLOWERS** are showy, yellow pea flowers measuring 12 mm (.5 in.) to 2.5 cm (1 in.) long. Flowers can cover the plants in the spring. **FRUITS** are densely hairy, egg-shaped legumes that measure 1 cm (.4 in.) to 2 cm (.75 in.) long. Pods open explosively. **TWIGS** are ribbed and green to grayish green, with spines up to 5 cm (2 in.) long.

HABITAT AND RANGE: Grows in coastal areas on disturbed sites such as roadsides and pastures, at elevations from sea level to 450 m (1,500 ft).

REMARKS: Gorse was introduced from western Europe and has become an invasive pest. Its densely arrayed branches contain volatile oils that present a serious fire hazard. Nitrogen-fixing bacteria form root nodules, adding considerable nitrogen to ecosystems via decaying plant parts. In time gorse creates a seed bank from which seedlings are recruited after a fire.

UMBELLULARIA (CALIFORNIA BAY)

The genus *Umbellularia* has 1 species that grows in California and Oregon.

Figure 201
California bay,
Umbellularia californica

CALIFORNIA BAY *Umbellularia californica*

(Fig. 201; Pl. 39)

DESCRIPTION: An erect, single- or multistemmed, medium-sized tree. When the species grows on serpentine soils, it can be a shrub. Mature plants are typically 9 m (30 ft) to 24 m (80 ft) tall and 30 cm (1 ft) to 90 cm (3 ft) in diameter. The largest tree grows in Siskiyou National Forest in Oregon and is 26.8 m (88 ft) tall and 4 m (13 ft) in diameter. Crowns are rounded to conical. Multistemmed tree crowns are multispired. This species sprouts vigorously. **LEAVES** are evergreen, simple, alternate, leathery, elliptical to lance shaped, and 5 cm (2 in.) to 15 cm (6 in.) long. Margins are entire. Upper surfaces are dark green, while lower surfaces are light green. Crushed leaves emit a strong, spicy, peppery aroma. **INFLORESCENCES** form flat-topped clusters of 4 to 10 flowers. **FLOWERS** are about 12 mm (.5 in.) in diameter and yellowish green. **FRUITS** are olivelike, fleshy, 2 cm (.75 in.) to 2.5 cm (1 in.) long, and attached to droopy stalks that resemble golf tees. Fruits are initially green, turning purple as they mature. **BARK** on mature trees is about 2.5 cm (1 in.) thick, brown, and broken into exfoliating scales.

HABITAT AND RANGE: Grows in forests and woodlands at elevations below 1,600 m (5,200 ft). It can be found on a wide variety of sites, from serpentine soils to exposed ridges to mountain slopes to valley bottoms. It also occurs in southwestern Oregon.
REMARKS: California bay is tolerant of shade and susceptible to fire. It sprouts from its base and along stems. Chemicals from this species' foliage are suspected of inhibiting germination of some herbaceous species. Its foliage is palatable to deer and birds, and small mammals eat its seeds. California bay wood is used in furniture, cabinets, and a variety of novelty items. Leaves of California bay are used in cooking but differ from *Laurus nobilis*—the bay tree whose leaves Europeans use to flavor food. California bay is planted as an ornamental. Native Americans used it as an insect repellent and to treat headaches and rheumatism. Other common names include pepperwood, Oregon myrtle, California laurel, bay laurel, myrtlewood, and spice-tree.

VACCINIUM (HUCKLEBERRY)

Vaccinium is a large genus, with at least 400 species, which grows throughout the Northern Hemisphere and in tropical mountains. Seven species are native to California.

Vaccinium is a collection of shrubs that have simple, deciduous or evergreen, thin or leathery, alternate leaves. Flowers are showy, pink to white blossoms, and each looks like an urn or bell with 4 or 5 lobes. Flowers are solitary or set in small clusters. Fruits are berries that develop from inferior ovaries.

Animals and people seek out *Vaccinium* fruits for their flavor and sugar content. Commercial blueberries and cranberries are members of this genus. Various native species have names like bilberries, grouseberries, huckleberries, and whortleberries.

1. Leaves are evergreen and leathery . . . . **evergreen huckleberry** (*V. ovatum*)
1. Leaves are deciduous and soft 2
   2. Stems are round or slightly angled in cross section . . . 3
   2. Stems are strongly angled and not round in cross section 4
3. Shrubs are less than 60 cm (2 ft) tall . . . . **dwarf huckleberry** (*V. caespitosum*)

3. Shrubs are over 60 cm (2 ft) tall . . . . **thinleaf huckleberry** (*V. membranaceum*)
   4. Shrubs are less than 30 cm (1 ft) tall . . . . **littleleaf huckleberry** (*V. scoparium*)
   4. Shrubs are over 30 cm (1 ft) tall . . . . **red huckleberry** (*V. parvifolium*)

DWARF HUCKLEBERRY *Vaccinium caespitosum*

DESCRIPTION: An expansive shrub less than 30 cm (1 ft) tall that spreads from underground, rooting stems. **LEAVES** are deciduous, green, thin, and mostly crowded at the ends of stems. Leaf blades have a broad oval or elliptical shape, or are widest above the middle. They are 12 mm (.5 in.) to 30 mm (1.25 in.) long and have fine-toothed margins. **INFLORESCENCES** consist of single flowers in the axils of the lowest leaves of the youngest shoots. **FLOWERS** are white to pink and urn shaped. **FRUITS** are less than 9 mm (.35 in.) in diameter, are blue, and may have a waxy covering. **TWIGS** are round or weakly angled and either green or reddish.

HABITAT AND RANGE: Grows in the mountains of northern California, from sea level to 3,600 m (12,000 ft). It is found in forests, along forest edges, in meadows, and even at alpine elevations. Outside of California the species ranges from Alaska eastward to New England.

REMARKS: At elevations below the tree line, *V. c.* var. *caespitosum* shrubs are taller and their berries lack the waxy covering. *V. c.* var. *paludicola,* which grows in alpine environments, produces flat mats with wax-covered berries. Another blue-fruited *Vaccinium* is the western blueberry (*V. uliginosum* ssp. *occidentale*). Its leaves have entire margins, and shrubs grow in montane and subalpine meadows in northern California.

THINLEAF HUCKLEBERRY *Vaccinium membranaceum*

DESCRIPTION: A spreading shrub that can grow to be 1.5 m (5 ft) tall. **LEAVES** are deciduous, green, and thin. Blades are oval to elliptical, often wider above the middle, and from 2.5 cm

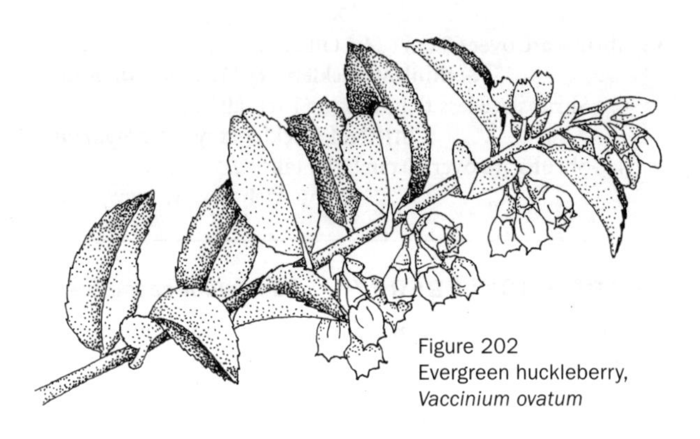

Figure 202
Evergreen huckleberry,
Vaccinium ovatum

(1 in.) to 5 cm (2 in.) long, with fine-toothed margins. **INFLORESCENCES** consist of single flowers in the axils of the leaves of the youngest shoots. **FLOWERS** are greenish white and urn shaped. **FRUITS** are about 10 mm (.4 in.) in diameter and either black or red. **TWIGS** are round to slightly angled and yellowish green.

HABITAT AND RANGE: Grows in forests, woodlands, and montane and subalpine chaparrals in the mountains of northern California, at elevations from 1,000 m (3,500 ft) to 2,100 m (7,000 ft). Outside of California it is found from Alaska to the Rocky Mountains.

REMARKS: Thinleaf huckleberry has large leaves with pointed tips, making it distinctive among California's deciduous huckleberries. The fruits are delicious and usually black, although some populations have red fruits. Some botanists call the red-fruited form *V. coccineum.*

EVERGREEN HUCKLEBERRY *Vaccinium ovatum*

(Fig. 202)

DESCRIPTION: A shrub that may grow to be 2.5 m (8 ft) tall. **LEAVES** are evergreen and leathery. Blades are thick, have a broad egg shape or are elliptical, and measure 12 mm (.5 in.) to 40 mm (1.5 in.) long. They have fine-toothed margins. **INFLO-**

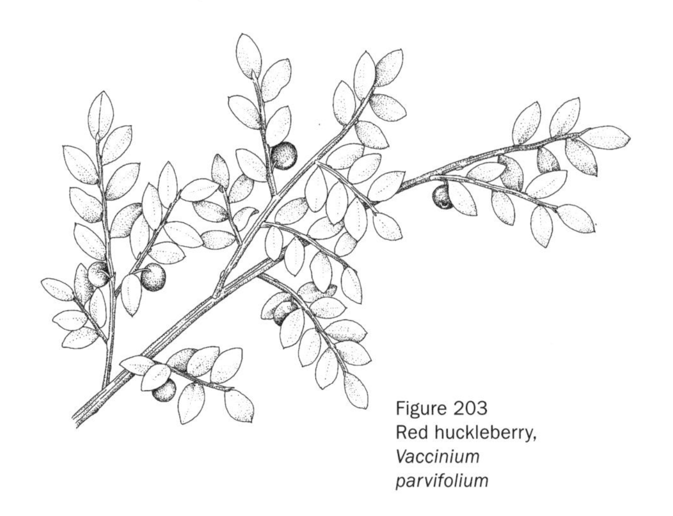

Figure 203
Red huckleberry,
Vaccinium parvifolium

RESCENCES consist of 1 to 5 flower clusters in leaf axils. **FLOWERS** are pink and bell shaped. **FRUITS** are 6 mm (.25 in.) to 9 mm (.35 in.) in diameter and black with a waxy white covering. **TWIGS** are round and gray.

HABITAT AND RANGE: Grows in and along the coastal scrubs and coastal forests of the state, at elevations below 750 m (2,500 ft). It ranges north to British Columbia.

REMARKS: Evergreen huckleberry may be the dominant shrub species in many of California's northern forests. The fruits are edible, and flavor depends on local site conditions. The foliage is used for Christmas decorations and in floral arrangements.

RED HUCKLEBERRY *Vaccinium parvifolium*

(Fig. 203)

DESCRIPTION: A straggly shrub that grows to be 1.5 m (5 ft) tall. **LEAVES** are deciduous, green, thin, and mostly crowded at the ends of stems. Blades are oval to elliptical and from 12 mm (.5 in.) to 20 mm (.75 in.) long, with entire or serrated margins. **INFLORESCENCES** consist of single flowers in the axils of leaves of the youngest shoots. **FLOWERS** are greenish or pink and urn shaped. **FRUITS** are 6 mm (.25 in.) to 10 mm (.4 in.) in diameter and red. **TWIGS** are strongly angled and green.

HABITAT AND RANGE: Grows from Alaska to the mountains of northern California in moist forests and woodlands, from sea level to 1,500 m (5,000 ft).

REMARKS: Red huckleberry often grows in habitats similar to those of evergreen huckleberry. Dwarf huckleberry may be confused with red huckleberry, but dwarf huckleberry has roundish twigs and blue berries. Red huckleberry leaf size can vary greatly.

LITTLELEAF HUCKLEBERRY *Vaccinium scoparium*

DESCRIPTION: A short, spreading shrub that grows from underground, rooting stems and is usually less than 30 cm (1 ft) tall. **LEAVES** are deciduous, green, thin, and distributed along the stems. Blades have a broad oval or elliptical shape, measure less than 12 mm (.5 in.) long, and have fine-toothed margins. **INFLORESCENCES** consist of single flowers in the axils of the lowest leaves of the youngest shoots. **FLOWERS** are pink and urn shaped. **FRUITS** are 3 mm (.12 in.) to 6 mm (.25 in.) in diameter and red. **TWIGS** are strongly angled and green.

HABITAT AND RANGE: The species is rare in California; it grows in the Klamath Mountains at elevations from 1,800 m (6,000 ft) to 3,600 m (12,000 ft). It can be found in the ground layer of subalpine forests. Outside of California it grows from Canada east to the northern Rocky Mountains.

REMARKS: Littleleaf huckleberry is easily distinguished among the *Vaccinium* species, as it has the smallest leaves outside of the introduced cranberry (*V. macrocarpon*). In California, cranberry grows only in an abandoned placer mine in Nevada County.

VIBURNUM (VIBURNUM)

The genus *Viburnum* has about 200 shrub species; 1 is native to California.

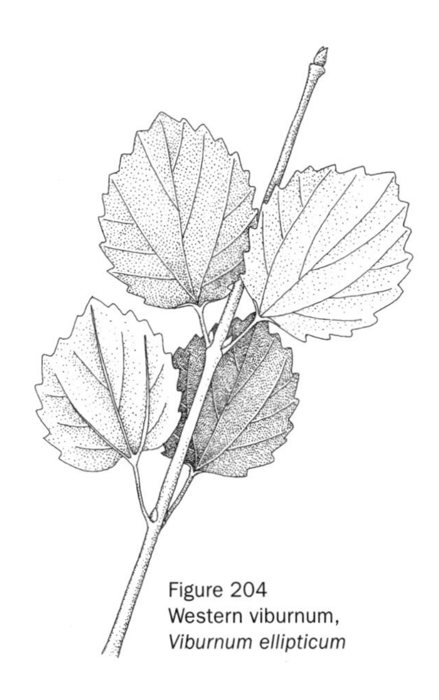

Figure 204
Western viburnum,
Viburnum ellipticum

WESTERN VIBURNUM ***Viburnum ellipticum***
(Fig. 204)

DESCRIPTION: A slender, loosely branched shrub that grows to be 1 m (3.3 ft) to 3.5 m (12 ft) tall. **LEAVES** are deciduous, opposite, simple, elliptical to round, 2.5 cm (1 in.) to 7.5 cm (3 in.) long, and 2 cm (.75 in.) to 5 cm (2 in.) wide. Margins are coarsely toothed except at the leaf bases. Leaf tips and bases are round. Upper surfaces are shiny dark green and generally hairless, while lower surfaces are pale green and slightly hairy. There are usually 3 (sometimes 5) veins originating from the base. **INFLORESCENCES** form branched, more or less flat-topped clusters of flowers. **FLOWERS** are white and from 6 mm (.25 in.) to 8 mm (.33 in.) wide. **FRUITS** are black, olivelike drupes about 12 mm (.5 in.) long. **TWIGS** are slender and reddish brown.

HABITAT AND RANGE: Grows from Washington south to the chaparrals and montane coniferous forests in central California. In California it is uncommon, and it occurs at elevations from 300 m (1,000 ft) to 1,400 m (4,500 ft), often on north slopes.

REMARKS: Flowers are malodorous.

WASHINGTONIA (FAN PALM)

The genus *Washingtonia* has 2 species growing in California, Arizona, and Baja California; 1 is native to Arizona and California, and the other is native to Baja California.

CALIFORNIA FAN PALM *Washingtonia filifera*

(Fig. 205; Pl. 40)

DESCRIPTION: A single-stemmed, small- to medium-sized tree. Mature trees are typically 9 m (30 ft) to 15 m (50 ft) tall and 30 cm (1 ft) to 60 cm (2 ft) in diameter. The largest grows in Sacramento and is 25.3 m (83 ft) tall and 96 cm (38 in.) in diameter. Crowns are rosettes that form at the tops of stout trunks. Unburned or unpruned trees have a skirt of dead leaves surrounding the trunk below the crown. Trees live to be about 200 years old. **LEAVES** form rosettes at the top of the trunk, and a persistent skirt of old leaves hangs beneath. Individual leaves are grayish green and grow in circular fans. They measure 1 m (3.3 ft) to 1.5 m (5 ft) in diameter and are palmately divided into between 40 and 60 folded segments. Each folded segment is usually ripped about one-half of its length. Margins have threadlike fibers. Petioles and segment bases have hooked spines on their margins. **FLOWERS** form long, droopy clusters within the crown. **FRUITS** are black drupes about 8 mm (.33 in.) in diameter. **BARK** is thick and rindlike.

HABITAT AND RANGE: Grows in disjunct desert oases along the San Andreas fault and in canyons. It is found in the Mojave and Colorado Deserts in California, as well as in Arizona and Baja California, at elevations below 1,200 m (4,000 ft).

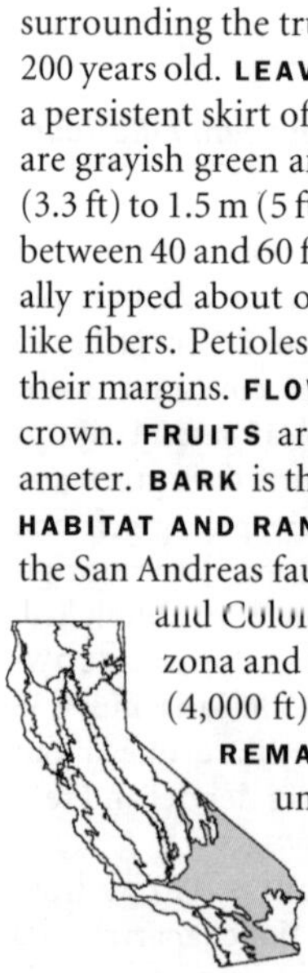

REMARKS: California fan palm is generally uncommon but is locally abundant. The species is intolerant of shade and is outcompeted by invading shrubs in the absence of fire or other disturbances. Coyotes

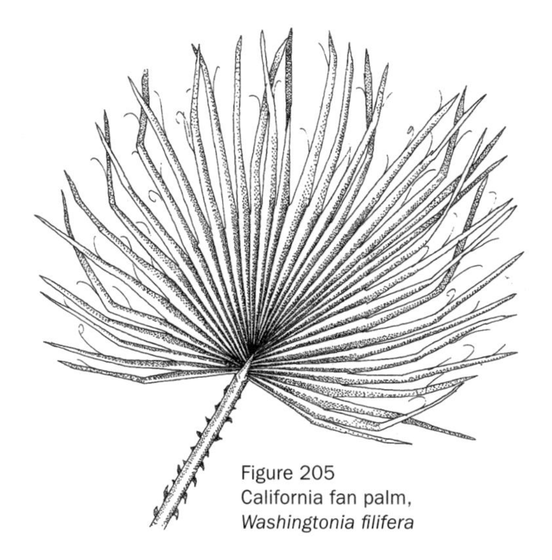

Figure 205
California fan palm,
Washingtonia filifera

facilitate seed dispersal by spreading it in their scat. Birds usually consume only the fleshy part of the fruit, not the seed. The palm provides critical habitat for many wildlife species. Native Americans used California fan palm oases for village sites. From the leaves they made shelters, food, clothing, and baskets. This species and its narrow-trunked relative, *W. robusta,* are widely planted as street trees. Water projects are impacting the California fan palm by drawing down water tables. Saltcedars are taking over some California fan palm groves.

YUCCA (YUCCA)

The genus *Yucca* has about 40 species of shrublike and treelike plants, which are native to arid North America. Four are native to California.

Leaves form elevated or basal rosettes. Individual leaves are evergreen, stiff, thick, swordlike, fibrous, and long. Margins are entire or serrated or have shredding fibers. Leaf tips have stout spines, and bases are broad. Flowers are clustered in a large, densely branched inflorescence. The inflorescences of California species range in length from 30 cm (1 ft) to 6.4 m (21 ft)! Individual flowers, which open at night, are whitish, fleshy, waxy, pendant, and 2 cm (.75 in.) to 12.5 cm (5 in.) long. Each flower has 3 sepals and 3 petals. Fruits are capsules.

1. Plants are treelike and have a definite trunk. Rosettes are elevated on branches 2
1. Plants are not treelike and lack a definite trunk. Rosettes are at ground level 3
   2. Leaves have shredding fibers on their margins . . . . **Mojave yucca** (*Y. schidigera*)
   2. Leaves do not have shredding fibers on their margins . . . . **Joshua tree** (*Y. brevifolia*)
3. Leaves have shredding fibers on their margins . . . . **banana yucca** (*Y. baccata*)
3. Leaves do not have shredding fibers on their margins . . . . **Our Lord's candle** (*Y. whipplei*)

BANANA YUCCA *Yucca baccata*

DESCRIPTION: A mostly stemless, shrublike plant with rosettes close to the ground. Rosettes grow to be 1.5 m (5 ft) tall and occur singly or occasionally as clumps. Clump diameters range between 1 m (3 ft) and 4.5 m (15 ft). **LEAVES** are evergreen, stiff, thick, swordlike, fibrous, yellowish green to bluish green, and 45 cm (1.5 ft) to 75 cm (2.5 ft) long. Margins have long, shredded, recurved fibers. Leaf tips have stout spines, while bases are flared and reddish. **INFLORESCENCES** are erect, branched, 60 cm (2 ft) to 90 cm (3 ft) long, and set on stalks less than 20 cm (8 in.) long. **FLOWERS** are bell shaped, 5 cm (2 in.) to 10 cm (4 in.) long, reddish brown on the outside, and whitish on the inside. **FRUITS** are capsules around 15 cm (6 in.) long. When young, they are fleshy, becoming pendant later.

HABITAT AND RANGE: Grows in dry woodlands on flats, foothills, and alluvial fans from southeastern California east to Utah and Texas. It is common in many low-elevation coniferous woodlands in the Southwest. In California it is uncommon, and it occurs in many kinds of woodlands in the eastern desert mountains, at elevations from 750 m (2,500 ft) to 1,300 m (4,200 ft).

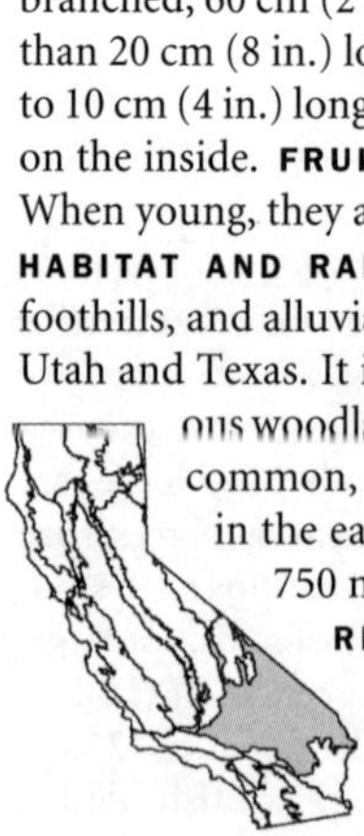

REMARKS: Banana yucca provides important shelter and food for many desert wildlife species, especially small mammals and small birds. Wood rats, deer, and big-

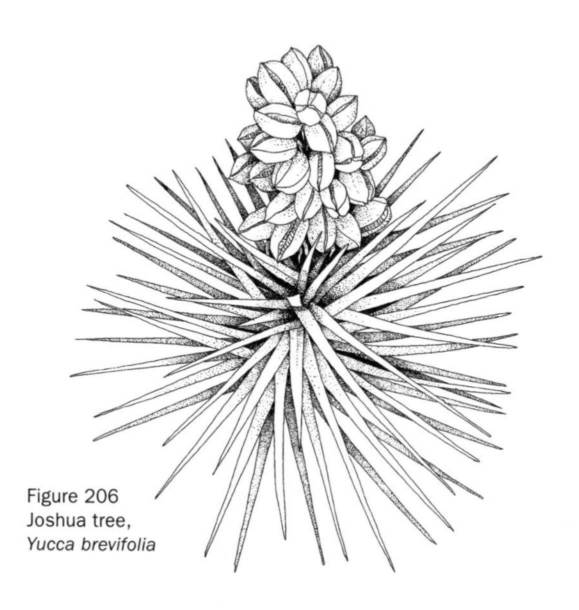

Figure 206
Joshua tree,
Yucca brevifolia

horn sheep eat its foliage, and rabbits, rodents, and other wildlife species eat its large fleshy fruits. Native Americans ate banana yucca fruits, used its roots to make shampoo, and made a variety of domestic products from its leaf fibers. Banana yucca hybridizes with Mojave yucca.

JOSHUA TREE (Fig. 206) *Yucca brevifolia*

DESCRIPTION: An erect, usually single-stemmed, treelike plant with rosettes at the ends of branches. Mature plants are typically 4.5 m (15 ft) to 12 m (40 ft) tall and 60 cm (2 ft) to 90 cm (3 ft) in diameter. The largest grows in San Bernardino County and is 11 m (36 ft) tall and 89 cm (35 in.) in diameter. Crowns are open and broadly rounded, with several to many loosely arrayed, crooked branches. **LEAVES** are evergreen, stiff, thick, swordlike, fibrous, dark green, and 20 cm (8 in.) to 30 cm (12 in.) long. Margins have small teeth. Leaves are broadest near their flared, whitish bases, then tapered toward spiny tips. **INFLORESCENCES** are erect, branched, 30 cm (1 ft) to 45 cm (1.5 ft) long, and set on stalks about 10 cm (4 in.) long. **FLOWERS** are bell shaped, 4 cm (1.5 in.) to 7 cm (2.75 in.) long, and greenish yellow to greenish white. **FRUITS** are light red to yellowish brown capsules that measure from 6 cm (2.5 in.) to 12.5 cm (5 in.) long. **TWIGS** are stout and usually have persistent dead leaves. **BARK** is soft, gray to reddish brown, rough, and fissured.

HABITAT AND RANGE: Grows in lowland woodlands, scrubs, and grasslands in the Mojave Desert in southeastern California and into Nevada, Utah, and Arizona. In California it is found on flats, foothills, and alluvial fans, at elevations from 600 m (2,000 ft) to 1,800 m (6,000 ft).

REMARKS: Joshua trees provide shelter and food for many desert wildlife species. Birds, small mammals, and reptiles nest in Joshua trees. New foliage, flowers, and fruit are eaten by wildlife, and the plants' shade brings relief from the desert heat. Joshua trees sprout following injury from fire. The wood from Joshua trees is used to make novelty items and veneer in limited quantities. In the past, the species was used to make fence posts, house beams, and paper. Native Americans fashioned a variety of domestic products from the leaves and roots. They ate the flower buds and seeds and made an alcoholic drink from the buds and flowers.

MOJAVE YUCCA (Fig. 207) *Yucca schidigera*

DESCRIPTION: An erect, usually single-stemmed, treelike or shrublike plant with rosettes at the ends of its branches. Plants occur singly or in clumps. Mature plants typically range in height from 1 m (3.3 ft) to 4.5 m (15 ft). The largest grows in Needles Resource Area, California, and is 7.3 m (24 ft) tall and 53 cm (21 in.) in diameter. Most plants are unbranched, although some have a few branches. Evergreen leaves nearly cover the trunks. The species is long-lived and may survive to be hundreds of years old. **LEAVES** are evergreen, stiff, thick, bayonetlike, fibrous, yellowish green to bluish green, and 30 cm (1 ft) to 1.5 m (5 ft) long. Margins have long, shredded, recurved fibers. Leaves are broadest near their flared, whitish bases and then constricted, becoming wide near the middle and then narrowing to a spiny tip. **INFLORESCENCES** are erect, branched, 60 cm (2 ft) to 120 cm (4 ft) long, and set on stalks measuring

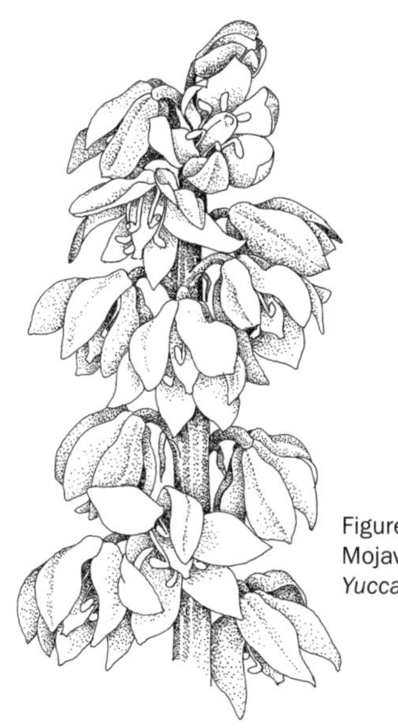

Figure 207
Mojave yucca,
Yucca schidigera

10 cm (4 in.) to 50 cm (20 in.) long. **FLOWERS** are spherical and whitish to cream colored, sometimes with purple streaks. **FRUITS** are fleshy, pendant capsules about 7.5 cm (3 in.) to 10 cm (4 in.) long and 4 cm (1.5 in.) wide. **BARK** is grayish brown, rough, and ridged.

HABITAT AND RANGE: Grows in scrubs, chaparrals, and lowland woodlands in southern California, southern Nevada, western Arizona, and northern Baja California. In California it is found near the coast in San Diego County, through the San Jacinto and Santa Rosa Mountains, and into the Mojave Desert and parts of the Colorado Desert, at elevations below 2,400 m (8,000 ft).

REMARKS: Small mammals and birds eat Mojave yucca fruits. With its tall stature, the plant provides shade, roosting sites, and nesting locations for wildlife. Native Americans ate Mojave yucca fruits and made soap, rope, shoes, and blankets from its leaves. Mojave yucca hybridizes with banana yucca.

OUR LORD'S CANDLE *Yucca whipplei*

DESCRIPTION: A stemless shrublike plant with dense rosettes close to the ground. Rosettes can occur singly or as clumps. The species is noted for its spectacular inflorescence, which ranges from 2 m (7 ft) to 6.4 m (21 ft) tall! **LEAVES** are evergreen, stiff, thick, daggerlike, fibrous, grayish green, and 45 cm (1.5 ft) to 90 cm (3 ft) long. Margins have small teeth. Leaves are broadest near their flared, white-to-greenish bases, then tapered toward brownish spiny tips. **INFLORESCENCES** are loosely branched, 20 cm (8 in.) to 4 m (13 ft) long, and set on a stalk measuring 1.5 m (5 ft) to 3.3 m (11 ft) long. Inflorescences can have between 100 and several thousand flowers. **FLOWERS** are spherical, creamy white, purplish tinged, and 2.5 cm (1 in.) to 5 cm (2 in.) long. **FRUITS** are spherical to egg-shaped capsules that measure 2.5 cm (1 in.) to 5 cm (2 in.) long.

HABITAT AND RANGE: Grows in chaparrals, scrubs, lowland woodlands, and lower montane forests. It is found from Monterey County and the southern Sierra Nevada to southwestern California and into Baja California, at elevations below 2,400 m (8,000 ft).

REMARKS: Five subspecies of Our Lord's candle are recognized: *Y. w.* ssp. *cespitosa, Y. w.* ssp. *intermedia, Y. w.* ssp. *percursa, Y. w.* ssp. *parishii,* and *Y. w.* ssp. *whipplei.* These subspecies vary in habit, inflorescence size, color, leaf shape, and flower shape. Some subspecies are fire adapted. Small mammals and birds eat the flowers and fruits. Native Americans ate the seeds and flower stalks and used the leaves to make rope and baskets.

APPENDIX A

GENERA GROUPED BY DISTINCTIVE MORPHOLOGICAL FEATURES

CONIFERS

SCALELIKE LEAVES

Calocedrus
Chamaecyparis
Cupressus
Juniperus
Thuja

AWL-LIKE LEAVES

Juniperus
Sequoiadendron

LINEAR LEAVES

Abies
Picea
Pseudotsuga
Sequoia
Taxus
Torreya
Tsuga

NEEDLELIKE LEAVES

Picea
Pinus

1 needle per bundle

Pinus monophylla

2 needles per bundle

Pinus contorta
P. monophylla
P. muricata
P. quadrifolia
P. radiata

3 needles per bundle

Pinus attenuata
P. coulteri
P. jeffreyi
P. ponderosa
P. quadrifolia
P. radiata
P. sabiniana
P. washoensis

4 needles per bundle

Pinus balfouriana
P. quadrifolia (2 to 5 needles per bundle)

5 needles per bundle

Pinus albicaulis
P. balfouriana
P. flexilis
P. lambertiana
P. longaeva
P. monticola
P. quadrifolia
P. torreyana

ERECT CONES

Abies
Thuja

BRACTS LONGER THAN CONE SCALES

Abies bracteata
A. magnifica var. *shastensis*
A. procera
Pseudotsuga macrocarpa
P. menziesii

DECIDUOUS CONE BRACTS AND SCALES

Abies

PELTATE CONE SCALES

Chamaecyparis
Cupressus
Sequoia
Sequoiadendron

BERRYLIKE CONES

Juniperus

ARILS

Taxus
Torreya

BROADLEAVED TREES AND SHRUBS

OPPOSITE LEAVES

Acer
Aesculus
Calycanthus
Cassiope
Ceanothus
Cephalanthus
Chilopsis
Coleogyne
Cornus
Erica
Euonymus
Fraxinus
Garrya
Kalmia
Keckiella
Larrea
Lonicera
Mimulus
Paxistima
Philadelphus
Salvia
Sambucus
Shepherdia
Staphylea
Symphoricarpos
Viburnum

COMPOUND LEAVES

Acacia
Acer
Aesculus
Ailanthus
Berberis
Cercidium
Chamaebatia
Chamaebatiaria
Cytisus
Fraxinus
Genista
Isomeris
Juglans
Larrea
Lupinus
Olneya
Pickeringia
Potentilla
Prosopis
Ptelea
Rhus
Robinia
Rosa
Rubus
Sambucus
Sorbus
Staphylea
Toxicodendron

PERSISTENT LEAVES

Adenostoma
Arbutus
Arctostaphylos
Artemisia
Atriplex
Baccharis
Berberis
Cassiope
Ceanothus
Cercocarpus
Chamaebatia
Chrysolepis
Chrysothamnus
Coleogyne
Cytisus
Dendromecon
Eastwoodia
Encelia
Erica
Eriodictyon
Eriogonum
Eucalyptus
Fremontodendron
Garrya
Gaultheria
Grayia
Heteromeles
Hymenoclea
Isomeris
Kalmia
Krascheninnikovia
Larrea
Ledum
Leucothoe
Lithocarpus
Lupinus
Malosma
Myrica
Nolina
Paxistima
Phyllodoce
Pickeringia
Pluchea
Potentilla
Prunus
Purshia
Quercus
Rhamnus
Rhododendron
Rhus
Rubus
Umbellularia
Vaccinium
Washingtonia
Yucca

LOBED LEAVES

Acer
Artemisia
Crataegus
Fremontodendron
Physocarpus
Platanus
Purshia
Quercus
Rhus
Ribes
Rubus
Symphoricarpos
Toxicodendron

ARMED

Acacia
Atriplex
Ceanothus
Cercidium
Coleogyne
Crataegus
Grayia
Lycium
Olneya
Prosopis
Prunus
Psorothamnus
Rhamnus
Ribes
Robinia
Rosa
Rubus
Shepherdia
Ulex

LEAF ROSETTES

Nolina
Washingtonia
Yucca

CATKINS

Alnus
Betula
Chrysolepis
Corylus
Garrya
Juglans
Lithocarpus
Populus
Quercus

ACHENES

Adenostoma
Artemisia
Baccharis
Calycanthus
Cercocarpus
Chamaebatia
Chrysothamnus
Coleogyne
Eastwoodia
Encelia
Ericameria
Eriogonum
Holodiscus
Hymenoclea
Lepidospartum
Neviusia
Platanus
Pluchea
Potentilla
Purshia
Rosa
Tetradymia

BERRIES

Arbutus
Berberis
Garrya
Lonicera
Lycium
Ribes
Symphoricarpos
Vaccinium

CAPSULES

Aesculus
Cassiope
Ceanothus
Cephalanthus
Chilopsis
Dendromecon
Erica
Eriodictyon
Eucalyptus
Euonymus
Fremontodendron
Gaultheria
Isomeris
Kalmia
Keckiella
Larrea
Ledum
Leucothoe
Menziesia
Mimulus
Nolina
Paxistima
Philadelphus
Phyllodoce
Populus
Rhododendron
Salix
Staphylea
Styrax
Tamarix
Yucca

DRUPES OR DRUPELETS

Arctostaphylos
Celtis
Cornus
Malosma
Oemleria
Prunus
Rhamnus
Rhus
Rubus
Sambucus
Shepherdia
Toxicodendron
Umbellularia
Viburnum
Washingtonia

FOLLICLES

Chamaebatiaria
Physocarpus
Spiraea

LEGUMES

Acacia
Cercidium
Cercis
Cytisus
Genista
Lupinus
Olneya
Pickeringia
Prosopis
Psorothamnus
Robinia
Spartium
Ulex

NUTS OR NUTLETS

Alnus
Betula
Chrysolepis
Corylus
Juglans
Larrea
Lithocarpus
Myrica
Quercus
Salvia

POMES

Amelanchier
Crataegus
Heteromeles
Malus
Sorbus

SAMARAS

Acer
Ailanthus
Fraxinus
Ptelea

UTRICLES

Atriplex
Grayia
Krascheninnikovia

APPENDIX B

CHECKLIST OF TREES AND SHRUBS

Species featured in this book are listed by taxonomic groups. The highest level distinguishes gnetophytes (Division Gnetophyta), conifers (Division Pinophyta), and broadleaved trees and shrubs (Division Magnoliophyta). Within each division, species are listed alphabetically by family and then by species. The name of the botanist who described the species is recorded next to the species name. Nomenclature follows Hickman 1993 unless noted by an asterisk. Common names appear in the right-hand column.

GNETOPHYTES

EPHEDRACEAE

Ephedra californica S. Watson	California ephedra
Ephedra viridis Cov.	green ephedra

CONIFERS

CUPRESSACEAE

Calocedrus decurrens (Torr.) Florin	incense-cedar
Chamaecyparis lawsoniana (A. Murr.) Parl.*	Port Orford–cedar
Chamaecyparis nootkatensis (D. Don) Spach*	Alaska yellow-cedar

Cupressus abramsiana C. B. Wolf	Santa Cruz cypress
Cupressus arizonica Greene	Arizona cypress
Cupressus bakeri Jepson	Baker cypress
Cupressus forbesii Jepson	Tecate cypress
Cupressus goveniana Gord.	Gowen cypress
Cupressus macnabiana A. Murr.	McNab cypress
Cupressus macrocarpa Gord.	Monterey cypress
Cupressus sargentii Jepson	Sargent cypress
Juniperus californica Carr.	California juniper
Juniperus communis L.	common juniper
Juniperus occidentalis Hook. var. *australis* (Vasek) A. & N. Holmgren	mountain juniper
Juniperus occidentalis Hook. var. *occidentalis*	western juniper
Juniperus osteosperma (Torr.) Little	Utah juniper
Sequoia sempervirens (D. Don) Endl.*	redwood
Sequoiadendron giganteum (Lindl.) Buchh.*	giant sequoia
Thuja plicata D. Don	western redcedar

PINACEAE

Abies amabilis (Dougl.) Forbes	Pacific silver fir
Abies bracteata (D. Don) Poit.	bristlecone fir
Abies concolor (Gordon & Glend.) Hildebr.	white fir
Abies grandis (Dougl.) Lindl.	grand fir
Abies lasiocarpa (Hook.) Nutt.	subalpine fir
Abies magnifica A. Murr. var. *magnifica*	red fir
Abies magnifica A. Murr. var. *shastensis* Lemmon	Shasta red fir
Abies procera Rehd.	noble fir
Picea breweriana S. Wats.	Brewer spruce
Picea engelmannii Engelm.	Engelmann spruce
Picea sitchensis (Bong.) Carr.	Sitka spruce
Pinus albicaulis Engelm.	whitebark pine
Pinus attenuata Lemmon	knobcone pine
Pinus balfouriana Grev. & Balf.	foxtail pine
Pinus contorta Dougl. ssp. *contorta*	beach pine
Pinus contorta Dougl. ssp. *murrayana* (Grev. & Balf.) Critchfield	lodgepole pine

Pinus coulteri D. Don	Coulter pine
Pinus flexilis James	limber pine
Pinus jeffreyi Grev. & Balf.	Jeffrey pine
Pinus lambertiana Dougl.	sugar pine
Pinus longaeva D. K. Bailey	Great Basin bristlecone pine
Pinus monophylla Torr. & Frém.	singleleaf pinyon
Pinus monticola Dougl.	western white pine
Pinus muricata D. Don	Bishop pine
Pinus ponderosa P. & C. Lawson	ponderosa pine
Pinus quadrifolia Sudworth	Parry pinyon
Pinus radiata D. Don	Monterey pine
Pinus sabiniana Dougl.	ghost pine
Pinus torreyana Carr.	Torrey pine
Pinus washoensis Mason & Stockwell	Washoe pine
Pseudotsuga macrocarpa (Vasey) Mayr	bigcone Douglas-fir
Pseudotsuga menziesii (Mirbel) Franco	Douglas-fir
Tsuga heterophylla (Raf.) Sarg.	western hemlock
Tsuga mertensiana (Bong.) Carr.	mountain hemlock

TAXACEAE

Taxus brevifolia Nutt.	Pacific yew
Torreya californica Torr.	California-nutmeg

BROADLEAVED TREES AND SHRUBS

ACERACEAE

Acer circinatum Pursh	vine maple
Acer glabrum Torr.	mountain maple
Acer macrophyllum Pursh	bigleaf maple
Acer negundo L.	box-elder

ANACARDIACEAE

Malosma laurina (Nutt.) Abrams	laurel sumac
Rhus integrifolia (Nutt.) Brewer & S. Watson	lemonadeberry
Rhus ovata S. Wats.	sugarbush
Rhus trilobata Nutt.	skunkbush
Toxicodendron diversilobum (Torr. & Gray) Greene	poison-oak

ARECACEAE

Washingtonia filifera (L. Linden) H. Wendl.	California fan palm

ASTERACEAE

Artemisia californica Less.	California sagebrush
Artemisia tridentata Nutt.	big sagebrush
Baccharis pilularis DC.	coyote brush
Baccharis salicifolia (Ruiz Lopez & Pavón) Pers.	mule fat
Chrysothamnus nauseosus (Pursh) Britt.	rubber rabbitbrush
Chrysothamnus parryi (Gray) Greene	Parry rabbitbrush
Chrysothamnus viscidiflorus (Hook.) Nutt.	yellow rabbitbrush
Eastwoodia elegans Brandegee	yellow-aster
Encelia californica Nutt.	California encelia
Encelia farinosa Torr. & Gray	brittlebush
Ericameria linearifolia (DC.) Urbatsch & Wussow	interior goldenbush
Hymenoclea salsola Gray	cheese bush
Lepidospartum squamatum (Gray) Gray	scalebroom
Pluchea sericea (Nutt.) Cov.	arrow-weed
Tetradymia canescens DC.	gray horsebrush

BERBERIDACEAE

Berberis aquifolium Pursh	Oregon-grape
Berberis nervosa Pursh	little Oregon-grape
Berberis pinnata Lag.	California barberry

BETULACEAE

Alnus incana (L.) Moench ssp. *tenuifolia* (Nutt.) Breitung	mountain alder
Alnus rhombifolia Nutt.	white alder
Alnus rubra Bong.	red alder
Alnus viridis (Vill.) Lam. & DC. ssp. *sinuata* (Regel) A. & D. Löve	Sitka alder
Betula glandulosa Michx.	resin birch
Betula occidentalis Hook.	water birch
Corylus cornuta Marsh. var. *californica* (A. DC.) Sharp	California hazel

BIGNONIACEAE

Chilopsis linearis (Cav.) Sweet	desert-willow

CALYCANTHACEAE

Calycanthus occidentalis Hook. & Arn.	spice bush

CAPPARACEAE

Isomeris arborea Nutt.	bladderpod

CAPRIFOLIACEAE

Lonicera conjugialis Kellogg	purple-flowered honeysuckle
Lonicera involucrata Spreng.	twinberry
Sambucus mexicana DC.	blue elderberry
Sambucus racemosa L.	red elderberry
Symphoricarpos albus (L.) Blake	common snowberry
Symphoricarpos mollis Nutt.	creeping snowberry
Symphoricarpos rotundifolius Gray	mountain snowberry
Viburnum ellipticum Hook.	western viburnum

CELASTRACEAE

Euonymus occidentalis Torr.	western burning bush
Paxistima myrsinites (Pursh) Raf.	Oregon boxwood

CHENOPODIACEAE

Atriplex canescens (Pursh) Nutt.	fourwing saltbush
Atriplex confertifolia (Torr. & Frém.) S. Wats.	shadscale
Grayia spinosa (Hook.) Moq.	hop-sage
Krascheninnikovia lanata (Pursh) A. D. J. Meeuse & Smit	winter fat

CORNACEAE

Cornus glabrata Benth.	brown dogwood
Cornus nuttallii Torr. & Gray	mountain dogwood
Cornus sericea L.	red osier
Cornus sessilis Dur.	blackfruit dogwood

ELAEAGNACEAE

Shepherdia argentea (Pursh) Nutt.	silver buffaloberry

ERICACEAE

Arbutus menziesii Pursh	Pacific madrone
Arctostaphylos canescens Eastw.	hoary manzanita
Arctostaphylos columbiana Piper	hairy manzanita
Arctostaphylos glandulosa Eastw.	Eastwood manzanita
Arctostaphylos glauca Lindl.	bigberry manzanita
Arctostaphylos manzanita Parry	common manzanita
Arctostaphylos nevadensis Gray	pinemat manzanita
Arctostaphylos nummularia Gray	glossyleaf manzanita
Arctostaphylos parryana Lemmon	Parry manzanita
Arctostaphylos patula Greene	greenleaf manzanita
Arctostaphylos pungens Kunth	Mexican manzanita
Arctostaphylos tomentosa (Pursh) Lindl.	woollyleaf manzanita
Arctostaphylos uva-ursi (L.) Spreng.	bearberry
Arctostaphylos viscida Parry	whiteleaf manzanita
Cassiope mertensiana (Bong.) D. Don	white heather
Erica lusitanica K. Rudolphi	heath
Gaultheria shallon Pursh	salal
Kalmia polifolia Wagenh.	laurel
Ledum glandulosum Nutt.	western Labrador-tea
Leucothoe davisiae Gray	black-laurel
Menziesia ferruginea Sm.	mock-azalea
Phyllodoce breweri (Gray) Heller	Brewer heather
Phyllodoce empetriformis (Smith) D. Don	Cascade heather
Rhododendron macrophyllum D. Don	Pacific rhododendron
Rhododendron occidentale (Torr.) Gray	western azalea
Vaccinium caespitosum Michx.	dwarf huckleberry

Vaccinium membranaceum Torr.	thinleaf huckleberry
Vaccinium ovatum Pursh	evergreen huckleberry
Vaccinium parvifolium Sm.	red huckleberry
Vaccinium scoparium Cov.	littleleaf huckleberry

FABACEAE

Acacia greggii Gray	catclaw acacia
Cercidium floridum Gray	blue palo verde
Cercis occidentalis Gray	western redbud
Cytisus scoparius (L.) Link	Scotch broom
Genista monspessulana (L.) L. Johnson	French broom
Lupinus arboreus Sims	yellow bush lupine
Olneya tesota Gray	ironwood
Pickeringia montana Nutt.	chaparral pea
Prosopis glandulosa Torr. var. *torreyana* (L. Benson) M. Johnston	honey mesquite
Psorothamnus spinosus (Gray) Barneby	smoke tree
Robinia pseudoacacia L.	black locust
Spartium junceum L.	Spanish broom
Ulex europaea L.	gorse

FAGACEAE

Chrysolepis chrysophylla (Hook.) Hjelmq.	golden chinquapin
Chrysolepis sempervirens (Kellogg) Hjelmq.	bush chinquapin
Lithocarpus densiflorus (Hook. & Arn.) Rehd.	tanoak
Quercus agrifolia Née	coast live oak
Quercus berberidifolia Liebm.	scrub oak
Quercus chrysolepis Liebm.	canyon live oak
Quercus cornelius-mulleri Nixon & Steele	Muller oak
Quercus douglasii Hook. & Arn.	blue oak
Quercus durata Jepson	leather oak
Quercus engelmannii Greene	Engelmann oak
Quercus garryana Hook.	Oregon white oak
Quercus john-tuckeri Nixon & Muller	Tucker oak

Quercus kelloggii Newberry	California black oak
Quercus lobata Née	valley oak
Quercus parvula Greene	island scrub oak
Quercus sadleriana R. Br. Campst.	Sadler oak
Quercus turbinella Greene	desert scrub oak
Quercus vacciniifolia Kellogg	huckleberry oak
Quercus wislizenii A. DC.	interior live oak

GARRYACEAE

Garrya elliptica Lindl.	coastal silk tassel
Garrya fremontii Torr.	Fremont silk tassel

GROSSULARIACEAE

Ribes aureum Pursh	golden currant
Ribes bracteosum Hook.	stink currant
Ribes californicum Hook. & Arn.	hillside gooseberry
Ribes cereum Dougl.	wax currant
Ribes divaricatum Dougl.	straggly gooseberry
Ribes lacustre (Pers.) Poir.	swamp currant
Ribes lobbii Gray	gummy gooseberry
Ribes menziesii Pursh	canyon gooseberry
Ribes montigenum McClatchie	mountain gooseberry
Ribes quercetorum Greene	oak gooseberry
Ribes roezlii Regel	Sierra gooseberry
Ribes sanguineum Pursh	red flowering currant
Ribes speciosum Pursh	fuchsia-flowered gooseberry

HIPPOCASTANACEAE

Aesculus californica (Spach) Nutt.	California buckeye

HYDRANGACEAE

Philadelphus lewisii Pursh	wild mock-orange

HYDROPHYLLACEAE

Eriodictyon californicum (Hook. & Arn.) Torr.	mountain balm

JUGLANDACEAE

Juglans californica S. Wats.	California black walnut

LAMIACEAE

Salvia apiana Jepson	white sage
Salvia dorrii (Kellogg) Abrams	desert sage
Salvia eremostachya Jepson	sand sage
Salvia leucophylla Greene	purple sage
Salvia mellifera Greene	black sage
Salvia mohavensis Greene	Mojave sage
Salvia pachyphylla Munz	mountain desert sage
Salvia sonomensis Greene	creeping sage

LAURACEAE

Umbellularia californica (Hook. & Arn.) Nutt.	California bay

LILIACEAE

Nolina parryi S. Wats.	Parry nolina
Yucca baccata Torr.	banana yucca
Yucca brevifolia Engelm.	Joshua tree
Yucca schidigera Ortgies	Mojave yucca
Yucca whipplei Torr.	Our Lord's candle

MYRICACEAE

Myrica californica Cham.	Pacific wax-myrtle
Myrica hartwegii S. Wats.	Sierra sweet-bay

MYRTACEAE

Eucalyptus globulus Labill.	blue gum

OLEACEAE

Fraxinus anomala S. Wats.	singleleaf ash
Fraxinus dipetala Hook. & Arn.	foothill ash
Fraxinus latifolia Benth.	Oregon ash
Fraxinus velutina Torr.	velvet ash

PAPAVERACEAE

Dendromecon rigida Benth.	bush poppy

PLATANACEAE

Platanus racemosa Nutt.	California sycamore

POLYGONACEAE

Eriogonum fasciculatum Benth.	California buckwheat

RHAMNACEAE

Ceanothus cordulatus Kellogg	mountain whitethorn
Ceanothus crassifolius Torr.	hoaryleaf ceanothus
Ceanothus cuneatus (Hook.) Nutt.	wedgeleaf ceanothus
Ceanothus diversifolius Kellogg	pine mat
Ceanothus gloriosus J. T. Howell	glory bush
Ceanothus greggii Gray	cupleaf ceanothus
Ceanothus incanus Torr. & Gray	coast whitethorn
Ceanothus integerrimus Hook. & Arn.	deerbrush
Ceanothus lemmonii Parry	Lemmon ceanothus
Ceanothus leucodermis Greene	chaparral whitethorn
Ceanothus megacarpus Nutt.	bigpod ceanothus
Ceanothus oliganthus Nutt.	hairy ceanothus
Ceanothus prostratus Benth.	Mahala mat
Ceanothus pumilus Greene	Siskiyou mat
Ceanothus thyrsiflorus Eschsch.	blue blossom
Ceanothus tomentosus C. Parry	woollyleaf ceanothus
Ceanothus velutinus Hook.	tobacco brush
Rhamnus californica Eschsch.	California coffeeberry
Rhamnus crocea Nutt.	spiny redberry
Rhamnus ilicifolia Kellogg	hollyleaf redberry
Rhamnus purshiana DC.	cascara

Rhamnus rubra Greene	Sierra coffeeberry
Rhamnus tomentella Benth.	hoary coffeeberry

ROSACEAE

Adenostoma fasciculatum Hook. & Arn.	chamise
Adenostoma sparsifolium Torr.	red shank
Amelanchier alnifolia (Nutt.) M. Roemer	western serviceberry
Cercocarpus betuloides Nutt.	birchleaf mountain-mahogany
Cercocarpus ledifolius Nutt.	curlleaf mountain-mahogany
Chamaebatia foliolosa Benth.	mountain misery
Chamaebatiaria millefolium (Torr.) Maxim.	fern bush
Coleogyne ramosissima Torr.	black bush
Crataegus suksdorfii (Sarg.) Kruschke	black hawthorn
Heteromeles arbutifolia (Lindl.) M. Roemer	toyon
Holodiscus discolor (Pursh) Maxim.	oceanspray
Holodiscus microphyllus Rydb.	rock-spiraea
Malus fusca (Raf.) Schneid.	Oregon crab apple
Neviusia cliftonii Shevock, Ertter, & Taylor	snow wreath
Oemleria cerasiformis (Hook. & Arn.) Landon	oso berry
Physocarpus capitatus (Pursh) Kuntze	ninebark
Potentilla fruticosa L.	shrub cinquefoil
Prunus andersonii Gray	desert peach
Prunus emarginata (Hook.) Walp.	bitter cherry
Prunus fasciculata (Torr.) Gray	desert almond
Prunus fremontii S. Wats.	desert apricot
Prunus ilicifolia (Hook. & Arn.) D. Dietr.	hollyleaf cherry
Prunus subcordata Benth.	Klamath plum
Prunus virginiana L.	choke cherry
Purshia mexicana (D. Don) Welsh var. stansburyana (Torr.) Welsh	cliffrose
Purshia tridentata (Pursh) DC.	bitterbrush
Rosa californica Cham. & Schlecht.	California rose
Rosa gymnocarpa Nutt.	wood rose

Rosa woodsii Lindl. var. *ultramontana* (S. Wats.) Jepson	interior rose
Rubus discolor Weihe & Nees	Himalayan blackberry
Rubus lasiococcus Gray	dwarf bramble
Rubus leucodermis Torr. & Gray	western raspberry
Rubus parviflorus Nutt.	thimbleberry
Rubus spectabilis Pursh	salmonberry
Rubus ursinus Cham. & Schlecht.	California blackberry
Sorbus scopulina Greene	western mountain-ash
Spiraea densiflora Greenm.	mountain spiraea
Spiraea douglasii Hook.	Douglas spiraea

RUBIACEAE

Cephalanthus occidentalis L.	button-willow

RUTACEAE

Ptelea crenulata Greene	California hop tree

SALICACEAE

Populus angustifolia James	narrowleaf cottonwood
Populus balsamifera L. ssp. *trichocarpa* (Torr. & Gray) Brayshaw	black cottonwood
Populus fremontii S. Wats. ssp. *fremontii*	Fremont cottonwood
Populus tremuloides Michx.	quaking aspen
Salix babylonica L.	weeping willow
Salix eastwoodiae Heller	Sierra willow
Salix exigua Nutt.	narrowleaf willow
Salix gooddingii Ball	black willow
Salix hookeriana Hook.	Hooker willow
Salix jepsonii Schneid.	Jepson willow
Salix laevigata Bebb	red willow
Salix lasiolepis Benth.	arroyo willow
Salix lucida Muhl.	shining willow
Salix lutea Nutt.	yellow willow
Salix melanopsis Nutt.	dusky willow
Salix scouleriana Hook.	Scouler willow
Salix sitchensis Bong.	Sitka willow

SCROPHULARIACEAE

Keckiella antirrhinoides (Benth.) Straw	yellow keckiella
Keckiella cordifolia (Benth.) Straw	straggly keckiella
Keckiella corymbosa (Benth.) Straw	red keckiella
Mimulus aurantiacus W. Curtis	bush monkeyflower

SIMAROUBACEAE

Ailanthus altissima (P. Mill.) Swingle	tree-of-heaven

SOLANACEAE

Lycium andersonii Gray	Anderson boxthorn
Lycium brevipes Benth.	desert-thorn

STAPHYLEACEAE

Staphylea bolanderi Gray	Sierra bladdernut

STERCULIACEAE

Fremontodendron californicum (Torr.) Cov.	flannelbush

STYRACACEAE

Styrax officinalis L.	snowdrop bush

TAMARICACEAE

Tamarix parviflora DC.	smallflower tamarisk

ULMACEAE

Celtis reticulata Torr.	netleaf hackberry

ZYGOPHYLLACEAE

Larrea tridentata (DC.) Cov.	creosote bush

GLOSSARY

Terms describing leaves, leaf shapes, leaf margins, twigs, flowers, and fruits and cones are illustrated at the end of the Glossary.

achene A small, dry, 1-seeded fruit that does not open readily, e.g., chamise, coyote brush, mountain-mahogany.

alternate A leaf arrangement characterized by 1 leaf at a node.

aril A fleshy appendage attached near the point of seed attachment, sometimes covering the seed, e.g., California-nutmeg, yew.

axil The interior angle formed by a stem and the petiole that it bears.

awl Shaped like the carpenter's tool for piercing leather or wood; characterized by a gradual taper from the base to a sharp point, as in certain conifers, e.g., giant sequoia, juniper.

berry A fleshy fruit with several to many seeds and no central pit or core, a product of 2 or more carpels making up the pistil, e.g., currant, gooseberry, madrone.

blade The flat, expanded portion of a leaf or petal.

bract A much-reduced leaf or leaflike structure at the base of a flower, inflorescence, or conifer cone scale.

bud An undeveloped flower or branch.

calyx The sepals collectively.

cambium A meristematic layer between the wood and inner bark.

capsule A dry, usually many-seeded fruit that opens along lines or pores, a product of 2 or more carpels making up the pistil, e.g., buckeye, ceanothus, yucca.

carpel The ovule-bearing structure of a flower.

catkin An inflorescence characterized by unisexual flowers arranged in a thin, drooping cluster, e.g., alder, birch.

complete flower A flower in which sepals, petals, stamens, and pistils are present.

compound leaf A leaf blade comprising 2 or more leaflets.

cone A reproductive structure made up of scales (often woody), papery bracts, and seeds (seed cones) or pollen sacs (pollen cones), e.g., spruce, pine, redwood.

corolla The petals collectively.

deciduous Falling from the plant within 1 growing season.

disk flower One of the central tubular flowers found in many inflorescences of plants in the sunflower family.

drupe A usually single-seeded fruit surrounded by a hard, stonelike covering (pit) that is covered with a pulpy or fleshy layer, e.g., cherry, dogwood, California bay.

drupelets A small drupe, often found in aggregations, e.g., blackberry.

elliptical Shaped like an ellipse, oval in general form.

endemic Native plants confined to a given region or locality.

entire A leaf edge that lacks features such as teeth or spines and is smooth.

epiphyte A plant that grows upon another plant for position or support, but that does not parasitize it.

evergreen Remaining green during the dormant season and persistent on the plant for 2 or more seasons.

exserted Protruding beyond other plant parts.

follicle A dry, usually many-seeded fruit that opens along a single line, e.g., fern bush, ninebark.

gland Used broadly for any warty protuberance.

hip A vaselike, leathery structure containing several seedlike achenes; the false fruit of the rose.

incomplete flower A flower that lacks at least one of the four floral series of a complete flower.

inflorescence The arrangement of flowers on a floral axis; also used to designate the sequence and location of flowers on a floral axis.

infructescence The arrangement of fruits on a floral axis.

internode Distance between two nodes.

involucre A series of bracts that subtends a flower or inflorescence.

keeled A raised or pronounced ridge.

lance shaped Narrow, tapering on both ends, and widest below the middle, as in a leaf shape.

layering A type of vegetative reproduction that occurs when a branch contacts the soil and roots develop.

leaflet A bladelike division of a compound leaf blade.

legume A dry, usually many-seeded fruit that opens along 2 lines, e.g., lupine, palo verde, redbud, black locust.

lenticel A lens-shaped pore on the surface of twigs.

linear Long and narrow with nearly parallel sides.

lobe Rounded leaf margins separated by indented sinuses.

margin The edge of a leaf.

montane Mid-elevation mountain zone between foothills and the subalpine zone.

needle A long, slender blade common in pines and some spruces.

node The point or region on a stem where buds and leaves are borne. The distance between nodes is called the internode.

nut A coarse, dry, hard, 1-seeded fruit that does not open readily. It is developed from 2 or more carpels making up the pistil, e.g., creosote bush, oak, walnut.

nutlet A small nut.

oblong Much longer than broad; the sides are more or less parallel.

open-grown Plants that grow with little to no competition from neighbors.

opposite A leaf arrangement characterized by 2 leaves at a node, these on opposing sides of the stem.

oval Having a broad elliptical shape.

palmate Originating from a common point; e.g., palmately compound leaves have leaflets radiating from the apex of the petiole.

pedicel The stalk that supports a single flower.

peltate Umbrella or shield shaped.

persistent Remaining attached.

petiole The portion of the leaf that supports the blade.

petiolule The stalk of a leaflet.

pinnate With leaflets arranged on both sides of the axis of a compound leaf.

pistil Female reproductive structure of a flower surrounding 1 or more developing seeds. At maturity, the pistil becomes a fruit.

pistillate Refers to a flower that has 1 or more pistils and lacks stamens.

pith The spongy, homogeneous, or chambered center of stems and roots.

podzol A soil type that is characterized by a whitish gray surface horizon.

pome A fruit with seeds in an inner core surrounded by a fleshy covering. The fruit does not open readily, e.g., apple, toyon, serviceberry.

prickle A sharp-pointed outgrowth on the epidermis of a twig or on a cone scale; the term is not synonymous with *spine* or *thorn*.

rachis The axis of an inflorescence or of a pinnately compound leaf.

ray flower One of the outer flowers in many inflorescences of plants in the sunflower family.

receptacle The more or less expanded tip of the pedicel upon which flower parts are borne.

reflexed Abruptly bent or turned downward or backward.

samara A dry, single-seeded fruit with a prominent wing. Fruits do not open readily, e.g., ash, tree-of-heaven.

scale A general term applied to any small, flat leaf or bract; often describes small, flat conifer leaves or the part of a conifer cone that supports seeds.

scalloped Having rounded teeth; usually describes a leaf margin.

sclerophyllous Plants with small, stiff, leathery leaves that often grow in dry summer climates.

sepal One of the component parts of the calyx that is typically green, although it may also be brightly colored and petal-like.

serotinous Refers to cones whose scales remain closed after seeds are mature.

serpentine A metamorphic rock with a high magnesium/calcium ratio. Soils formed from serpentine are relatively infertile.

serrated Having sharp, pointed teeth; usually describes a leaf margin.

sessile Without a stalk.

simple leaf A leaf made up of a blade, petiole, and stipules. The blade is not divided into leaflets.

spine A sharp-pointed, modified leaf or portion of a modified leaf, as in the stipular spines of acacia; used by some authors instead of *thorn*, and distinguished by its leaf origin.

spiny Having spines.

spur shoot A short, lateral branch lacking internode elongation.

stamen The pollen-producing organ of the flower.

staminate Refers to a flower that has 1 or more stamens and lacks pistils.

stipule An appendage, usually seen as paired structures, inserted at the base of the petiole; stipules are part of a leaf and not bracts subtending it.

stomatal bloom A waxy white coating around stomata; on some conifer leaves it can take the form of conspicuous lines (see Plate 9).

stomate A pore (usually in a leaf) that allows gas exchange. The plural is *stomata.*

style The elongated, usually narrow, part of the pistil between the ovary and the stigma.

subalpine High-elevation zone between the montane and alpine zone. The subalpine zone is a zone of forests and woodlands, while the alpine is above forest line.

thorn A sharp-pointed, simple or branched stem that develops from a bud.

toothed Having coarse, sharp teeth that are roughly perpendicular to the margins.

ultramafic Igneous and metamorphic rocks that are rich in magnesium and iron. Soils formed from ultramafic rocks are relatively infertile.

utricle A 1-seeded, more or less bladdery fruit, e.g., saltbush, hop-sage.

whorled A circular arrangement involving 3 or more leaves, flowers, or cones at a node.

LEAVES

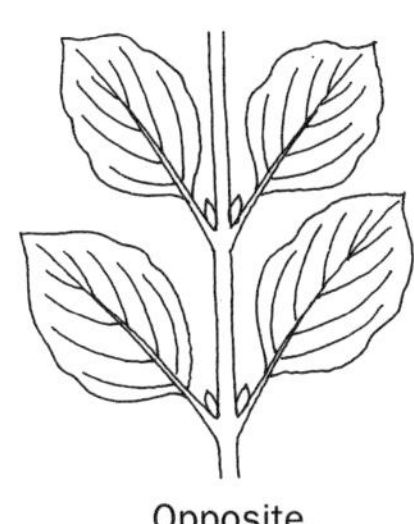

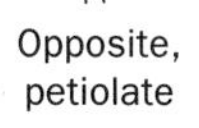

Opposite, petiolate

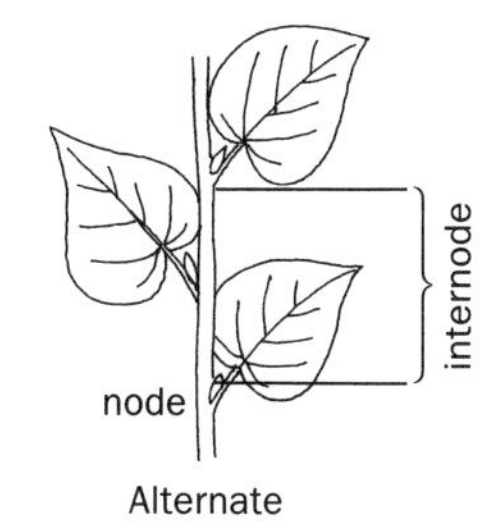

Alternate

Whorled

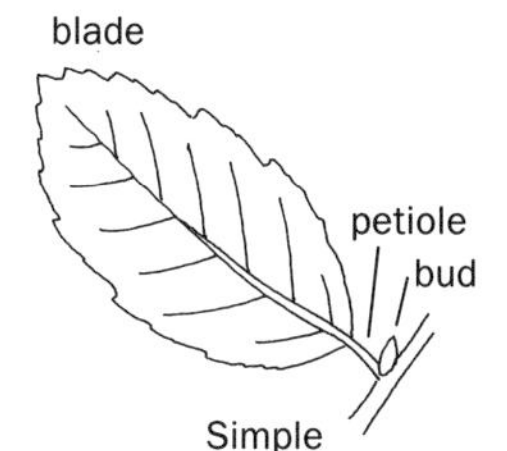

Simple

Pinnately compound

Palmately compound

LEAF SHAPES

Elliptical Oval Round

Linear Oblong Lanceolate (lance shaped)

Egg shaped Triangular

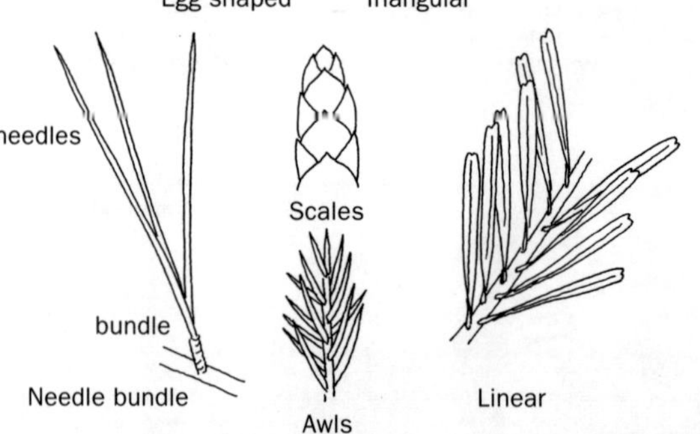

LEAF MARGINS

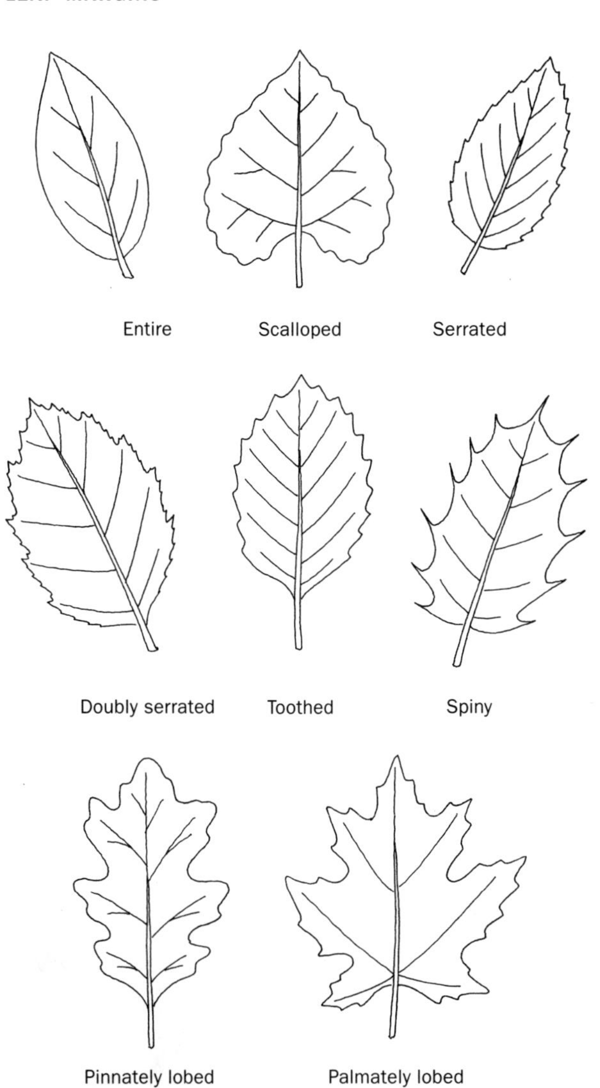

TWIGS

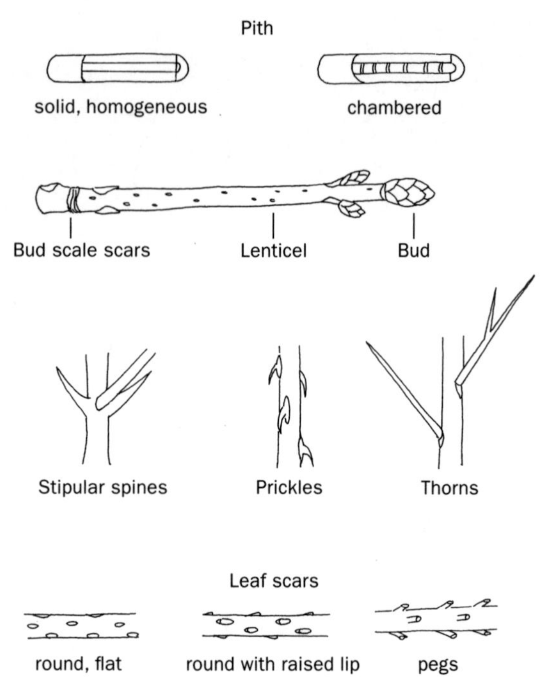
Pith
solid, homogeneous
chambered
Bud scale scars
Lenticel
Bud
Stipular spines
Prickles
Thorns
Leaf scars
round, flat
round with raised lip
pegs

FLOWERS

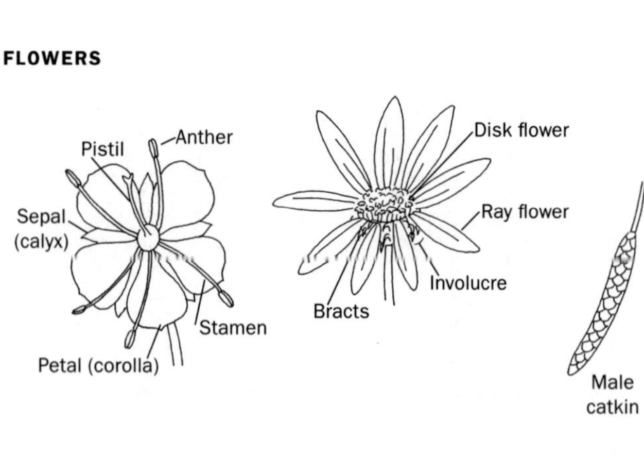
Pistil
Anther
Sepal
(calyx)
Stamen
Petal (corolla)
Disk flower
Ray flower
Involucre
Bracts
Male
catkin

FRUITS AND CONES

Samara

Achene

Follicle

Legume

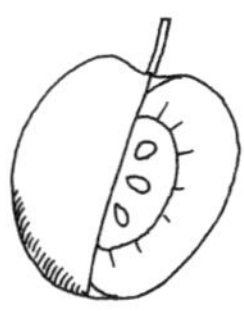
Pome

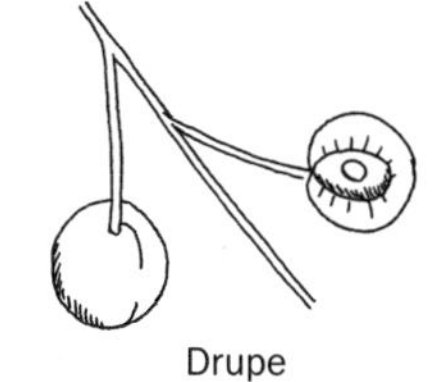
Drupe

Nut

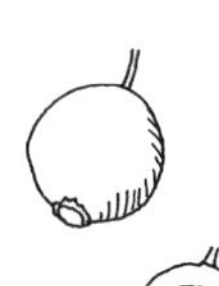
Berry

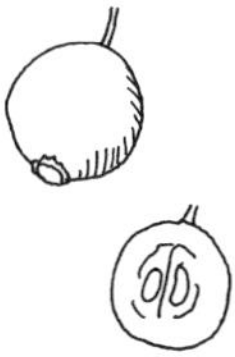

Capsule

Utricle

Aril

Seed-bearing catkin

Seed-bearing cone

REFERENCES

Bailey, R. G. 1995. *Description of the ecoregions of the United States.* Misc. Publ. No. 1391. 2d ed. Washington, D.C.: USDA Forest Service.

Barbour, M. G., and W. D. Billings. 1988. *North American terrestrial vegetation.* New York: Cambridge University Press.

Barbour, M. G., and J. Major, eds. 1988. *Terrestrial vegetation of California.* New expanded ed. New York: Wiley-Interscience, 1977. Reprint, Sacramento: California Native Plant Society.

Burns, R. M., and B. H. Honkala. 1990. *Silvics of North America.* Vol. 1, *Conifers.* Agriculture Handbook 654. Washington, D.C.: USDA Forest Service.

———. 1990. *Silvics of North America.* Vol. 2, *Hardwoods.* Agriculture Handbook 654. Washington, D.C.: USDA Forest Service.

Cannon, B. 1998. National register of big trees: 1998–99. *American Forests* 104 (1): 24–46.

ECOMAP. 1993. National hierarchical framework of ecological units. Unpublished administrative paper. Washington, D.C.: USDA Forest Service.

Elias, T. S. 1980. *The complete trees of North America: Field guide and natural history.* New York: Van Nostrand Reinhold.

Farrar, J. L. 1995. *Trees of the northern United States and Canada.* Ames: Iowa State University Press.

Gadek, P. A., D. L. Alpers, M. M. Heslewood, and C. J. Quinn.

2000. Relationships within Cupressaceae sensu lato: A combined morphological and molecular approach. *American Journal of Botany* 87 (7): 1044–1057.

Goudey, C. B., and D. W. Smith, comps. and eds. 1994. *Ecological units of California: Subsections* [ecological map]. San Francisco: Pacific Southwest Region, USDA Forest Service.

Greuter, W., chair of the editorial committee, with F. R. Barrie, H. M. Burdet, W. G. Chaloner, V. Demoulin, D. L. Hawksworth, P. M. Jørgensen, D. H. Nicolson, P. C. Silva, P. Trehane, and J. McNeill. 1994. *International code of botanical nomenclature, adopted by the Fifteenth International Botanical Congress, Yokohama, August-September 1993.* Regnum Vegetabile 131. Königstein, Germany: Koeltz Scientific Books.

Griffin, J. R., and W. B. Critchfield. 1972. *The distribution of forest trees in California.* Research paper PSW-82. Berkeley: USDA Forest Service.

Harlow, W. M., E. S. Harrar, J. W. Hardin, and F. M. White. 1996. *Textbook of dendrology.* 8th ed. New York: McGraw-Hill.

Hickman, J. C., ed. 1993. *The Jepson manual: Higher plants of California.* Berkeley and Los Angeles: University of California Press.

Holland, V. L., and D. J. Keil. 1995. *California vegetation.* Dubuque, Iowa: Kendall/Hunt.

Kartesz, J. T. 1994. *A synonymized checklist of the vascular flora of the United States, Canada, and Greenland.* Vol. 1, checklist. 2d ed. Portland, Ore.: Timber Press.

———. 1994. *A synonymized checklist of the vascular flora of the United States, Canada, and Greenland.* Vol. 2, thesaurus. 2d ed. Portland, Ore.: Timber Press.

Little, E. L. 1979. *Checklist of United States trees (native and naturalized).* Agriculture Handbook 541. Washington, D.C.: USDA Forest Service.

Little, E. L., and W. B. Critchfield. 1969. *Subdivisions of the genus* Pinus *(pines).* Miscellaneous Publication No. 1144. Washington, D.C.: USDA Forest Service.

McMinn, H. E. 1939. *An illustrated manual of California shrubs.* Berkeley and Los Angeles: University of California Press.

McNab, W. H., and P. I. Avers, comps. 1994. *Ecological subregions of the United States: Section descriptions.* WO-WSA-5. Washington, D.C.: USDA Forest Service.

Miles, S. R., and C. B. Goudey. 1997. *Ecological subregions of California: Section and subsection descriptions.* R5-EM-TP-005. San Francisco: Pacific Southwest Region, USDA Forest Service.

Munz, P. A. 1959. *A California flora.* Berkeley and Los Angeles: University of California Press.

———. 1968. *Supplement to a California flora.* Berkeley and Los Angeles: University of California Press.

Newsholme, C. 1992. *Willows: The genus* Salix. Portland, Ore.: Timber Press.

Peattie, D. C. 1950. *A natural history of western trees.* Boston: Houghton Mifflin.

Randall, W. R., R. F. Keniston, D. N. Bever, and E. C. Jensen. 1988. *Manual of Oregon trees and shrubs.* Corvallis, Ore.: O.S.U. Bookstores.

Sawyer, J. O., and T. Keeler-Wolf. 1995. *A manual of California vegetation.* Sacramento: California Native Plant Society.

Skinner, M. W., and B. M. Pavlik. 1994. *California Native Plant Society's inventory of rare and endangered vascular plants of California.* 5th ed. Sacramento: California Native Plant Society Press.

Smith, J. P., and J. O. Sawyer. 1993. *Keys to the families and genera of vascular plants in northwest California.* Eureka, Calif.: Mad River Press.

Storer, T. I., and R. L. Usinger. 1963. *Sierra Nevada natural history: An illustrated handbook.* Berkeley and Los Angeles: University of California Press.

INDEX

The index lists scientific names for the genus only, followed by the range of pages on which the species accounts for that genus will be found in alphabetical order. In addition, the index lists both genera and all individual species under their common names, with cross-references included for common names not used in this book. Boldface page numbers denote illustrations.

Abies (fir), 22–35
Acacia (acacia), 130
acacia, catclaw, 130, **131**
Acer (maple), 130–137
Adenostoma (chamise), 137–140
Aesculus (buckeye), 140–141
Ailanthus (ailanthus), 141–142
alder (*Alnus*), 142–148
alder
- mountain, 143–144
- red, 146–147, **147**
- Sitka, 147–148, **148;** pl. 13
- thinleaf. *See* alder, Sitka
- white, 144–146, **145**

almond, Nevada wild. *See* desert peach
Alnus (alder), 142–148
Amelanchier (serviceberry), 149–150
antelope brush. *See* bitterbrush (species)
apple (*Malus*), 274–275
arborvitae (*Thuja*), 102–104
Arbutus (madrone), 150–152
Arctostaphylos (manzanita), 152–167
arrow-weed (*Pluchea*), 291–292
arrow-weed, 292
Artemisia (sagebrush), 167–169
ash (*Fraxinus*), 237–241
ash
- California. *See* ash, foothill
- dwarf. *See* ash, singleleaf
- flowering. *See* ash, foothill
- foothill, 238–239, **239**
- Oregon, 239–240
- singleleaf, 237–238

ash (*continued*)
velvet, 240–241, **241**
aspen
quaking, 297–298, **297**
trembling. *See* aspen, quaking
Atriplex (saltbush), 169–172
azalea
California. *See* azalea, western
western, 342–343, **342**

Baccharis (coyote brush), 172–174
barberry, California, 177–178
bay, California (*Umbellularia*), 402–404
bay, California (species), 403–404, **403;** pl. 39
bearberry, 164–165, **165;** pl. 14
Berberis (Oregon-grape), 174–178
Betula (birch), 178–180
bigcone-spruce. *See* Douglas-fir, bigcone
bilberry, dwarf. *See* huckleberry, dwarf
birch (*Betula*), 178–180
birch
bog. *See* birch, resin
red. *See* birch, water
resin, 178–179
scrub. *See* birch, resin
water, 179–180, **180**
bitterbrush (*Purshia*), 311–313
bitterbrush (species), 312–313, **312**
black bush (*Coleogyne*), 216–217
black bush (species), 216–217, **216**
blackberry (*Rubus*), 361–367
blackberry
California, 366–367, **367**
Himalayan, 362–363, **363**
roughfruit. *See* roughfruit-berry
wild. *See* blackberry, California
blackcap. *See* raspberry, western
black-laurel (*Leucothoe*), 264–265
black-laurel (species), 265, **265**
bladdernut (*Staphylea*), 392–393
bladdernut, Sierra, 392–393, **393**
bladderpod (*Isomeris*), 253–254
bladderpod (species), 253–254, **254**
blue blossom, 196–197, **196;** pl. 18
boxelder, 136–137, **137**
boxthorn (*Lycium*), 270–272
boxthorn
Anderson, 271–272, **271**
desert. *See* desert-thorn
boxwood (*Paxistima*), 283–284
boxwood, mountain. *See* mountain lover
brittlebush, 228–229, **228**
broom (*Cytisus*), 224–225
broom
chaparral. *See* coyote brush (species)
French (species), 246–247, **247**
Scotch, 224–225, **225**
Spanish (species), 390
broom, French (*Genista*), 246–247
broom, Spanish (*Spartium*), 390
buckbrush. *See* ceanothus, wedgeleaf
buckeye (*Aesculus*), 140–141
buckeye, California, 140–141, **141;** pl. 12

buckthorn (*Rhamnus*), 334–340
buckthorn
California. *See* coffeeberry, California
cascara. *See* cascara
hoary. *See* coffeeberry, hoary
hollyleaf. *See* redberry, hollyleaf
Sierra. *See* coffeeberry, Sierra
spiny. *See* redberry, spiny
buckwheat, California, 233–234, **233**
buckwheat, wild (*Eriogonum*), 232–234
buffaloberry (*Shepherdia*), 387–388
buffaloberry, silver, 387–388, **388**
burning bush (*Euonymus*), 235–237
burning bush, western, 235–237, **236**
burrobrush. *See* cheese bush (species)
burrobush, white. *See* cheese bush (species)
button-willow (*Cephalanthus*), 199–201
button-willow (species), 200–201, **200**

California-lilac (*Ceanothus*), 182–198
California-nutmeg, 104–105, **105**
Calocedrus (incense-cedar), 35–36
Calycanthus (spice bush), 180–181
cascara, 337–338, **337;** pl. 32
Cassiope (white heather), 181–182
Ceanothus (California-lilac), 182–198
ceanothus
bigpod, 193
blue blossom. *See* blue blossom
chaparral whitethorn. *See* whitethorn, chaparral
coast whitethorn. *See* whitethorn, coast
cupleaf, 189–190, **189**
deerbrush. *See* deerbrush
desert. *See* ceanothus, cupleaf
Gregg's. *See* ceanothus, cupleaf
glory bush. *See* glory bush
hairy, 193–194
hoaryleaf, 186
Lemmon, 192
Mahala mat. *See* Mahala mat
Mojave. *See* ceanothus, cupleaf
mountain whitethorn. *See* whitethorn, mountain
pine mat. *See* pine mat
Point Reyes. *See* glory bush
Siskiyou mat. *See* Siskiyou mat
tobacco brush. *See* tobacco brush
trailing. *See* pine mat
varnish leaf. *See* tobacco brush
wedgeleaf, 186–187, **187**
woollyleaf, 197
Celtis (netleaf hackberry), 198–199
Cephalanthus (button-willow), 199–201
Cercidium (palo verde), 201–202
Cercis (redbud), 202–204
Cercocarpus (mountain-mahogany), 204–207

Chamaebatia (mountain misery), 207–208
Chamaebatiaria (fern bush), 208–209
Chamaecyparis (whitecedar), 36–40
chamise (*Adenostoma*), 137–140
chamise (species), 138, **139**
chaparral-pea (*Pickeringia*), 289–290
chaparral-pea (species), 289–290, **289**
cheese bush (*Hymenoclea*), 252–253
cheese bush (species), 252–253, **252**
cherry (*Prunus*), 301–308
cherry
  bitter, 303–304, **304**
  choke, 307–308, **308**
  evergreen. *See* cherry, hollyleaf
  hollyleaf, 305–306, **306**
  western choke. *See* cherry, choke
Chilopsis (desert-willow), 209–210
chinquapin (*Chrysolepis*), 210–213
chinquapin
  bush, 212–213, **212**
  giant. *See* chinquapin, golden
  golden, 211–212
  Sierra. *See* chinquapin, bush
Christmas berry. *See* toyon (species)
Chrysolepis (chinquapin), 210–213
Chrysothamnus (rabbitbrush), 213–215
cinquefoil (*Potentilla*), 298–299
cinquefoil, shrub, 298–299, **299**
cliffrose, 311–312; pl. 28
coffeeberry
  California, 335
  chaparral. *See* coffeeberry, hoary
  coast. *See* coffeeberry, California
  hoary, 339–340, **339**
  Sierra, 338–339, **338**
Coleogyne (black bush), 216–217
Cornus (dogwood), 217–221
Corylus (hazel), 221–223
cottonwood (*Populus*), 292–298
cottonwood
  Alamo. *See* cottonwood, Fremont
  black, 294–295, **295;** pl. 27
  Fremont, 295–297, **296**
  narrowleaf, 293–294
coyote brush (*Baccharis*), 172–174
coyote brush (species), 172–173, **173**
crab apple
  Oregon, 274–275, **274**
  western. *See* crab apple, Oregon
Crataegus (hawthorn), 223–224
cream bush (*Holodiscus*), 250–252
cream bush (species). *See* oceanspray
creosote bush (*Larrea*), 261–262
creosote bush (species), 261–262, **262**
Cupressus (cypress), 40–48
currant and gooseberry (*Ribes*), 346–356
currant
  golden, 348
  red flowering, 355–356, **355;** pl. 36
  stink, 348–349

swamp, 350–351
wax, 349–350
cypress (*Cupressus*), 40–48
cypress
Arizona, 42–43
Baker, 43–44
Cuyamaca. *See* cypress, Arizona
Forbes. *See* cypress, Tecate
Gowen, 44–45
McNab, 45–46, **46**
Mendocino. *See* cypress, Gowen
Monterey, 47–48, **47**
Piute. *See* cypress, Arizona
pygmy. *See* cypress, Gowen
Santa Cruz, 42
Sargent, 48
Tecate, 44
Cytisus (broom), 224–225

deerbrush, 190–191, **191**
Dendromecon (bush poppy), 225–226
desert almond, 304–305
desert apricot, 305
desert peach, 302, **303**
desert sweet. *See* fern bush (species)
desert tea. *See* ephedra, California
desert-thorn, 272
desert-willow (*Chilopsis*), 209–210
desert-willow (species), 209–210, **210**
dogwood (*Cornus*), 217–221
dogwood
blackfruit, 221
brown, 218
mountain, 218–219, **219;** pl. 21
osier, red. *See* red osier
Pacific. *See* dogwood, mountain
smooth. *See* dogwood, brown
Douglas-fir (*Pseudotsuga*), 94–97
Douglas-fir
— (species), 96–97, **97**
bigcone, 95–96, **95**

Eastwoodia (yellow-aster), 226–227
elderberry (*Sambucus*), 385–387
elderberry
blue, 385–386
red, 386–387, **386**
Encelia (encelia), 227–229
encelia, California, 227–228
Ephedra (Mormon-tea), 49–50
ephedra
California, 49
green, 50, **50;** pl. 2
Erica (heath), 229–230
Ericameria (goldenbush), 230–231
Eriodictyon (yerba santa), 231–232
Eriogonum (wild buckwheat), 232–234
Eucalyptus (gum tree), 234–235
Euonymus (burning bush), 235–237

fan palm (*Washingtonia*), 410–411
fan palm, California, 410–411, **411;** pl. 40
fern bush (*Chamaebatiaria*), 208–209
fern bush (species), 208–209, **209**
fir (*Abies*), 22–35
fir
bristlecone, 25–26, **26;** pl. 1
grand, 28–29, **29**
lovely. *See* fir, Pacific Silver
noble, 33–35, **34**
Pacific silver, 24–25, **24**
red, 31–32, **31**

fir (*continued*)
Santa Lucia. *See* fir, bristlecone
Shasta red, 32–33, **33**
silvertip. *See* fir, red
subalpine, 29–30, **30**
white, 26–28, **27**
flannelbush (*Fremontodendron*), 241–243
flannelbush (species), 241–243, **242**
Fraxinus (ash), 237–241
Fremontia. *See* flannelbush (species)
Fremontodendron (flannelbush), 241–243
furze. *See* gorse

Garrya (silk tassel), 243–245
Gaultheria (wintergreen), 245–246
Genista (French broom), 246–247
giant sequoia (*Sequoiadendron*), 99–100
giant sequoia (species), 99–100, **100;** pl. 10
glory bush, 188
glory mat. *See* glory bush
goldenbush (*Ericameria*), 230–231
goldenbush, interior, 230–231, **231**
gooseberry and currant (*Ribes*), 346–356
gooseberry
canyon, 352
fuchsia-flowered, 356
gummy, 351; pl. 35
hillside, 349
mountain, 352–353, **353**
oak, 353–354
Sierra, 354–355, **354**
straggly, 350
gorse (*Ulex*), 401–402
gorse (species), 401–402, **402**
Grayia (hop-sage), 247–248
gum, blue, 234–235, **235**
gum tree (*Eucalyptus*), 234–235

hard-tack. *See* mountain-mahogany, birchleaf
hawthorn (*Crataegus*), 223–224
hawthorn, black, 223–224, **223**
hazel (*Corylus*), 221–223
hazel, California, 221–223, **222;** pl. 22
heath (*Erica*), 229–230
heath (species), **229,** 230
heather
Brewer, 286
Cascade, 286–287, **287**
love. *See* heather, white (species)
northern mountain. *See* heather, Cascade
red-mountain. *See* heather, Brewer
white (species), 181–182, **182**
heather, mountain (*Phyllodoce*), 285–287
heather, white (*Cassiope*), 181–182
hemlock (*Tsuga*), 105–109
hemlock
mountain, 107–109, **108;** pl. 11
western, 106–107, **107**
Heteromeles (toyon), 248–249
holly, California. *See* toyon (species)
Holodiscus (cream bush), 250–252
honeysuckle, purpleflower, 268

hop tree (*Ptelea*), 309–311
hop tree, California, 310–311, **310**
hop-sage (*Grayia*), 247–248
hop-sage (species), 247–248, **248**
hop-sage, spiny. *See* hop-sage (species)
horsebrush (*Tetradymia*), 399–400
horsebrush, gray, 399–400
huckleberry (*Vaccinium*), 404–408
huckleberry
  California. *See* huckleberry, evergreen
  dwarf, 405
  evergreen, 406–407, **406**
  littleleaf, 408
  red, 407–408, **407**
  thinleaf, 405–406
Hymenoclea (cheese bush), 252–253

incense-cedar (*Calocedrus*), 35–36
incense-cedar (species), 35–36, **36**
ironwood (*Olneya*), 282–283
ironwood (species), 282–283, **283**
Isomeris (bladderpod), 253–254

Jim brush. *See* ceanothus, hairy
Joshua-tree, 413–414, **413**
Juglans (walnut), 254–256
juniper (*Juniperus*), 51–58
juniper
  California, 52–53, **53**
  common, 53–54, **54**
  mountain, 54–55; pl. 3
  Sierra. *See* juniper, mountain
  Utah, 57–58, **57**
  western, 55–57, **56**
Juniperus (juniper), 51–58

Kalmia (laurel), 256–257
Keckiella (keckiella), 257–260
keckiella
  red, 259–260, **259**
  straggly, 258–259
  yellow, 258
ket-ket-dizze. *See* mountain misery (species)
kinnikinnick. *See* bearberry
Krascheninnikovia (winter fat), 260–261

Labrador-tea (*Ledum*), 262–264
Labrador-tea, western, 263–264, **263**
Larrea (creosote bush), 261–262
laurel (*Kalmia*), 256–257
laurel
  — (species), 256–257, **257**
  California. *See* bay, California (species)
  Sierra. *See* black-laurel (species)
Ledum (Labrador-tea), 262–264
lemonadeberry, 343–344
Lepidospartum (scalebroom), 264
Leucothoe (black-laurel), 264–265
Lithocarpus (tanoak), 266–267
locust (*Robinia*), 356–358
locust, black, 356–358, **357**
Lonicera (twinberry), 267–269
lupine (*Lupinus*), 269–270
lupine, yellow bush, 269–270, **270**
Lupinus (lupine), 269–270
Lycium (boxthorn), 270–272

madrone (*Arbutus*), 150–152
madrone, Pacific, 150–152, **151**
Mahala mat, 194–195, **194**
mahonia
hollyleaf. *See* Oregon-grape (species)
longleaf. *See* Oregon-grape, little
shinyleaf. *See* barberry, California
Malosma (laurel sumac), 272–273
Malus (apple), 274–275
manzanita (*Arctostaphylos*), 152–167
manzanita
bearberry. *See* bearberry
bigberry, 156–157, **157**
common, 158–159, **158**
crown. *See* manzanita, Eastwood
Eastwood, 155–156, **156**
explorer's. *See* manzanita, woollyleaf
fire. *See* manzanita, glossyleaf
glossyleaf, 160–161, **160**
greenleaf, 161–163, **162**
hairy, 154–155
hoary, 154
Mexican, 163
Parry, 161
pinemat, 159
pointleaf. *See* manzanita, Mexican
whiteleaf, 165–167, **166;** pl. 15
woollyleaf, 163–164
maple (*Acer*), 130–137
maple
bigleaf, 135–136, **136**
box elder. *See* boxelder
mountain, 133–135, **134**
vine, 132–133, **133**
Menziesia (mock-azalea), 275–276
mesquite (*Prosopis*), 299–301
mesquite, honey, 299–301, **300**
Mimulus (monkeyflower), 276–277
mock-azalea (*Menziesia*), 275–276
mock-azalea (species), 275–276, **276**
mock-orange (*Philadelphus*), 284–285
mock-orange, wild, 284–285, **285**
monkeyflower (*Mimulus*), 276–277
monkeyflower, bush, 276–277, **277**
Mormon-tea (*Ephedra*), 49–50
mountain balm, 231–232, **232**
mountain lover, 283–284, **284**
mountain misery (*Chamaebatia*), 207–208
mountain misery
— (species), 207–208, **207**
Sierra. *See* mountain misery (species)
mountain-ash (*Sorbus*), 388–390
mountain-ash, western, 388–390, **389**
mountain-laurel. *See* azalea, western
mountain-mahogany (*Cercocarpus*), 204–207
mountain-mahogany
birchleaf, 204–205, **205**
curlleaf, 205–207, **206;** pl. 20
desert. *See* mountain-mahogany, curlleaf
western. *See* mountain-mahogany, birchleaf
mule fat, 173–174
Myrica (wax-myrtle), 277–279

netleaf hackberry (*Celtis*), 198–199

netleaf hackberry (species), 198–199, **199**
Neviusia (snow wreath), 279–280
ninebark (*Physocarpus*), 287–288
ninebark (species), 288, **288**
Nolina (nolina), 280–281
nolina, Parry, 280–281

oak (*Quercus*), 313–334
oak
- blue, 320–322, **321;** pl. 29
- California black, 326–328, **327;** pls. 30, 31
- canyon live, 318–320, **319**
- coast live, 315–317, **316**
- deer. *See* oak, Sadler
- desert scrub, 331–332
- Engelmann, 322–324, **324**
- Garry. *See* oak, Oregon white
- huckleberry, 332–333, **332**
- interior live, 333–334, **334**
- island scrub, 329–330
- leather, 322, **323**
- maul. *See* oak, canyon live
- mesa. *See* oak, Engelmann
- Muller, 320
- Oregon. *See* oak, Oregon white
- Oregon white, 324–325, **325**
- Sadler, 330–331, **331**
- scrub, 317–318, **317**
- shrub live. *See* oak, desert scrub
- Tucker, 326
- valley, 328–329, **329**

oceanspray, 250–251, **250**
Oemleria (oso berry), 281–282
Olneya (ironwood), 282–283
Oregon-grape (*Berberis*), 174–178
Oregon-grape
- —(species), 175–176, **175;** pl. 16
- dwarf. *See* Oregon-grape, little
- little, 176–177, **177**

Oregon-myrtle. *See* bay, California (species)
oso berry (*Oemleria*), 281–282
oso berry (species), 281–282, **281**
Our Lord's candle, 416

palo verde (*Cercidium*), 201–202
palo verde, blue, 201–202, **201**
Paxistima (boxwood), 283–284
pearl bush. *See* cheese bush (species)
Philadelphus (mock-orange), 284–285
Phyllodoce (mountain heather), 285–287
Physocarpus (ninebark), 287–288
pepperwood. *See* bay, California (species)
Picea (spruce), 58–63
Pickeringia (chaparral-pea), 289–290
pigeonberry, mountain. *See* coffeeberry, Sierra
pine (*Pinus*), 63–94
pine
- beach, 71–72, **72**
- Bishop, 84–85, **85**
- bull. *See* pine, ghost
- Coulter, 74–75, **75**
- digger. *See* pine, ghost
- foothill. *See* pine, ghost
- fourleaf pinyon. *See* pine, Parry pinyon
- foxtail, 70–71, **70;** pl. 5
- ghost, 90–91, **91;** pl. 8
- gray. *See* pine, ghost
- Great Basin bristlecone, 80–81, **80**

pine (*continued*)
Jeffrey, 77–78, **77**
knobcone, 68–69, **69**
limber, 75–76, **76**
lodgepole, 72–74, **73**
Monterey, 88–90, **89;** pl. 7
Parry pinyon, 87–88, **88**
ponderosa, 85–87, **86**
radiata. *See* pine, Monterey
shore. *See* pine, beach
Sierra lodgepole. *See* pine, lodgepole
silver. *See* pine, western white
singleleaf pinyon, 81–82, **82**
sugar, 78–79, **79**
tamarack. *See* pine, lodgepole
Torrey, 91–92, **92**
Washoe, 93–94, **93**
western white, 83–84, **83;** pl. 6
western yellow. *See* pine, ponderosa
whitebark, 67–68, **68**
pine mat, 187–188
Pinus (pine), 63–94
Platanus (sycamore), 290–291
Pluchea (arrow-weed), 291–292
plum
Indian. *See* oso berry (species)
Klamath, 307
Sierra. *See* plum, Klamath
poison-oak (*Toxicodendron*), 400–401
poison-oak (species), 400–401, **401**
poppy, bush (*Dendromecon*), 225–226
poppy, bush (species), 225–226, **226**
Populus (cottonwood), 292–298
Port Orford–cedar, 37–39, **38**
Potentilla (cinquefoil), 298–299
Prosopis (mesquite), 299–301
Prunus (cherry), 301–308
Pseudotsuga (Douglas-fir), 94–97
Psorothamnus (smoke tree), 308–309
Ptelea (hop tree), 309–311
Purshia (bitterbrush), 311–313

Quercus (oak), 313–334

rabbitbrush (*Chrysothamnus*), 213–215
rabbitbrush
Parry, 215
rubber, 214, **214**
yellow, 215
raspberry, western, 364–365, **364**
red osier, 220–221, **220**
red shank, 138–140
redberry
hollyleaf, 336
spiny, 335–336
redbud (*Cercis*), 202–204
redbud, western, 202–204, **203;** pl. 19
redcedar, western, 102–104, **103**
redwood (*Sequoia*), 97–99
redwood
— (species), 98–99, **98;** pl. 9
coast. *See* redwood (species)
Sierra. *See* giant sequoia (species)
Rhamnus (buckthorn), 334–340
Rhododendron (rhododendron), 340–343
rhododendron
coast. *See* rhododendron, Pacific
Pacific, 340–341, **341;** pl. 33

Rhus (sumac), 343–346
ribbon wood. *See* red shank
Ribes (currant and gooseberry), 346–356
Robinia (locust), 356–358
rock-spiraea, 251–252
Rosa (rose), 358–360
rose (*Rosa*), 358–360
rose
  California, 359
  interior, 360, **361**
  wood, 359–360, **360**
rose-bay, California. *See* rhododendron, Pacific
roughfruit-berry, 363–364
Rubus (blackberry), 361–367

sage (*Salvia*), 378–384
sage
  black, 382–383, **383;** pl. 37
  creeping, 384
  desert, 380–381, **381**
  Mojave, 383
  mountain desert, 384
  purple, 382
  sand, 381–382
  white, 379–380
sagebrush (*Artemisia*), 167–169
sagebrush
  big, 168–169, **169**
  California, 167–168, **168**
salal, 245–246, **246**
Salix (willow), 368–378
salmonberry, 366
saltbush (*Atriplex*), 169–172
saltbush
  fourwing, 170–171
  spiny. *See* shadscale
Salvia (sage), 378–384
Sambucus (elderberry), 385–387
scalebroom (*Lepidospartum*), 264
scalebroom (species), 264
seep-willow. *See* mule fat
Sequoia (redwood), 97–99
sequoia (species). *See* giant sequoia (species)
Sequoiadendron (giant sequoia), 99–100
serviceberry (*Amelanchier*), 149–150
serviceberry, western, 149–150, **149**
shadscale, 171–172, **171**
sheep-laurel. *See* azalea, western
Shepherdia (buffaloberry), 387–388
silk tassel (*Garrya*), 243–245
silk tassel
  coastal, 243–244, **244**
  Fremont, 244–245; pl. 23
Siskiyou mat, 195–196
skunkbush, 345–346, **345;** pl. 34
smoke tree (*Psorothamnus*), 308–309
smoke tree (species), 308–309, **309**
snow wreath (*Neviusia*), 279–280
snow wreath (species), 279–280
snowberry (*Symphoricarpos*), 395–398
snowberry
  common, 395–396
  creeping, 396–397, **397**
  mountain, 397–398
snowbrush. *See* tobacco brush
snowdrop bush, 394–395, **394**
Sorbus (mountain-ash), 388–390
Spanish bayonet. *See* yucca, banana
Spanish dagger. *See* yucca, Mojave
Spartium (Spanish broom), 390
spice bush (*Calycanthus*), 180–181

spice bush (species), 180–181, **181;** pl. 17
spice tree. *See* bay, California (species)
Spiraea (spiraea), 390–392
spiraea
  Douglas, 392; pl. 38
  mountain, 391–392, **391**
spruce (*Picea*), 58–63
spruce
  Brewer, 59–60, **60;** pl. 4
  Engelmann, 60–62, **61**
  Sitka, 62–63, **62**
squaw carpet. *See* Mahala mat
Staphylea (bladdernut), 392–393
sticky-laurel. *See* tobacco brush
stinking-cedar. *See* California-nutmeg
storax (*Styrax*), 394–395
Styrax (storax), 394–395
styrax, California. *See* snow-drop bush
sugarbush, 344–345, **344**
sumac (*Rhus*), 343–346
sumac
  laurel (species), 272–273, **273**
  lemon. *See* lemonadeberry
  mahogany. *See* lemonade-berry
  skunk. *See* skunkbush
  sugar. *See* sugarbush
sumac, laurel (*Malosma*), 272–273
sweet-bay, Sierra, 278–279
sycamore (*Platanus*), 290–291
sycamore, California, 290–291, **291**
Symphoricarpos (snowberry), 395–398

tamarisk (*Tamarix*), 398–399
tamarisk, smallflower, 398–399, **399**
Tamarix (tamarisk), 398–399
tanoak (*Lithocarpus*), 266–267
tanoak (species), 266–267, **267;** pl. 25
Taxus (yew), 101–102
Tetradymia (horsebrush), 399–400
thimbleberry, 365–366, **365**
Thuja (arborvitae), 102–104
tobacco brush, 197–198
Torreya (Torreya), 104–105
Toxicodendron (poison-oak), 400–401
toyon (*Heteromeles*), 248–249
toyon (species), 249, **249**
trapper's tea. *See* Labrador-tea, western
tree-of-heaven, 141–142, **142**
Tsuga (hemlock), 105–109
twinberry (*Lonicera*), 267–269
twinberry
  — (species), 268, **269;** pl. 26
  purpleflower. *See* honey-suckle, purpleflower

Ulex (gorse), 401–402
Umbellularia (California bay), 402–404

Vaccinium (huckleberry), 404–408
Viburnum (viburnum), 408–410
viburnum, western, 409–410, **409**

walnut (*Juglans*), 254–256
walnut, California black, 254–256, **255;** pl. 24
Washingtonia (fan palm), 410–411
water-wally. *See* mule fat
wax-myrtle (*Myrica*), 277–279

wax-myrtle
- Pacific, 278, **279**
- Sierra. *See* sweet-bay, Sierra

whitecedar (*Chamaecyparis*), 36–40
whitethorn
- chaparral, 192–193
- coast, 190
- mountain, 185–186, **185**

whortleberry, grouse. *See* huckleberry, littleleaf
willow (*Salix*), 368–378
willow
- arroyo, 374–375
- black, 372–373, **372**
- dusky, 377
- Hooker, 373
- Jepson, 373–374
- narrowleaf, 370–371, **371**
- red, 374
- Scouler, 377–378
- shining, 375–376, **375**
- Sierra, 370
- Sitka, 378
- weeping, 369–370
- yellow, 376–377

winter fat (*Krascheninnikovia*), 260–261
winter fat (species), 260–261, **261**
wintergreen (*Gaultheria*), 245–246

yellow-aster (*Eastwoodia*), 226–227
yellow-aster (species), 226–227
yellow-cedar, Alaska, 39–40, **40**
yerba santa (*Eriodictyon*), 231–232
yerba santa (species). *See* mountain balm
yew (*Taxus*), 101–102
yew, Pacific, 101–102, **102**
Yucca (yucca), 411–416
yucca
- banana, 412–413
- chaparral. *See* Our Lord's candle
- Mojave, 414–415, **415**

ABOUT THE AUTHORS

John D. Stuart is Professor of Forestry at Humboldt State University. John O. Sawyer is Professor of Botany at Humboldt State University and coauthor, with T. Keeler-Wolf, of *A Manual of California Vegetation* (1995), among other books. Illustrator Andrea J. Pickart is an ecologist with the U.S. Fish and Wildlife Service.

Text: 9/10.5 Minion
Display: Franklin Gothic Demi and Book
Composition: Integrated Composition Systems
Printing and binding: Everbest Printing Company